A Visual Guide to Stata Graphics

Third Edition

A Visual Guide to Stata Graphics

Third Edition

MICHAEL N. MITCHELL

A Stata Press Publication
StataCorp LP
College Station, Texas

Copyright © 2004, 2008, 2012 by StataCorp LP
All rights reserved. First edition 2004
Second edition 2008
Third edition 2012

Published by Stata Press, 4905 Lakeway Drive, College Station, Texas 77845
Typeset in LaTeX 2_ε
Printed in the United States of America

10 9 8 7 6 5 4 3 2 1

ISBN-10: 1-59718-106-4
ISBN-13: 978-1-59718-106-8

Library of Congress Control Number: 2011942526

Dedication

I dedicate this book to the teachers of the world. I have been fortunate to have been touched by many special teachers, and I will always be grateful for what they kindly gave to me. I thank (in order of appearance) Larry Grossman, Fred Perske, Rosemary Sheridan, Donald Butler, Jim Torcivia, Richard O'Connell, Linda Fidell, and Jim Sidanius. These teachers all left me gifts of knowledge and life lessons that help me every day. Even if they do not all remember me, I will always remember them.

Acknowledgments

Although there is one name on the cover of this book, many people have helped to make this book possible. Without them, this book would have remained a dream, and I could have never shared it with you. I thank those people who helped that dream become the book you are now holding.

I thank the warm people at Stata, who were generous in their assistance and who always find a way to be friendly and helpful. In particular, I thank Vince Wiggins for his generosity of time, insightful advice, boundless enthusiasm, whimsical sense of humor, and commitment to help make this book the best that it could be. His insight and support has been priceless. I am also grateful to Jeff Pitblado, who created the LaTeX tools that made the layout of this book possible. Without the benefit of his time and talent, I would still be learning LaTeX instead of writing these acknowledgments. Also, I would like to thank the Stata technical support team, especially Derek Wagner, for patiently working with me on my numerous questions. My thanks go to Lisa Gilmore for her kind and patient support when I got stuck. I also thank Deirdre Patterson for her very helpful, detailed, and thorough editing. Finally, I wish to express my deep thanks to Annette Fett both for her delightful cover design and for the terrific and innovative screen captures she created for chapter 2.

I also thank, in alphabetical order, Xiao Chen, Phil Ender, Frauke Kreuter, and Christine Wells for their support and extremely helpful suggestions.

Contents

Preface to the Third Edition

This third edition updates the second edition of this book, reflecting new features available in Stata version 12. Since version 10, Stata has added several new graphical features, including a command for creating contour plots, options that give you greater control over the display of text, and the ability to create graphs from the results of the `margins` command. Additional sections have been added to this third edition that illustrate these new features.

A new section has been added that illustrates the use of the `twoway contour` command; see Twoway : Contour (141). You can see Options : Text Display (388) for information about how to specify symbols, subscripts, and superscripts, as well as how to display text in bold or italics; this section also describes how you can display text using different fonts. A new section has also been added that describes how you can customize graphs created using the `marginsplot` command; see Appendix : Marginsplot (444).

This third edition also includes minor updates here and there to bring the text up to date for use with Stata version 12.

Simi Valley, California
December 2011

Preface to the Second Edition

I cannot believe that it has been over three years since the release of the first edition of this book. A lot has changed since then, and that includes the way that Stata graphics have evolved. Although the core features remain the same, there have been many enhancements and more features added, the most notable being the addition of the interactive point-and-click Stata Graph Editor.

The second edition of this book has been thoroughly revised to address these new features, especially the Graph Editor. This edition has an entire chapter devoted to the use of the Graph Editor; see chapter 2. Also, almost every example in this book has been augmented to include descriptions of how the Graph Editor can be used to create the customizations being illustrated via commands.

The Stata Graph Editor and Stata `graph` commands offer powerful tools for customizing your graphs, and I hope that the coverage of both side by side helps you to use each to their fullest capacity. To emphasize this point, I wrote a section that describes certain areas where I feel that commands are especially superior to the Graph Editor, and areas where I feel the Graph Editor is especially superior to commands; see section 2.9. Although I still feel that commands provide a primary mode of creating graphs, you need to use the Graph Editor for only a short amount of time to see what a smart and powerful tool it is. Whereas commands offer the power of repeatability, the Graph Editor provides a nimble interface that permits you to tangibly modify graphs like a potter directly handling clay. I hope that this book helps you to integrate the effective use of both of these tools into your graph-making toolkit.

As with the first edition, updating this second edition has been both a challenge and a delight. I have endeavored to make this book a tool that you would find friendly, logical, intuitive, and above all, useful. I really hope you like it!

Simi Valley, California
April 2008

Preface to the First Edition

It is obvious to say that graphics are a visual medium for communication. This book takes a visual approach to help you learn about how to use Stata graphics. While you can read this book in a linear fashion or use the table of contents to find what you are seeking, it is designed to be "thumbed through" and visually scanned. For example, the right margin of each right page has what I call a *Visual Table of Contents* to guide you through the chapters and sections of the book. Generally, each page has three graphs on it, allowing you to see and compare as many as six graphs at a time on facing pages. For a given graph, you can see the command that produced it, and next to each graph is some commentary. But don't feel compelled to read the commentary; often, it may be sufficient just to see the graph and the command that made it.

This is an informal book and is written in an informal style. As I write this, I picture myself sitting at the computer with you, and I am showing you examples that illustrate how to use Stata graphics. The comments are written very much as if we were sitting down together and I had a couple of points to make about the graph that I thought you might find useful. Sometimes, the comments might seem obvious, but because I am not there to hear your questions, I hope it is comforting to have the obvious stated just in case there was a bit of doubt.

While this book does not spend much time discussing the syntax of the graph commands (because you will be able to infer the rules for yourself after seeing a number of examples), the Intro : Options (20) section discusses some of the unique ways that options are used in Stata graph commands and compares them with the way that options are used in other Stata commands.

I strived to find a balance to make this book comprehensive but not overwhelming. As a result, I have omitted some options I thought would be seldom used. So, just because a feature is not illustrated in this book, this does not mean that Stata cannot do that task, and I would refer to [G-2] **graph** for more details. I try to include frequent cross-references to [G-2] **graph**; for example, see also [G-3] *axis_options*. I view this book as a complement to the *Stata Graphics Reference Manual*, and I hope that these cross-references will help you use these two books in a complementary manner. Note that, whenever you see references to [G-2] **xyz**, you can either find "xyz" in the *Stata Graphics Reference Manual* or type `whelp xyz` within Stata. The manual and the help have the same information, although the help may be more up to date and allows hyperlinking to related topics.

Each chapter is broken into a number of sections showing different features and options for the particular kind of graph being discussed in the chapter. The examples illustrate how these options or features can be used, focusing on examples that isolate these features so you are not distracted by irrelevant aspects of the Stata command or graph. While this approach improves the clarity of presentation, it does sacrifice some realism because graphs frequently have many options used together. To address this, there is a section addressing strategies for building up more complicated graphs, Intro : Building graphs (29), and a section giving tips on creating more complicated graphs, Appendix : More examples (466). These sections are geared to help you see how you can combine options to make more complex and feature-rich graphs.

While this book is printed in color, this does not mean that it ignores how to create monochrome (black & white) graphs. Some of the examples are shown using monochrome graphs illustrating how you can vary colors using multiple shades of gray and how you can vary other attributes, such as marker symbol and size, line width, and pattern, and so forth. I have tried to show options that would appeal to those creating color or monochrome graphs.

The graphs in this book were created using a set of schemes specifically created for this book. Despite differences in their appearance, all the schemes increase the size of textual and other elements in the graphs (e.g., titles) to make them more readable, given the small size of the graphs in this book. You can see more about the schemes in Intro : Schemes (15) and how to obtain them in Appendix : Online supplements (482). While one purpose of the different schemes is to aid in your visual enjoyment of the book, they are also used to illustrate the utility of schemes for setting up the look and default settings for your graphs. See Appendix : Online supplements (482) for information about how you can obtain these schemes.

Stata has a number of graph commands for producing special-purpose statistical graphs. Examples include graphs for examining the distributions of variables (e.g., `kdensity`, `pnorm`, or `gladder`), regression diagnostic plots (e.g., `rvfplot` or `lvr2plot`), survival plots (e.g., `sts` or `ltable`), time series plots (e.g., `ac` or `pac`), and ROC plots (e.g., `roctab` or `lsens`). To cover these graphs in enough detail to add something worthwhile would have expanded the scope and size of this book and detracted from its utility. Instead, I have included a section, Appendix : Stat graphs (431), that illustrates a number of these kinds of graphs to help you see the kinds of graphs these commands create. This is followed by Appendix : Stat graph options (438), which illustrates how you can customize these kinds of graphs using the options illustrated in this book.

If I may close on a more personal note, writing this book has been very rewarding and exciting. While writing, I kept thinking about the kind of book you would want to help you take full advantage of the powerful, but surprisingly easy to use, features of Stata graphics. I hope you like it!

Simi Valley, California
February 2004

1 Introduction

This chapter begins by briefly telling you about the organization of this book and giving you tips to help you use it most effectively. The next section gives a short overview of the different kinds of Stata graphs that will be examined in this book, and that section is followed by an overview of the different kinds of schemes that will be used for showing the graphs in this book. The fourth section illustrates the structure of options in Stata graph commands. In a sense, the second, third, and fourth sections of this chapter are a thumbnail preview of the entire book, showing the types of graphs covered, how you can control their overall look, and the general structure of options used within those graphs. The final section is about the process of creating graphs.

1.1 Using this book

I hope that you are eager to start reading this book but will take just a couple of minutes to read this section to get some suggestions that will make the book more useful to you. First, there are many ways you might read this book, but perhaps I can suggest some tips:

- Read this chapter before reading the other chapters, as it provides key information that will make the rest of the book more understandable.

- Although you might read a traditional book cover to cover, this book has been written so that the chapters stand on their own. You should feel free to dive into any chapter or section of any chapter.

- Sometimes you might find it useful to visually scan the graphs rather than to read. I think this is a good way to familiarize yourself with the kinds of features available in Stata graphs. If a certain feature catches your eye, you can stop and see the command that made the graph and even read the text explaining the command.

- Likewise, you might scan a chapter just by looking at the graphs and the part of the command in red, which is the part of the command highlighted in that graph. For example, scanning the chapter on bar charts in this way would quickly familiarize you with the kinds of features available for bar graphs and would show you how to obtain those features.

The right margin contains what I call the *Visual Table of Contents*. It is a useful tool for quickly finding the information you seek. I frequently use the *Visual Table of Contents* to cross-reference information within the book. By design, Stata graphs share many common features. For example, you use the same kinds of options to control legends across different types of graphs. It would be repetitive to go into detail about legends for bar charts, box

1

plots, and so on. Within each kind of graph, legends are briefly described and illustrated, but the details are described in the *Options* chapter in the section titled *Legend*. This is cross-referenced in the book by saying something like "for more details, see Options : Legend (361)", indicating that you should look to the *Visual Table of Contents* and thumb to the *Options* chapter and then to the *Legend* section, which begins on page 361.

Sometimes it may take an extra cross-reference to get the information you need. Say that you want to make the *y*-axis title large for a bar chart by using the `ytitle()` option, so you first consult Bar : Y-axis (213). This gives you some information about using `ytitle()`, but then that section refers you to Options : Axis titles (327), where more details about axis titles are described. This section then refers you to Options : Textboxes (379) for more complete details about options to control the display of text. That section shows more details but then refers to Styles : Textsize (428), where all the possible text sizes are described. I know this sounds like a lot of jumping around, but I hope that it feels more like drilling down for more detail, that you feel you are in control of the level of detail that you want, and that the *Visual Table of Contents* eases the process of getting the additional details.

Most pages of this book have three graphs per page, with each graph being composed of the graph itself, the command that produced it, and some descriptive text. An example is shown below, followed by some points to note.

```
graph twoway scatter propval100 ownhome, msymbol(Sh)
```

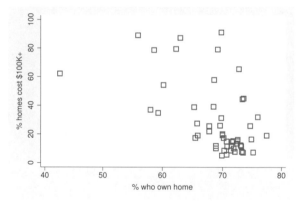

Here we use the `msymbol()` (marker symbol) option to make the symbols large hollow squares; see Options : Markers (307) for more details. The `graph twoway` portion of the command is optional.

Double-click on any of the markers and change the **Symbol** to `Hollow square`.

Uses allstates.dta & scheme vg_s2c

- The command itself is displayed in a `typewriter font`, and the salient part of the command (i.e., `msymbol(Sh)`) is in `this color`—both in the command and when referenced in the descriptive text.

- When commands or parts of commands are given in the descriptive text (e.g., `graph twoway`), they are displayed in the `typewriter font`.

- Many of the descriptions contain cross-references, for example, Options : Markers (307), which means to flip to the *Options* chapter and then to the section *Markers*. Equivalently, go to page 307.

- The names of some options are shorthand for two or more words that are sometimes explained; for instance, "we use the `msymbol()` (marker symbol) option to make ...".

- Many examples include more instructions describing how the Stata Graph Editor can be used to accomplish the same customization as illustrated in the command (in this example, how the Graph Editor can be used to obtain the equivalent of `msymbol(Sh)`). The ⬛ icon indicates that the instructions that follow apply to the use of the Graph Editor. The instructions assume that you have run the command *omitting* the highlighted portion of the command (e.g., omitting `msymbol(Sh)`) and that you have started the Graph Editor. The Graph Editor can be started in one of three ways: 1) by selecting **File** and then **Start Graph Editor** from the Graph window menu, 2) by clicking the Start Graph Editor ⬛ icon in the Graph window toolbar, or 3) by right-clicking on the graph and selecting **Start Graph Editor**. Once the Graph Editor is started, you can follow the instructions given (e.g., you can double-click on any of the markers, and in the dialog box that appears, you can then change the setting for the *Symbol* option to *Hollow square*). See Editor (35) for more details about using the Graph Editor.

- The descriptive text always concludes by telling you the name of the data file and scheme used for making the graph. Here the data file was *allstates.dta*, and the scheme was *vg_s2c.scheme*. You can read the data file over the Internet by using the **vguse** command, which is added to Stata when you install the online supplements; see Appendix : Online supplements (482). If you are connected to the Internet and your Stata is fully up to date, you can simply type **vguse allstates** to use that file over the Internet, and you can run the graph command shown to create the graph.

- Sometimes there is not enough space to describe the command as well as describe how to use the Graph Editor to accomplish the customization illustrated. In such cases, the description will conclude with " ⬛ See the next graph". The descriptive text for the next example will begin with the ⬛ icon and will be dedicated to illustrating how to use the Graph Editor for that particular customization.

If you want your graphs to look like the ones in the book, you can display them using the same schemes. See Appendix : Online supplements (482) for information about how to download the schemes used in this book. Once you have downloaded the schemes, you can then type the following commands in the Stata Command window:

```
. set scheme vg_s2c
. vguse allstates
. graph twoway scatter propval100 ownhome, msymbol(Sh)
```

After you issue the **set scheme vg_s2c** command, subsequent graph commands will show graphs with the **vg_s2c** scheme. You could also add the **scheme(vg_sc2)** option to the graph command to specify that the scheme be used just for that graph; for example,

```
. graph twoway scatter propval100 ownhome, msymbol(Sh) scheme(vg_s2c)
```

Generally, all commands and options are provided in their complete form. Commands and options are usually not abbreviated. However, for purposes of typing, you may want to use abbreviations. The previous example could have been abbreviated to

```
. gr tw sc propval100 ownhome, m(Sh)
```

The **gr** could have been omitted, leaving

```
. tw sc propval100 ownhome, m(Sh)
```

The `tw` also could have been omitted, leaving

```
. sc propval100 ownhome, m(Sh)
```

For guidance on appropriate abbreviations, consult [G-2] **graph**.

This book has been written based on the features available in Stata version 12.0. In the future, Stata may evolve to make the behavior of some of these commands change. If this happens, you can use the `version` command to make Stata run the graph commands as though they were run under version 12.0. For example, if you were running Stata version 13.0 but wanted a graph command to run as though you were running Stata 12.0, you could type

```
. version 12.0: graph twoway scatter propval100 ownhome
```

and the command would be executed as if you were running version 12.0. Or, perhaps you want a command to run as it did under Stata 11.2, you would then type

```
. version 11.2: graph twoway scatter propval100 ownhome
```

This book has a number of associated online resources to complement the book. Appendix : Online supplements (482) has more information about these online resources and how to access them. I strongly suggest that you install the online supplements, which make it easier to run the examples from the book. To install the supplemental programs, schemes, and help files, type from within Stata

```
. net from http://www.stata-press.com/data/vgsg
. net install vgsg
```

For an overview of what you have installed, type `help vgsg` within Stata. Then, with the `vguse` command, you can use any dataset from the book. Likewise, all the custom schemes used in the book will be installed into your copy of Stata, and you can use them to display the graphs, as described earlier in this section.

Finally, I would like to emphasize that the goal of this book is to help you learn and use the Stata graph commands and the Graph Editor for the purposes of creating graphs in Stata. I assume that you know the kind of graph you want to create and that you are turning to this book for advice on how to make that graph. I don't provide guidance on how to select the right kind of graph for visualizing your data or the merits of one graphical method over another. For such guidance, I would refer readers to books such as *The Visual Display of Quantitative Information, Second Edition* by Edward R. Tufte and *Visualizing Data* by William S. Cleveland, as well as your favorite statistical book.

1.2   Types of Stata graphs

Stata has a wide variety of graph types. This section introduces the types of graphs Stata produces, and it covers twoway plots (including scatterplots, line plots, fit plots, fit plots with confidence intervals, area plots, bar plots, range plots, and distribution plots), scatterplot matrices, bar charts, box plots, dot plots, and pie charts. Let's begin by exploring the variety of twoway plots that can be created with `graph twoway`. For this introduction, they are combined into six families of related plots: scatterplots and fit plots, line plots, area plots, bar plots, range plots, and distribution plots. Now let's turn to scatterplots and fit plots.

`graph twoway scatter propval100 popden`

Here is a basic scatterplot. The variable `propval100` is placed on the *y* axis, and `popden` is placed on the *x* axis. See Twoway : Scatter (89) for more details about these kinds of plots.
Uses allstates.dta & scheme vg_s2c

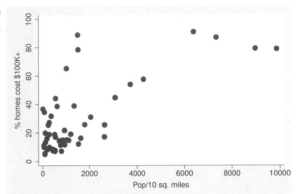

`twoway scatter propval100 popden`

We can start the previous command with just `twoway`, and Stata understands that this is shorthand for `graph twoway`.
Uses allstates.dta & scheme vg_s2c

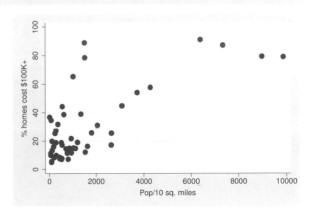

`twoway lfit propval100 popden`

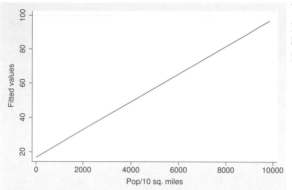

We now make a linear fit (`lfit`) line predicting `propval100` from `popden`. See Twoway : Fit (106) for more information about these kinds of plots.
Uses allstates.dta & scheme vg_s2c

`twoway (scatter propval100 popden) (lfit propval100 popden)`

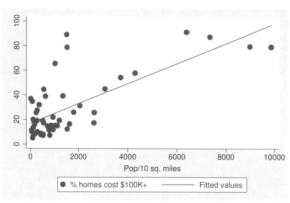

Stata allows us to overlay `twoway` graphs. In this example, we make a classic plot showing a scatterplot overlaid with a fit line by using the `scatter` and `lfit` commands. For more details about overlaying graphs, see Twoway : Overlaying (152).
Uses allstates.dta & scheme vg_s2c

`twoway (scatter propval100 popden) (lfit propval100 popden)`
`    (qfit propval100 popden)`

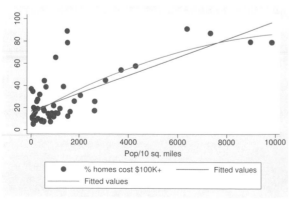

The ability to combine `twoway` plots is not limited to overlaying just two plots; we can overlay multiple plots. Here we overlay a scatterplot (`scatter`) with a linear fit (`lfit`) line and a quadratic fit (`qfit`) line.
Uses allstates.dta & scheme vg_s2c

```
twoway (scatter propval100 popden) (mspline propval100 popden)
    (fpfit propval100 popden) (mband propval100 popden)
    (lowess propval100 popden)
```

Stata has other kinds of fit methods in addition to linear and quadratic fits. This example includes a median spline (`mspline`), fractional polynomial fit (`fpfit`), median band (`mband`), and lowess (`lowess`). For more details, see Twoway : Fit (106).

Uses allstates.dta & scheme vg_s2c

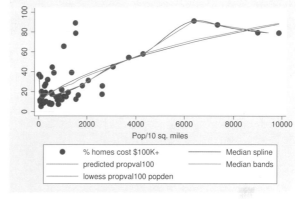

```
twoway (lfitci propval100 popden) (scatter propval100 popden)
```

In addition to being able to plot a fit line, we can plot a linear fit line with a confidence interval by using the `lfitci` command. We also overlay the linear fit and confidence interval with a scatterplot. See Twoway : CI fit (108) for more information about fit lines with confidence intervals.

Uses allstates.dta & scheme vg_s2c

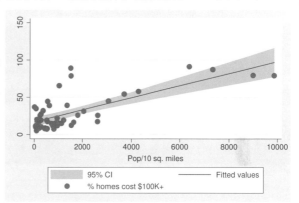

```
twoway dropline close tradeday
```

This `dropline` graph shows the closing prices of the S&P 500 by trading day for the first 40 days of 2001. A `dropline` graph is like a `scatter`plot because each data point is shown with a marker, but a dropline for each marker is shown as well. For more details, see Twoway : Scatter (89).

Uses spjanfeb2001.dta & scheme vg_s2c

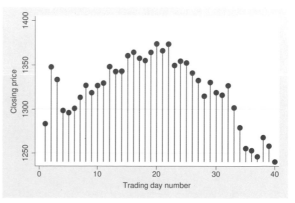

Using this book   Types of Stata graphs   Schemes   Options   Building graphs

Introduction   Editor   Twoway   Matrix   Bar   Box   Dot   Pie   Options   Standard options   Styles   Appendix

`twoway spike close tradeday`

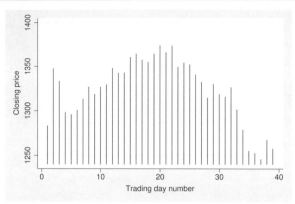

Here we use a `spike` plot to show the same graph as the previous one. It is like the `dropline` plot, but no markers are put on the top. For more details, see Twoway : Scatter (89).
Uses spjanfeb2001.dta & scheme vg_s2c

`twoway dot close tradeday`

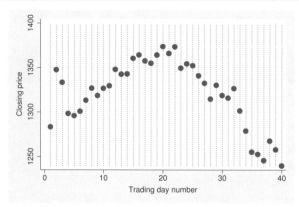

The `dot` plot, like the `scatter`plot, shows markers for each data point but also adds a dotted line for each of the x values. For more details, see Twoway : Scatter (89).
Uses spjanfeb2001.dta & scheme vg_s2c

`twoway line close tradeday, sort`

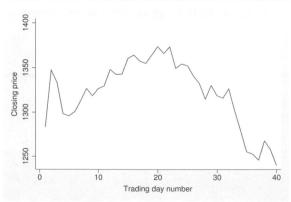

We use the `line` command in this example to make a simple line graph. See Twoway : Line (112) for more details about line graphs.
Uses spjanfeb2001.dta & scheme vg_s2c

`twoway connected close tradeday, sort`

This `twoway connected` graph is
similar to the `twoway line` graph,
except that a symbol is shown for each
data point. For more information, see
Twoway : Line (112).
Uses spjanfeb2001.dta & scheme vg_s2c

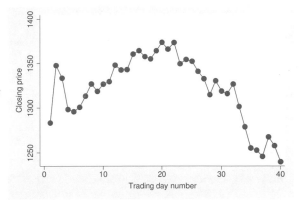

`twoway tsline close, sort`

The `tsline` (time-series line) command
makes a line graph where the x variable
is a date variable that has been
previously declared by using `tsset`; see
[TS] **tsset**. This example shows the
closing price of the S&P 500 by trading
date. For more information, see
Twoway : Line (112).
Uses sp2001ts.dta & scheme vg_s2c

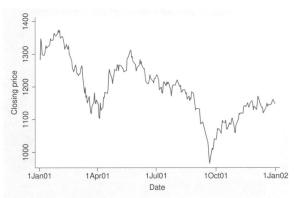

`twoway tsrline high low, sort`

The `tsrline` (time-series range line)
command makes a line graph showing
the high and low prices of the S&P 500
by trading date. For more information,
see Twoway : Line (112).
Uses sp2001ts.dta & scheme vg_s2c

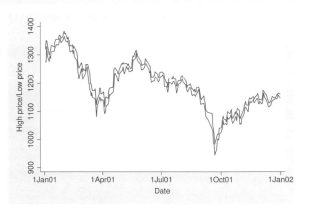

Introduction   Editor   Twoway   Matrix   Bar   Box   Dot   Pie   Options   Standard options   Styles   Appendix

Using this book   Types of Stata graphs   Schemes   Options   Building graphs

`twoway area close tradeday, sort`

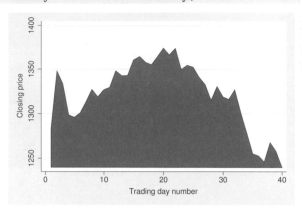

An `area` plot is similar to a `line` plot, but the area under the line is shaded. See Twoway : Area (119) for more information about area plots.
Uses spjanfeb2001.dta & scheme vg_s2c

`twoway bar close tradeday`

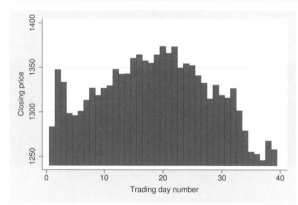

Here is an example of a `twoway bar` plot. For each x value, a bar is shown corresponding to the height of the y variable. This command shows a continuous x variable as compared with the `graph bar` command, which would be useful when you have a categorical x variable. See Twoway : Bar (121) for more details about bar plots.
Uses spjanfeb2001.dta & scheme vg_s2c

`twoway rarea high low tradeday, sort`

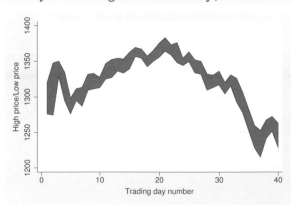

This example illustrates the use of `rarea` (range area) to graph the high and low prices with the area filled. If you used `rline` (range line), the area would not be filled. See Twoway : Range (123) for more details.
Uses spjanfeb2001.dta & scheme vg_s2c

twoway rconnected high low tradeday, sort

The `rconnected` (range connected) command makes a graph similar to the previous one, except that a marker is shown at each value of the x variable and the area between is not filled. If you instead used `rscatter` (range scatter), the points would not be connected. See Twoway : Range (123) for more details.

Uses spjanfeb2001.dta & scheme vg_s2c

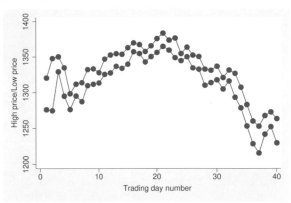

twoway rcap high low tradeday, sort

Here we use `rcap` (range cap) to graph the high and low prices with a spike and a cap at each value of the x variable. If you used `rspike` instead, spikes would be displayed but not caps. If you used `rcapsym`, the caps would be symbols that could be modified. See Twoway : Range (123) for more details.

Uses spjanfeb2001.dta & scheme vg_s2c

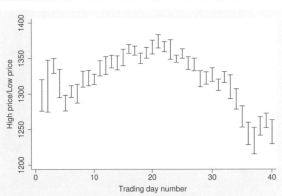

twoway rbar high low tradeday, sort

The `rbar` command graphs the high and low prices with bars at each value of the x variable. See Twoway : Range (123) for more details.

Uses spjanfeb2001.dta & scheme vg_s2c

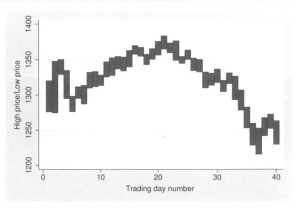

`twoway histogram popk, freq`

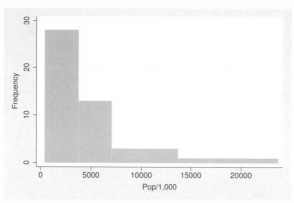

The `twoway histogram` command shows the distribution of one variable. It is often useful when overlaid with other twoway plots; otherwise, the `histogram` command would be preferable. See Twoway : Distribution (133) for more details.
Uses allstates.dta & scheme vg_s2c

`twoway kdensity popk`

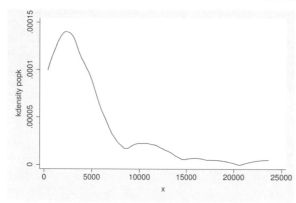

The `twoway kdensity` command shows a kernel-density plot and is useful for examining the distribution of one variable. It can be overlaid with other twoway plots; otherwise, the `kdensity` command would be preferable. See Twoway : Distribution (133) for more details.
Uses allstates.dta & scheme vg_s2c

`twoway function y=normden(x), range(-4 4)`

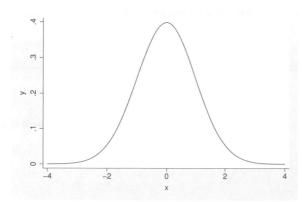

The `twoway function` command allows an arbitrary function to be drawn over a range of specified values. See Twoway : Distribution (133) for more details.
Uses allstates.dta & scheme vg_s2c

twoway contour depth northing easting

The twoway contour command creates contour plots representing three-dimensional data in two dimensions. See Twoway : Contour (141) for more details.

Uses sandstone.dta & scheme vg_s2c

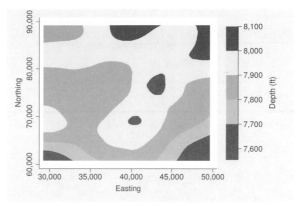

graph matrix propval100 rent700 popden

The graph matrix command shows a scatterplot matrix. See Matrix (161) for more details.

Uses allstates.dta & scheme vg_s2c

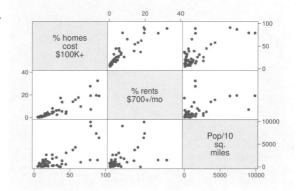

graph hbar popk, over(division)

This example shows how the graph hbar (horizontal bar) command is often used to show the values of a continuous variable broken down by one or more categorical variables. graph hbar is merely a rotated version of **graph bar**. See Bar (173) for more details.

Uses allstates.dta & scheme vg_s2c

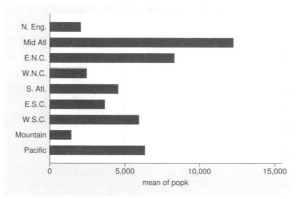

`graph hbox popk, over(division)`

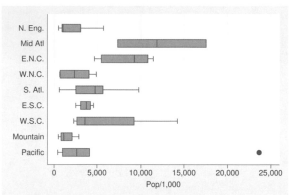

Here is the previous graph as a box plot by using the `graph hbox` (horizontal box) command, which is commonly used for showing the distribution of one or more continuous variables, broken down by one or more categorical variables. `graph hbox` is merely a rotated version of **graph box**. See Box (227) for more details.

Uses allstates.dta & scheme vg_s2c

`graph dot popk, over(division)`

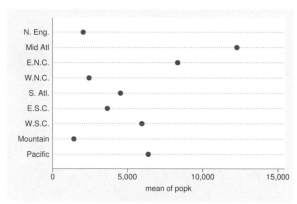

Here the previous plot is shown as a dot plot by using `graph dot`. Dot plots are often used to show one or more summary statistics for one or more continuous variables, broken down by one or more categorical variables. See Dot (263) for more details.

Uses allstates.dta & scheme vg_s2c

`graph pie popk, over(region)`

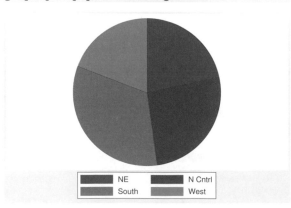

The `graph pie` command creates a pie chart. See Pie (289) for more details.

Uses allstates.dta & scheme vg_s2c

1.3 Schemes

Whereas the previous section was about the different types of graphs Stata can make, this section is about the different kinds of looks that you can have for Stata graphs. The basic starting point for the look of a graph is a scheme, which controls just about every aspect of the look of the graph. A scheme sets the stage for the graph, but you can use options to override the settings in a scheme. As you might surmise, if you choose (or develop) a scheme that produces graphs similar to the final graph you want to make, you can reduce the need to customize your graphs using options. This section gives you a basic idea of what schemes can do and introduces you to the schemes used throughout the book. See Intro : Using this book (1) for more details about how to select and use schemes and Appendix : Online supplements (482) for more information about how to download them.

`twoway scatter propval100 rent700 ownhome, scheme(vg_s1c)`

This scatterplot illustrates the vg_s1c scheme. It is based on the s1color scheme but increases the sizes of elements in the graph to make them more readable. This scheme is in color and has a white background, both inside the plot region and in the surrounding area.

Uses allstates.dta & scheme vg_s1c

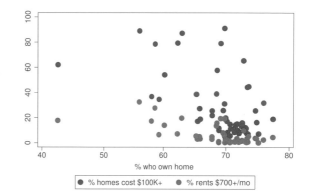

`twoway scatter propval100 rent700 ownhome, scheme(vg_s1m)`

This scatterplot is similar to the last one but uses the vg_s1m scheme, the monochrome equivalent of the vg_s1c scheme. It is based on the s1mono scheme but increases the sizes of elements in the graph to make them more readable. This scheme is in black and white and has a white background, both inside the plot region and in the surrounding area.

Uses allstates.dta & scheme vg_s1m

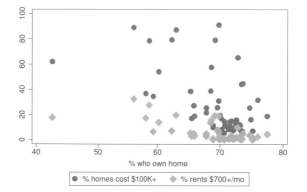

```
graph hbox wage, over(grade) asyvar nooutsides legend(rows(2))
    scheme(vg_s2c)
```

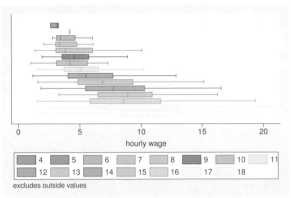

This box plot shows an example of the vg_s2c scheme. It is based on the s2color scheme but increases the sizes of elements in the graph to make them more readable. In this scheme, the plot region has a white background, but the surrounding area (the graph region) is light blue.

Uses nlsw.dta & scheme vg_s2c

```
graph hbox wage, over(grade) asyvar nooutsides legend(rows(2))
    scheme(vg_s2m)
```

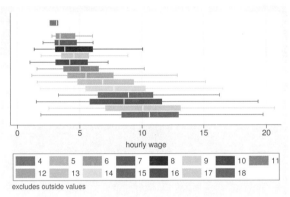

This box plot is similar to the previous one but uses the vg_s2m scheme, the monochrome equivalent of the vg_s2c scheme. This scheme is based on the s2mono scheme but increases the sizes of elements in the graph to make them more readable. This scheme is in black and white, and it has a white background in the plot region but is light gray in the surrounding graph region.

Uses nlsw.dta & scheme vg_s2m

```
graph hbar wage, over(occ7, label(nolabels)) blabel(group, position(base))
    scheme(vg_palec)
```

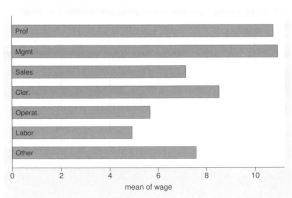

This horizontal bar chart shows an example of the vg_palec scheme. It is based on the s2color scheme but makes the colors of the bars/boxes/markers paler by decreasing the intensity of the colors. As shown in this example, one use of this scheme is to make the colors of the bars pale enough to include text labels inside bars.

Uses nlsw.dta & scheme vg_palec

```
graph hbar wage, over(occ7, label(nolabels)) blabel(group, position(base))
    scheme(vg_palem)
```

This example is the same as the last one but uses the `vg_palem` scheme, the monochrome equivalent of the `vg_palec` scheme. This scheme is based on the `s2mono` scheme but makes the colors of the bars/boxes/markers paler by decreasing the intensity of the colors.

Uses nlsw.dta & scheme vg_palem

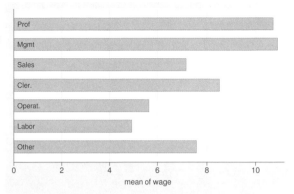

```
scatter propval100 rent700 ownhome, scheme(vg_outc)
```

This scatterplot illustrates the `vg_outc` scheme. It is based on the `s2color` scheme but makes the fill color of the bars/boxes/markers white, so they appear hollow. The plot region is a light blue to contrast with the white fill color. This scheme is useful to see the number of markers present where numerous markers are close or partially overlapping.

Uses allstates.dta & scheme vg_outc

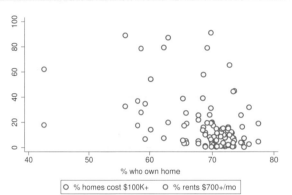

```
scatter propval100 rent700 ownhome, scheme(vg_outm)
```

This example is similar to the previous one but illustrates the `vg_outm` scheme, the monochrome equivalent of the `vg_outc` scheme. It is based on the `s2mono` scheme but makes the fill color of the bars/boxes/markers white, so they appear hollow.

Uses allstates.dta & scheme vg_outm

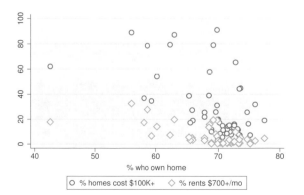

Introduction   Editor   Twoway   Matrix   Bar   Box   Dot   Pie   Options   Standard options   Styles   Appendix

Using this book   Types of Stata graphs   Schemes   Options   Building graphs

```
twoway (scatter ownhome borninstate if stateab=="DC", mlabel(stateab))
     (scatter ownhome borninstate), legend(off) scheme(vg_samec)
```

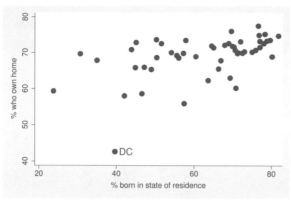

This is an example of the vg_samec scheme, which is based on the s2color scheme and makes all the markers, lines, bars, etc., the same color, shape, and pattern. Here the second scatter command labels Washington DC, which normally would be shown in a different color; with this scheme, the marker is the same. This scheme has a monochrome equivalent called vg_samem, which is not illustrated.
Uses allstates.dta & scheme vg_samec

```
graph hbar commute, over(division) asyvar scheme(vg_lgndc)
```

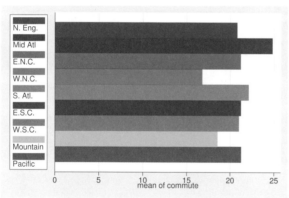

This horizontal bar chart shows an example of the vg_lgndc scheme. It is based on the s2color scheme but changes the default attributes of the legend, namely, showing the legend in one column to the left of the plot region, with the key and symbols placed atop each other. It can be efficient to place the legend to the left of the graph. This scheme has a monochrome equivalent called vg_lgndm, which is not illustrated here.
Uses allstates.dta & scheme vg_lgndc

```
graph bar commute, over(division) asyvar legend(rows(3)) scheme(vg_past)
```

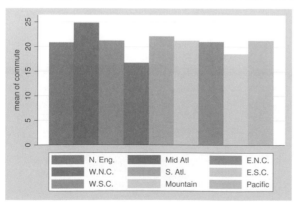

This bar chart shows an example of the vg_past scheme. It is based on the s2color scheme but selects subdued pastel colors and provides a sand background for the surrounding graph region and an eggshell color for the inner plot region and legend area.
Uses allstates.dta & scheme vg_past

`twoway scatter rent700 propval100,` `scheme(vg_rose)`

This bar chart shows an example of the
vg_rose scheme. It is based on the
s2color scheme but uses a different set
of colors for the background (eggshell)
and for the plot area (a light rose
color). By default, the grid lines are
omitted and the labels for the y axis
are horizontal.

Uses allstates.dta & scheme vg_rose

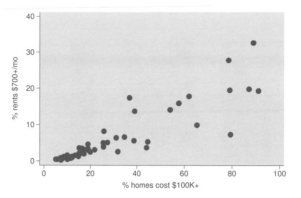

`graph bar commute, over(division) asyvar legend(rows(3))` `scheme(vg_blue)`

This bar chart shows an example of the
vg_blue scheme. It is based on the
s2color scheme but uses a set of blue
colors, with a light blue background
and a light blue-gray color for the plot
area. By default, the grid lines are
omitted and the labels for the y axis
are horizontal.

Uses allstates.dta & scheme vg_blue

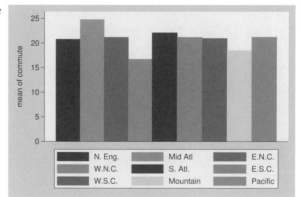

`graph bar commute, over(division) asyvar legend(rows(3))` `scheme(vg_teal)`

This is an example of the vg_teal
scheme. This scheme is also based on
the **s2color** scheme but uses an
olive-teal background. It also
suppresses the display of grid lines and
makes the labels for the y axis display
horizontally by default.

Uses allstates.dta & scheme vg_teal

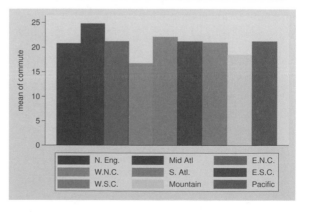

```
graph bar commute, over(division) asyvar legend(rows(3)) scheme(vg_brite)
```

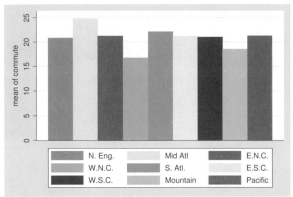

This bar chart shows an example of the vg_brite scheme. It is based on the s2color scheme but selects a bright set of colors and changes the background to light khaki.

Uses allstates.dta & scheme vg_brite

This section has just scratched the surface of all there is to know about schemes in Stata. I hope that it helps you see how schemes create a starting point for your graph and that, by choosing a scheme that is most similar to the look you want, you can save time and effort in customizing your graphs.

1.4   Options

Learning to create effective Stata graphs is ultimately about using options to customize the look of a graph until you are pleased with it. This section illustrates the general rules and syntax for Stata graph commands, starting with their basic structure and followed by illustrations showing how options work in the same way across different kinds of commands. Stata graph options work much like other options in Stata; however, there are more features that extend their power and functionality. Although these examples will use the twoway scatter command for illustration, most of the principles illustrated extend to all kinds of Stata graph commands. Although the emphasis of this section is on the use of Stata commands for customizing graphs, you could accomplish many of these customizations by using the Graph Editor; see Editor (35) for more information.

`twoway scatter propval100 rent700`

Consider this basic scatterplot. To add
a title to this graph, we can use the
`title()` option as illustrated in the
next example.
Uses allstates.dta & scheme vg_s2c

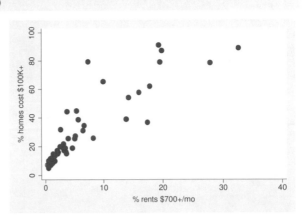

`twoway scatter propval100 rent700,`
    `title("This is a title for the graph")`

Just as with any Stata command, the
`title()` option comes after a comma,
and here it contains a quoted string
that becomes the title of the graph.
Uses allstates.dta & scheme vg_s2c

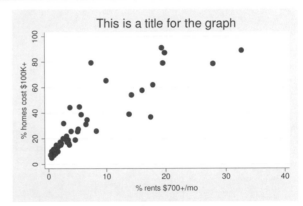

`twoway scatter propval100 rent700,`
    `title("This is a title for the graph", box)`

In this example, we add `box` as an
option within `title()` to place a box
around the title. If the default for the
current scheme had included a box,
then you could have used the `nobox`
option to suppress it.
Uses allstates.dta & scheme vg_s2c

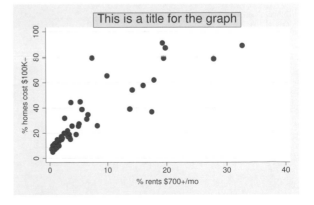

```
twoway scatter propval100 rent700,
   title("This is a title for the graph", box size(small))
```

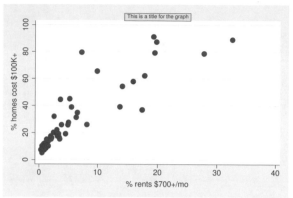

Let's take the last graph and modify the title to make it small. We add `size(small)` to the `title()` option to change the title's size, as well as the box size, to small.
Uses allstates.dta & scheme vg_s2c

```
twoway scatter propval100 rent700,
   title("This is a title for the graph", box size(small))
   msymbol(S)
```

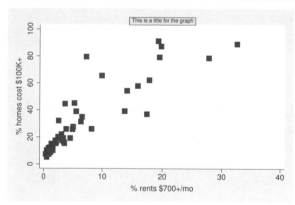

Say that we want the symbols to be displayed as squares. We add another option, `msymbol(S)`, to indicate that we want the marker symbol to be displayed as a square (`S` for square). Adding one option at a time is a common way to build a Stata graph. In the next graph, we will change gears and start building a new graph to show other aspects of options.
Uses allstates.dta & scheme vg_s2c

```
twoway scatter propval100 rent700
```

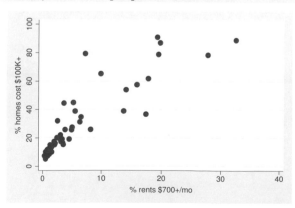

Let's return to this simple scatterplot. Say that we want the labels for the x axis to change from 0 10 20 30 40 to 0 5 10 15 20 25 30 35 40.
Uses allstates.dta & scheme vg_s2c

`twoway scatter propval100 rent700, xlabel(0(5)40)`

Here we add the `xlabel()` option to label the *x* axis from 0 to 40, incrementing by 5. But say that we want the labels to be displayed larger.
Uses allstates.dta & scheme vg_s2c

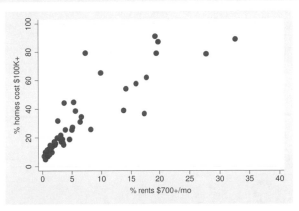

`twoway scatter propval100 rent700, xlabel(0(5)40, labsize(huge))`

Here we add the `labsize()` (label size) option to increase the size of the labels for the *x* axis. Now say that we were happy with the original numbering (0 10 20 30 40) but wanted the labels to be huge.
Uses allstates.dta & scheme vg_s2c

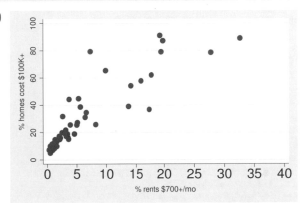

`twoway scatter propval100 rent700, xlabel(, labsize(huge))`

The `xlabel()` option we use here indicates that we are content with the numbers chosen for the label of the *x* axis because we have nothing before the comma. After the comma, we add the `labsize()` option to increase the size of the labels for the *x* axis.
Uses allstates.dta & scheme vg_s2c

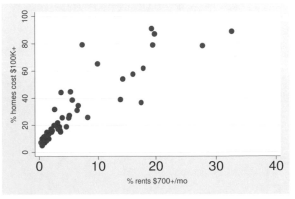

Introduction

Using this book   Editor

Types of Stata graphs   Twoway

Schemes   Matrix

Options   Bar

Building graphs   Box

Dot

Pie

Options

Standard options

Styles

Appendix

Now let's consider some examples using the `legend()` option to show that some options do not require or permit the use of commas within them. Also, the following examples show where we might properly specify an option repeatedly.

`twoway scatter propval100 rent700 popden`

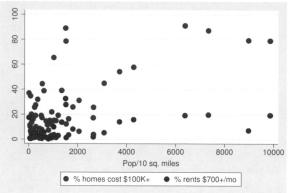

Here we show two y variables, `propval100` and `rent700`, graphed against population density, `popden`. Stata has created a legend, helping us see which symbols correspond to which variables. We can use the `legend()` option to customize the legend.

Uses allstates.dta & scheme vg_s2c

`twoway scatter propval100 rent700 popden, legend(cols(1))`

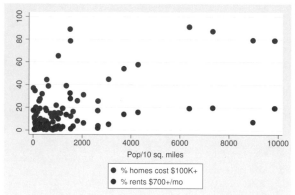

By using the `legend(cols(1))` option, we make the legend display in one column. We did not use a comma because, with the `legend()` option, there is no natural default argument. If we had included a comma within the `legend()` option, Stata would have reported this as an error.

Uses allstates.dta & scheme vg_s2c

```
twoway scatter propval100 rent700 popden,
   legend(cols(1) label(1 "Property Value"))
```

This example adds another option,
`label()`, within the `legend()` option to
change the label for the first variable.
Uses allstates.dta & scheme vg_s2c

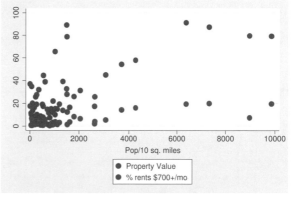

```
twoway scatter propval100 rent700 popden,
   legend(cols(1) label(1 "Property Value") label(2 "Rent"))
```

Here we add another `label()` option
within the `legend()` option; this option
changes the label for the second
variable. We can use the `label()`
option repeatedly to change the labels
for the different variables.
Uses allstates.dta & scheme vg_s2c

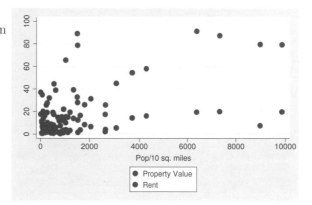

Finally, let's consider an example that shows how to use the `twoway` command to overlay
two plots. The following examples show how each graph can have its own options and how
options can apply to the overall graph.

Introduction   Editor   Twoway   Matrix   Bar   Box   Dot   Pie   Options   Standard options   Styles   Appendix

Using this book   Types of Stata graphs   Schemes   Options   Building graphs

```
twoway (scatter propval100 popden)
       (lfit propval100 popden)
```

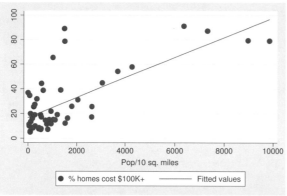

This graph shows a scatterplot predicting property value from population density and shows a linear fit between these two variables. Say that we wanted to change the symbol displayed in the scatterplot and the thickness of the line for the linear fit. *Uses allstates.dta & scheme vg_s2c*

```
twoway (scatter propval100 popden, msymbol(S))
       (lfit propval100 popden, lwidth(vthick))
```

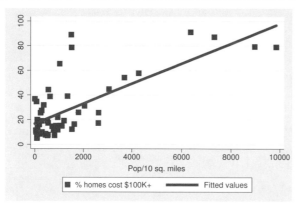

We add the `msymbol()` option to the `scatter` command to change the symbol to a square, and we add the `lwidth()` (line width) option to the `lfit` command to make the line very thick. When we overlay two plots, each plot can have its own options that operate on its respective parts of the graph. However, some parts of the graph are shared, for example, the title. *Uses allstates.dta & scheme vg_s2c*

```
twoway (scatter propval100 popden, msymbol(S))
       (lfit propval100 popden, lwidth(vthick)),
       title("This is the title of the graph")
```

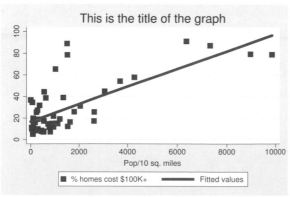

We add the `title()` option to the end of the command, placed after a comma. That final comma signals that options concerning the overall graph are to follow, here, the `title()` option. *Uses allstates.dta & scheme vg_s2c*

One of the beauties of Stata graph commands is the way that different graph commands share common options. If you want to customize the display of a legend, you do it by using the same options, whether you are using a bar graph, a box plot, a scatterplot, or any other kind of Stata graph. Once you learn how to control legends with one type of graph, you have learned how to control legends for all types of graphs. This is illustrated with a couple of examples.

`twoway scatter propval100 rent700 popden, legend(position(1))`

Consider this scatterplot. We add the `legend()` option to make the legend display in the one o'clock position on the graph, putting the legend in the top right corner.

Uses allstates.dta & scheme vg_s2c

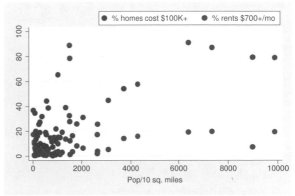

`graph bar propval100 rent700, over(nsw) legend(position(1))`

Here we use the **graph bar** command, which is a completely different command from the previous one. Even though the graphs are different, the `legend()` option we supply is the same and has the same effect. Many (but not all) options function in this way, sharing a common syntax and having common effects.

Uses allstates.dta & scheme vg_s2c

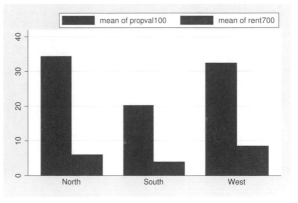

```
graph matrix propval100 rent700 popden, legend(position(1))
```

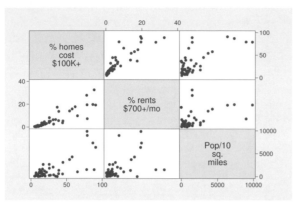

Compare this example with the previous two. The **graph matrix** command does not support the **legend()** option because this graph does not need or produce legends. In the Matrix (161) chapter, for example, there are no references to legends, an indication that this is not a relevant option for this kind of graph. Even though we included this irrelevant option, Stata ignored it and produced an appropriate graph anyway.

Uses allstates.dta & scheme vg_s2c

Because legends work the same way with different types of Stata graph commands, only one place discusses legends in detail: Options : Legend (361). However, it is useful to see examples of legends for each type of graph that uses them. Each chapter, therefore, includes a brief section describing legends for each type of graph discussed in that chapter. Likewise, Options (307) describes most options in detail, with a brief section in every chapter discussing how each option works for that specific type of graph. As shown for legends, some options are not appropriate for some types of graphs, so those options are not discussed with the commands that do not support them.

Although you can use an option like **legend()** with many, but not all, kinds of Stata graph commands, you can use other kinds of options with almost every kind of Stata graph. These are called *standard options*. To help you differentiate these kinds of options, they are discussed in their own chapter, Standard options (395). Because these options can be used with most types of graph commands, they are generally not discussed in the chapters about the different types of graphs, except when their usage interacts with the options illustrated. For example, **subtitle()** is a standard option, but its behavior takes on a special meaning when used with the **legend()** option, so the **subtitle()** option is discussed in the context of legends. Consistent with what was previously shown, the syntax of standard options follows the same kinds of rules that have been illustrated, and their usage and behavior are uniform across the many types of Stata graph commands.

1.5 Building graphs

I have three agendas in writing this section. First, I wish to show the process of building complex graphs a little bit at a time. At the same time, I illustrate how to use the resources of this book to get the bits of information needed to build these graphs. Finally, I hope to show that, even though a complete Stata graph command might look complicated and overwhelming, the process of building the graph slowly is actually straightforward and logical.

Let's first build a bar chart that looks at property values broken down by region of the country. Then we will modify the legend and bar characteristics, add titles, and so forth.

Although this section emphasizes the process of building and customizing graphs with Stata commands, you could accomplish many of these customizations by using the Graph Editor; see Editor (35) for more information.

graph display

Say that we want to create this graph. For now, the syntax is concealed, just showing the **graph display** command to show the previously drawn graph. It might be overwhelming at first to determine all the options needed to make this graph. To ease our task, we will build it a bit at a time, refining the graph and fixing any problems we find.
Uses allstates.dta & scheme vg_past

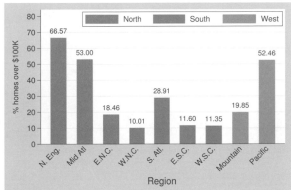

graph bar propval100, over(nsw) over(division)

We begin by seeing that this is a bar chart and look at Bar : Y-variables (173) and Bar : Over (177). We take our first step toward making this graph by making a bar chart showing **propval100** and adding over(nsw) and over(division) to break down the means by **nsw** and **division**.
Uses allstates.dta & scheme vg_past

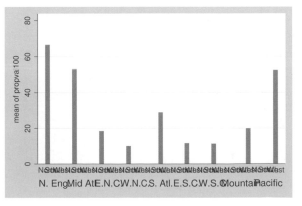

`graph bar propval100, over(nsw) over(division) nofill`

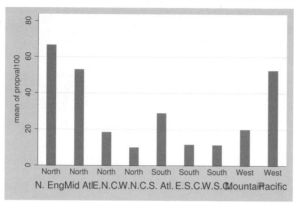

The previous graph is not quite what we want because we see every `division` shown with every `nsw`, but for example, the Pacific region only appears in the West. In Bar : Over (177), we see that we can add the `nofill` option to show only the combinations of `nsw` and `division` that exist in the data file. Next we will look at the colors of the bars.

Uses allstates.dta & scheme vg_past

`graph bar propval100, over(nsw) over(division) nofill asyvars`

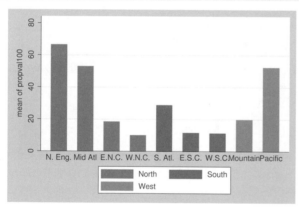

The last graph is getting closer, but we want the bars for North, South, and West displayed in different colors and labeled with a legend. In Bar : Y-variables (173), we see that the `asyvars` option will accomplish this. Next we will change the title for the y axis.

Uses allstates.dta & scheme vg_past

`graph bar propval100, over(nsw) over(division) nofill asyvars`
`    ytitle("% homes over $100K")`

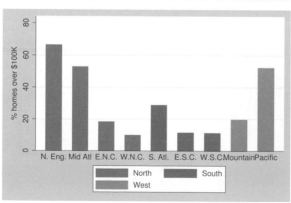

Now we want to put a title on the y axis. In Bar : Y-axis (213), we see examples illustrating the use of `ytitle()` for putting a title on the y axis. Here we put a title on the y axis, but now we want to change the labels for the y axis to go from 0 to 80, incrementing by 10.

Uses allstates.dta & scheme vg_past

```
graph bar propval100, over(nsw) over(division) nofill asyvars
    ytitle("% homes over $100K") ylabel(0(10)80, angle(0))
```

The Bar : Y-axis (213) section also tells us about the ylabel() option. In addition to changing the labels, we want to change the angle of the labels. From the Bar : Y-axis (213) section, we see that we can use the angle() option to change the angle of the labels. Now that we have the y axis labeled as we want, let's next look at the title for the x axis.

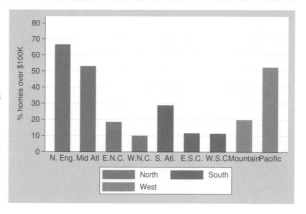

Uses allstates.dta & scheme vg_past

```
graph bar propval100, over(nsw) over(division) nofill asyvars
    ytitle("% homes over $100K") ylabel(0(10)80, angle(0)) b1title(Region)
```

After having used the ytitle() option to label the y axis, we might be tempted to use the xtitle() option to label the x axis, but this axis is a categorical variable. In Bar : Cat axis (193), we see that this axis is treated differently because of that. To put a title below the graph, we use the b1title() option. Now let's turn our attention to formatting the legend.

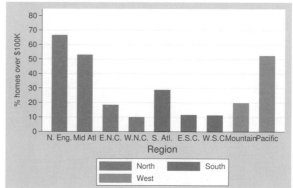

Uses allstates.dta & scheme vg_past

```
graph bar propval100, over(nsw) over(division) nofill asyvars
    ytitle("% homes over $100K") ylabel(0(10)80, angle(0)) b1title(Region)
    legend(rows(1) position(1) ring(0))
```

Here we want to use the legend() option to make the legend have one row in the top right corner within the plot area. In Bar : Legend (201), we see that the rows(1) option makes the legend appear in one row and that the position(1) option puts the legend in the one o'clock position. The ring(0) option puts the legend inside the plot region. Next let's label the bars.

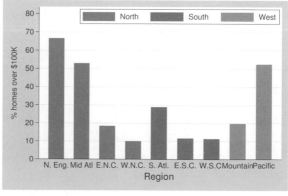

Uses allstates.dta & scheme vg_past

```
graph bar propval100, over(nsw) over(division) nofill asyvars
    ytitle("% homes over $100K") ylabel(0(10)80, angle(0)) b1title(Region)
    legend(rows(1) position(1) ring(0)) blabel(bar)
```

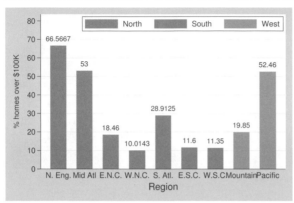

We want each bar labeled with the height of the bar. Bar: Legend (201) shows how we can do this by using the blabel() (bar label) option to label the bars in lieu of legends. blabel(bar) labels the bars with their height.
Uses allstates.dta & scheme vg_past

```
graph bar propval100, over(nsw) over(division) nofill asyvars
    ytitle("% homes over $100K") ylabel(0(10)80, angle(0)) b1title(Region)
    legend(rows(1) position(1) ring(0)) blabel(bar, format(%4.2f))
```

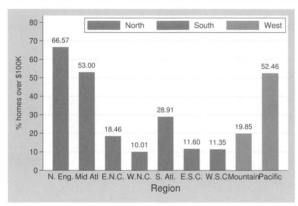

We want the label for each bar to end in two decimal places, and we see in Bar: Legend (201) that we can use the format() option to format these numbers as we want.
Uses allstates.dta & scheme vg_past

```
graph bar propval100, over(nsw) over(division, label(angle(45))) nofill
    ytitle("% homes over $100K") ylabel(0(10)80, angle(0)) b1title(Region)
    legend(rows(1) position(1) ring(0)) blabel(bar, format(%4.2f)) asyvars
```

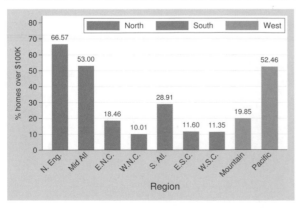

Finally, in Bar: Cat axis (193), we see that we can add the label(angle(45)) option to the over() option to specify that labels for that variable be shown at a 45-degree angle so they do not overlap each other.
Uses allstates.dta & scheme vg_past

I hope this section has shown that it is not that difficult to create complex graphs by building them one step at a time. You can use the resources in this book to seek out each piece of information you need and then put those pieces together the way you want to create your own graphs. For more information about how to integrate options to create complex Stata graphs, see Appendix : More examples (466).

2   Editor

The Graph Editor gives you a variety of tools and techniques for modifying your graphs. Even if you are a seasoned veteran of using Stata `graph` commands, you will find times where the Graph Editor offers a more elegant solution and even times when the Graph Editor offers you options not available with `graph` commands. In fact, this chapter concludes with a section that compares the Graph Editor with `graph` commands to help identify situations where one tool may be preferable to the other; see Editor : Editor vs. commands (82).

This chapter starts with a quick overview of the Graph Editor, followed by a section on the Object Browser [see Editor : Browser (43)], one of the powerful tools within the Graph Editor. Then the heart of this chapter covers the types of tasks that you can perform using the Graph Editor:

- Modifying the way objects such as markers, titles, axes, and colors look in the graph; see Editor : Modifying (47).
- Adding objects such as text, lines, and markers to the graph; see Editor : Adding (54).
- Moving objects such as titles, added text, added lines, legends, and axes; see Editor : Moving (62).
- Hiding or showing objects such as titles, added text, added lines, legends, and axes; see Editor : Hiding/Showing (73).
- Locking objects to prevent yourself from accidentally moving or modifying them; see Editor : Locking/Unlocking (76).
- Using the Graph Recorder to record and save customizations, permitting you to automate the process of applying such customizations by simply *playing back* the actions you recorded; see Editor : Graph Recorder (78).

As mentioned before, this chapter concludes with a section that compares the Graph Editor with Stata graph commands; see Editor : Editor vs. commands (82).

For this chapter, I recommend that you read each section from beginning to end (rather than jumping around), and I strongly encourage you to replicate the examples yourself in Stata. The examples are designed to touch upon each of the major features of the Graph Editor, so by replicating the examples, you will see each feature in action for yourself.

Finally, I would like to note that many examples used in this chapter are trivial or silly because they focus solely on showing how the Graph Editor works.

2.1   Overview of the Graph Editor

Figure 2.1 shows the Graph Editor annotated to point out its major elements and tools.

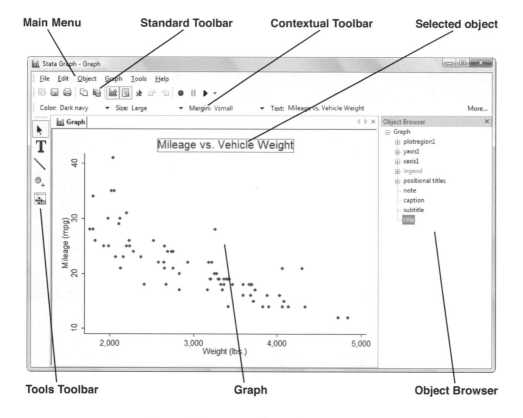

Figure 2.1: Stata Graph Editor

This section illustrates how these tools work by showing them in action. Let's begin by typing the following commands to make a graph, and then we will learn how to start the Graph Editor.

```
. vguse allstates
. graph twoway scatter propval100 ownhome
```

Starting and stopping the Graph Editor

Once you have created a graph, you can start the Graph Editor in one of three ways: 1) by selecting **File** and then **Start Graph Editor** from the Graph window menu, 2) by clicking on the Start Graph Editor ⌧ icon in the Graph window toolbar, or 3) by right-clicking on the graph and selecting **Start Graph Editor**. Once you have started the Graph Editor, you can stop it in the same three ways: by selecting **File** and then **Stop Graph Editor** from the Graph window menu, by clicking on the Stop Graph Editor ⌧ icon in the Graph window toolbar, or by right-clicking on the graph and selecting **Stop Graph Editor**.

Contextual Toolbar

One of the most handy features of the Graph Editor is the Contextual Toolbar (see figure 2.1). If you have not already done so, run the commands from the previous subsection to create the scatterplot, and then start the Graph Editor. Using your mouse, click on the title of the y axis (% homes cost $100K+). Notice the toolbar that appears above the graph with options to modify the **Color**, **Size**, **Margin**, and **Text** of the selected object (here, the title of the y axis). This toolbar is called the Contextual Toolbar because it changes based on the context of what you have selected. Now click on a y-axis label (e.g., 20). Notice the Contextual Toolbar changes, offering choices for modifying the labels for the y axis. Try clicking on one of the markers in the scatterplot, and note how the Contextual Toolbar now offers you choices appropriate for modifying the markers. In fact, every time you click on a different object within the graph, the Contextual Toolbar will update with options for modifying that object. Let's try this in the following examples.

`scatter propval100 ownhome`

Example 1 of 2. We can modify the title of the x axis by clicking on it. Within the Contextual Toolbar change the **Size** to Huge. The size of the axis label immediately becomes huge.
Uses allstates.dta & scheme vg_s2c

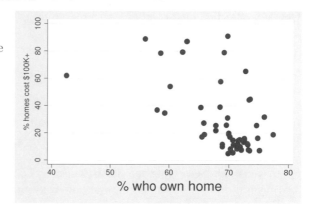

`scatter propval100 ownhome`

Example 2 of 2. Let's place a box around the title. We do not see any such choice in the Contextual Toolbar, but note the **More...** button at the far right of that toolbar. Click the **More...** button and a dialog box is displayed offering the full compliment of options for modifying this object (the x-axis title). From the *Box* tab, check **Place box around text** and then click on **Apply**. The change happens immediately, and the dialog box remains open so other modifications can be made. Click on **OK**.
Uses allstates.dta & scheme vg_s2c

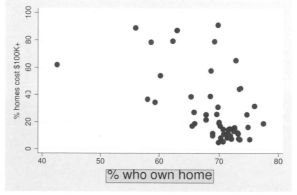

Introduction   Editor   Twoway   Matrix   Bar   Box   Dot   Pie   Options   Standard options   Styles   Appendix

Overview   Browser   Modifying   Adding   Moving   Hiding/Showing   Locking/Unlocking   Graph Recorder   Editor vs. commands

Standard Toolbar

Now I would like to call your attention to the Standard Toolbar:

The Standard Toolbar (see figure 2.1) includes many familiar icons common to many applications, namely, ⊞ to open a graph, ⊟ to save a graph, ⊜ to print a graph, ⧉ to copy a graph to the clipboard (for pasting into other applications), ☞ to undo the last modification, and ↻ to redo the last undo operation.

Start/Stop Graph Editor. The Start/Stop Graph Editor ⊮ icon starts the Graph Editor if it is currently stopped and stops the Graph Editor if it is currently started.

Deselect object. Once you have selected an object, you can click on the Deselect object ✶ icon to deselect the currently selected object.

Save and Open graphs. You might like the changes you made to the previous graph. These changes simply live in memory and will be lost if you do not save them. You can save the changes permanently by using the Save ⊟ icon or by selecting **File** and then **Save As...** from the Main Menu. To retrieve a saved graph, close the Graph Editor, and either use the Open ⊞ icon or select **File** and then **Open** from the Graph window or **Open...** from the main Stata window menu. You can edit any *live* graph, including those created in previous Stata versions. See Appendix : Save/Redisplay/Combine (456) for more details about saving and using graphs, as well as more about live graphs.

Undo and Redo. Let's illustrate the use of these icons by clicking on a marker and then from within the Contextual Toolbar changing the **Color** to Red, the **Size** to Small and the **Symbol** to Large X (in that order). Now you can click on the Undo ☞ icon to undo the **Symbol** change, click on Undo ☞ again to undo the change in **Size**, and click on Undo ☞ yet again to undo the change in **Color**. If you preferred the color change before the Undo process, you can click on Redo ↻ and the color will revert back to the way it was before the Undo operation. The Undo/Redo process can take a while because Stata must redraw the entire graph. Sometimes it is faster to simply change the setting back to the way it was.

Recorder. The Standard Toolbar includes icons that allow you to play ▶ a graph recording and to begin ● a graph recording; when a recording is in progress, the ● icon ends a recording. The pause ‖ icon permits you to pause a graph recording, and then clicking on the same icon again resumes the recording process. These are discussed in more detail in Editor : Graph Recorder (78).

Graph preferences. Let's take a brief detour to look at preferences that can be set for graphs. From the Main Menu of the Graph Editor, select **Edit** and then **Preferences...**. In the *General* tab, you can set the **Scheme** and **Font**. You can also check **Create multiple graphs as tabs in a single window**, which as the title implies will display all your graphs in one window with the ability to tab among the graphs (rather than the default of creating a separate window for every graph). In the *Printer* tab, there are options to change the **Font**, **Color mapping**, **Magnifications**, and **Margins**. You can also check **Print logo** to display the Stata logo on your graphs. In the *Clipboard* tab, there are similar preferences to control the image that is copied from Stata and pasted into another application. These preferences are saved permanently, but you can change them at any time.

Print and Copy. We might want to print or copy and paste the previous graph into another application. Before doing so, I suggest checking the printer and/or clipboard preferences; see previous paragraph. Once you have set your preferences, you can use the Print ⊖ icon to print your graph or the Copy ⧉ icon to copy your graph to the clipboard for pasting into other applications.

Schemes. Before saving, printing, or copying a graph to the clipboard, you may want to select a different scheme. To change the scheme, you must first stop the Graph Editor (if started), and then from the Graph window menu you can select **Edit** and then **Apply New Scheme**. From the dialog box that appears, select the **New scheme** for the graph and click on **OK**. You can then display, save, print, or copy the graph with the new scheme applied.

Others. The Standard Toolbar also includes a Rename 🖫 icon to rename your graph (the equivalent of the `graph rename` command). The Hide Object Browser 🖻 icon hides/shows the Object Browser. (I suggest you always leave the Object Browser displayed.)

Tools Toolbar

At the left of the graph, you can find the Tools Toolbar (see figure 2.1), which contains the Pointer ⬉ icon, Add Text **T** icon, Add Line ＼ icon, Add Marker •+ icon, and Grid Edit 🖫 icon. By default, the Pointer ⬉ icon is selected, which allows you to select and move objects. Let's illustrate these tools using the following examples.

Introduction Overview Browser Editor Modifying Twoway Adding Matrix Moving Bar Hiding/Showing Box Locking/Unlocking Dot Graph Recorder Pie Editor vs. commands Options Standard options Styles Appendix

`scatter propval100 ownhome, title(My Title)`

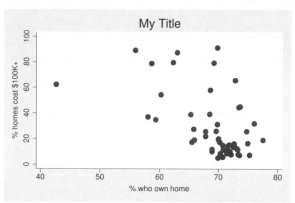

Example 1 of 5. We return to this simple scatterplot.
Uses allstates.dta & scheme vg_s2c

`scatter propval100 ownhome, title(My Title)`

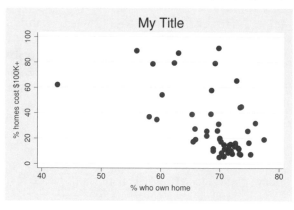

Example 2 of 5. With the Pointer ⬉ tool selected, we can select objects by clicking on them. When we do so, the Contextual Toolbar appears, allowing us to modify the properties of the object. We can also double-click on objects to modify their properties. For example, double-click on the title and then change the **Size** to Huge. This brings up the same dialog box that you would get if you chose **More...** from the Contextual Toolbar.
Uses allstates.dta & scheme vg_s2c

`scatter propval100 ownhome, title(My Title)`

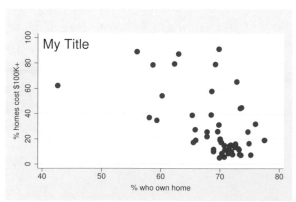

Example 3 of 5. The Pointer ⬉ icon can be used to move an object. Position the cursor over the title (i.e., My Title), and click and hold and drag the title to a new location. Drag the title to the top left corner inside the graph. There is a big gap where the title was. Let's try this another way.
Uses allstates.dta & scheme vg_s2c

`scatter propval100 ownhome, title(My Title)`

Example 4 of 5. Click on the Undo ☞ icon to put the title back where it was. Now select the Grid Edit 🔲 icon from the Tools Toolbar. Click and hold the title and drag it into the center of the plot region until the plot region is shown in light red with thick red lines bordering it. Release the mouse and this places the title in the center of the plot region and removes the space used in the title section.
Uses allstates.dta & scheme vg_s2c

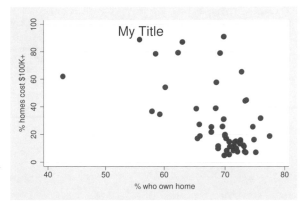

`scatter propval100 ownhome, title(My Title)`

Example 5 of 5. Now return to the Pointer ➤ icon. Click and hold the title and drag it to the top left corner of the plot region. In fact, we can drag the title anywhere we like within the plot region. More details about moving objects (including more about the use of the Grid Edit 🔲 tool) is described in Editor : Moving (62).
Uses allstates.dta & scheme vg_s2c

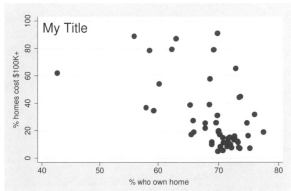

`twoway (scatter ownhome borninstate) (lfit ownhome borninstate)`

Example 1 of 4. Let's consider some examples that illustrate adding lines and text to a graph. This graph shows a scatterplot with a fitted line, and the fitted line seems highly influenced by the observation in the lower left corner. Inspection of the data shows that point to be DC. Let's call attention to this point.
Uses allstates.dta & scheme vg_s2c

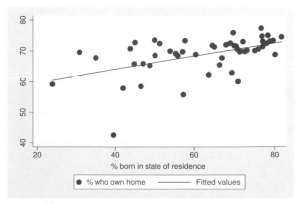

```
twoway (scatter ownhome borninstate) (lfit ownhome borninstate)
```

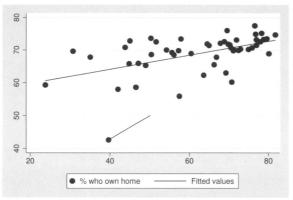

Example 2 of 4. Select the Add Line ⬉ tool and point (roughly) to the coordinates (50,50). Click and hold and extend the line toward the marker for DC.
Uses allstates.dta & scheme vg_s2c

```
twoway (scatter ownhome borninstate) (lfit ownhome borninstate)
```

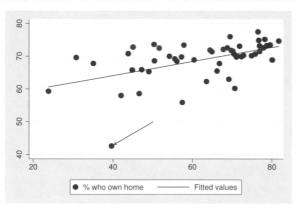

Example 3 of 4. We can convert this line into an arrow. Select the Pointer ⬉ tool and click on the line to select it. Then, in the Contextual Toolbar, change **Arrowhead** to Head.
Uses allstates.dta & scheme vg_s2c

```
twoway (scatter ownhome borninstate) (lfit ownhome borninstate)
```

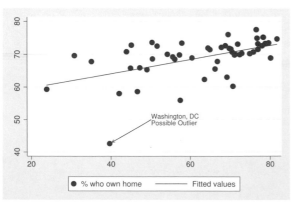

Example 4 of 4. Now let's select the Add Text **T** tool. Click at the start of the arrow, and a textbox dialog will appear. For the **text** enter `"Washington, DC" "Possible Outlier"`. Learn more about adding lines, text, and markers in Editor : Adding (54).
Uses allstates.dta & scheme vg_s2c

2.2   Object Browser

The Object Browser (see figure 2.1, page 36) contains a nested listing of all the objects in the graph. Some objects are nested within other (container) objects; for example, the object named **legend** is actually a container object that contains the elements within the legend. The Object Browser is a powerful tool for editing graphs. Once you have created a graph, consider some of the ways that you can use the Object Browser.

Some objects can be difficult to select with the Pointer ⬉ tool. For example, it can be difficult to select the entire legend because you usually end up selecting an object within the legend. Using the Object Browser, you can select the legend by clicking on **legend**.

You can show hidden objects or hide shown objects. Although it is possible to right-click on an object to hide it, the Object Browser is the only means of selecting a hidden object in order to show it. For example, you can show the legend by right-clicking on **legend** in the Object Browser and selecting **show**. See Editor : Hiding/Showing (73) for more details.

You can modify objects that are currently empty. For example, the title of your graph might be empty. Because the title is empty, you cannot double-click on the title to change its contents. However, from the Object Browser, you can double-click on **title** and add a title to the graph. See Editor : Modifying (47) for more details.

You can lock the contents of an object to prevent it from being modified (or moved), or you can unlock an object to permit it to be modified. For example, right-click on **legend** in the Object Browser and select **lock object**; now the legend cannot be modified until it is unlocked. See Editor : Locking/Unlocking (76) for more details.

Most things can be done in the Graph Editor without recourse to the Object Browser. Selecting an object in the Browser is the same as selecting it on the graph itself (and vice versa). Even right-clicking works in the Object Browser. Even so, I believe that many things are easier with the Object Browser, and a little investment in understanding the Browser makes Stata graphs easier to understand and easier to manipulate.

This section focuses on understanding the contents of the Object Browser to help you relate the objects displayed in a graph with how they are named and shown within the Object Browser. The later sections of this chapter illustrate how to use the Object Browser to modify, hide/show, and lock/unlock objects; see Editor : Modifying (47), Editor : Hiding/Showing (73), and Editor : Locking/Unlocking (76) for more details.

Introduction   Editor   Twoway   Matrix   Bar   Box   Dot   Pie   Options   Standard options   Styles   Appendix

Overview   Browser   Modifying   Adding   Moving   Hiding/Showing   Locking/Unlocking   Graph Recorder   Editor vs. commands

```
graph twoway scatter propval100 urban
```

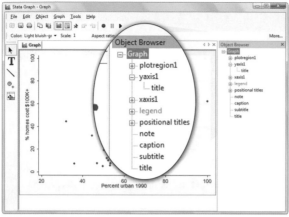

The top-level object in the Browser is the name of the graph (which, by default, is **Graph**). Within **Graph**, there is **plotregion1**, **yaxis1**, **xaxis1**, **legend**, **positional titles**, and then **note**, **caption**, **subtitle**, and **title**. The **legend** is dimmed because the legend is not currently visible. If we click on the plus sign in front of an object, we can see the objects nested within it. For example, within **yaxis1** is an object named **title** that refers to the title for the y axis.
Uses allstates.dta & scheme vg_s2c

```
graph twoway (scatter propval100 urban) (lfit propval100 urban),
    name(scatfit)
```

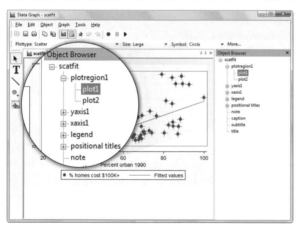

This graph, named `scatfit`, contains objects similar to the previous graph. The legend is now visible and is no longer dimmed. The **plotregion1** contains two objects: **plot1**, which refers to the scatterplot, and **plot2**, which refers to the fitted line. Click on **plot1** and the markers are selected. Try clicking on other objects (such as the objects within the **legend**), and see what becomes selected in the graph.
Uses allstates.dta & scheme vg_s2c

```
graph twoway scatter propval100 rent700 urban, yaxis(1 2)
    name(scat2)
```

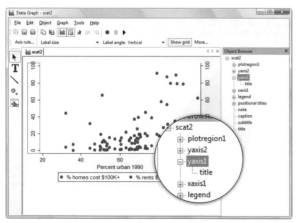

This graph, named `scat2`, shows similar objects to the previous graph, but this graph has two y axes, reflected in the Object Browser as **yaxis1** and **yaxis2**. Click on each to see which is which. Click on the **title** within **yaxis1**, and nothing appears to be shown. This is because the title is blank; hence, there is nothing to show.
Uses allstates.dta & scheme vg_s2c

graph combine scatfit scat2, altshrink name(both)

We now combine the two previous graphs into a graph named **both**. The objects within the first graph can be found within **plotregion1** under **graph1**, whereas **graph2** holds the objects for the second graph. This nesting is why objects from the first graph cannot be moved into the second graph (and vice versa).

Uses allstates.dta & scheme vg_s2c

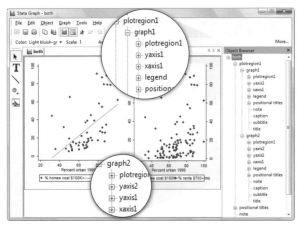

graph twoway (scatter propval100 urban rent700, yaxis(1 2)), by(north)

Below **plotregion1** is **plotregion1[1]**, which is the plot region for the first (left) graph, and **plotregion1[2]**, which is the plot region for the second (right) graph. Likewise, all the objects for the first graph carry the suffix **[1]** and the objects for the second graph have the suffix **[2]**. All these objects for both graphs are nested within **plotregion1**. Because of this structure, you can move objects among the two graphs because they all reside at the same level of nesting.

Uses allstates.dta & scheme vg_s2c

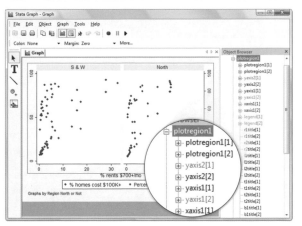

graph matrix propval100 rent700 urban

The scatterplot matrix contains a **plotregion1**, which contains the objects of the scatterplot matrix. There are nine plot regions, corresponding to the nine cells of the matrix named **plotregion1_1** (for row 1, column 1) up to **plotregion3_3**. The objects **titlegrid1** to **titlegrid3** are the titles on the diagonal. The axes are named **xaxis1** to **xaxis3** and **yaxis1** to **yaxis3**.

Uses allstates.dta & scheme vg_s2c

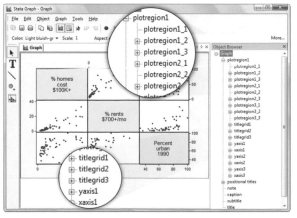

`graph bar wage tenure, over(union) over(married)`

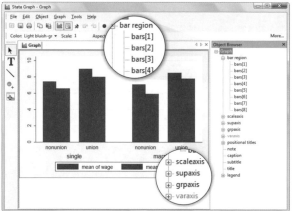

A bar chart has a **bar region** that contains objects for each bar. The **scaleaxis** refers to the continuous axis (here, the y axis), the **supaxis** refers to *single/married*, the **grpaxis** refers to *union/nonunion*, and the **varaxis** is hidden and generally not useful. Other familiar objects include **positional titles**, **note**, **caption**, **subtitle**, **title**, and **legend**.

Uses nlsw.dta & scheme vg_s2c

`graph box wage tenure, over(union) over(married)`

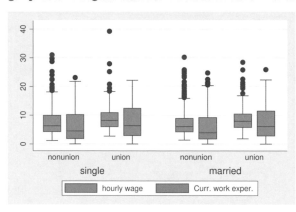

Given the similarities between `graph bar` and `graph box`, it is no surprise that the structure of the objects for this graph is similar to the previous graph. The main difference is that this graph contains a **box region**, which contains eight **boxes** (one for each box in the graph), as well as several objects named **outsides**, corresponding to each of the markers representing outside values.

Uses nlsw.dta & scheme vg_s2c

`graph dot wage tenure, over(union) over(married)`

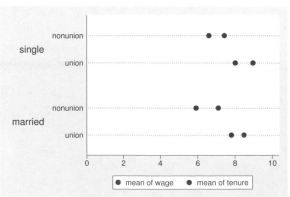

Again these objects are similar to the objects in the previous graph. Within the **dot region** there are eight **points** that refer to the eight markers in the graph. Also, there are four **bars** that refer to the horizontal dotted lines in the graph.

Uses nlsw.dta & scheme vg_s2c

This section has focused on understanding the structure of the objects within the Object Browser. For further details on how to modify, move, hide, or lock these objects, see Editor : Modifying (47), Editor : Moving (62), Editor : Hiding/Showing (73), and Editor : Locking/Unlocking (76).

2.3   Modifying objects

This section will focus on how you can modify objects within your graph. It will show you both the different tools and strategies that you can use with the Graph Editor for modifying objects within your graphs, as well as expose you to the variety of objects that can be modified using the Graph Editor. The purpose is not to cover all the kinds of modifications that you can make to objects but to emphasize the techniques that you can use for modifying objects. The rest of the chapters of this book explore the wide variety of modifications that you can make to objects, and most examples describe how to use the Graph Editor to do so. Let's start by creating a graph.

`scatter propval100 ownhome`

Example 1 of 6. Let's modify the title of the x axis. Click on the title of the x axis. Then, using the Contextual Toolbar, change the **Size** to Huge. The graph reflects the effect of that change. *Uses allstates.dta & scheme vg_s2c*

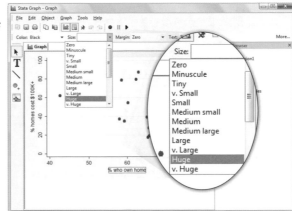

`scatter propval100 ownhome`

Example 2 of 6. Let's modify the title of the x axis. Click on the title of the x axis. Then, using the Contextual Toolbar, change the **Text** to Percent who own their home and press *Enter*. The graph reflects the new title. *Uses allstates.dta & scheme vg_s2c*

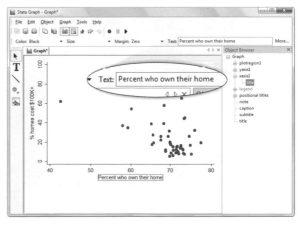

`scatter propval100 ownhome`

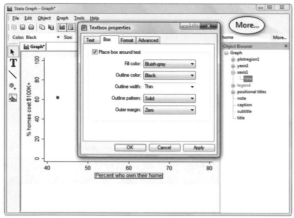

Example 3 of 6. Continuing with the previous example, let's place a box around the title. There is no such choice in the Contextual Toolbar, but note the **More...** button at the far right of that toolbar. Click it once and the full compliment of options for modifying this object (the x title) are now available. From the *Box* tab, check **Place box around text** and then click on **Apply**. The change happens immediately, and the dialog box remains open.
Uses allstates.dta & scheme vg_s2c

`scatter propval100 ownhome`

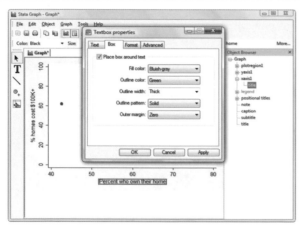

Example 4 of 6. Because we used the **Apply** button in the previous example, we can continue to make other modifications from the dialog box. Let's make the line around the box thick and green by changing the **Outline width** to Thick and the **Outline color** to Green and then click on **OK**.
Uses allstates.dta & scheme vg_s2c

`scatter propval100 ownhome`

Example 5 of 6. Let's make similar modifications to the y title. This time let's double-click on the y title, and the same dialog box appears as though we had chosen **More...** from the Contextual Toolbar. In the *Text* tab, change the **Size** to Huge. In the *Box* tab, check **Place box around text**; change the **Outline width** to Thick and the **Outline color** to Green. Click on **OK**.
Uses allstates.dta & scheme vg_s2c

`scatter propval100 ownhome`

Example 6 of 6. Now let's modify the
y-axis labels by first clicking on one
(e.g., 20). Then, in the Contextual
Toolbar, change the **Label angle** to
`Horizontal` and click on **Show grid** to
hide the grid. Finally, click on **Axis
Rule...** and select **Range/Delta**.
Enter 0 for the **Minimum value**, 100
for the **Maximum value**, and 10 for
the **Delta**.
Uses allstates.dta & scheme vg_s2c

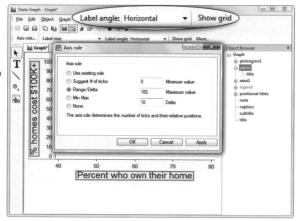

I encourage you to double-click on either the *x*-axis or the *y*-axis title and play around
with the settings within the **Text**, **Box**, **Format**, and **Advanced** tabs. To immediately
see the effect of your change, you can click on the **Apply** button. You can learn more about
controlling the display of text in Options : Textboxes (379). Also, you can learn more about
how to modify axis labels in Options : Axis labels (330) and Options : Axis scales (339).

Now let's create a new example and explore other ways to make modifications to your
graphs.

`scatter ownhome borninstate`

Example 1 of 4. Consider this graph. It
is a nice graph, but it could use some
titles to better explain the graph.
Uses allstates.dta & scheme vg_s2c

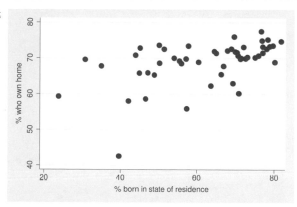

Introduction   Editor   Twoway   Matrix   Bar   Box   Dot   Pie   Options   Standard options   Styles   Appendix

Overview   Browser   Modifying   Adding   Moving   Hiding/Showing   Locking/Unlocking   Graph Recorder   Editor vs. commands

`scatter ownhome borninstate`

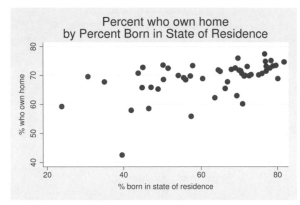

Example 2 of 4. There are many ways to add a title to the graph. Within the Object Browser, double-click on **title**. For the text, enter `"Percent who own home" "by Percent Born in State of Residence"` and click on **OK**. The two pairs of double quotes allow us to make the title display as two lines.
Uses allstates.dta & scheme vg_s2c

`scatter ownhome borninstate`

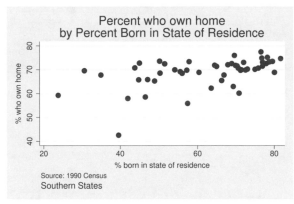

Example 3 of 4. We can further annotate this graph by adding a note and a caption. Instead of using the Object Browser, we can use the Main Menu to select **Graph** and then **Titles**. Change the **Caption** to `Southern States` and the **Note** to `Source: 1990 Census` and click on **OK**.
Uses allstates.dta & scheme vg_s2c

`scatter ownhome borninstate`

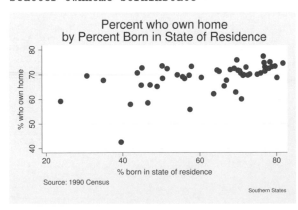

Example 4 of 4. Let's make the caption (i.e., `Southern States`) much smaller and move it to the right corner. Now that the caption is part of the graph, we can simply double-click on it and change the **Size** to `Small`. In the *Format* tab, change the **Position** to `East`. See Standard options: Titles (395) for more information on modifying titles, notes, and captions.
Uses allstates.dta & scheme vg_s2c

```
scatter ownhome borninstate, mlabel(stateab) xlabel(25(5)85)
```

Example 1 of 4. In this series of
examples, we will examine modifying
markers and marker labels. We have
switched to the *allstates3* data file,
which contains just the states in the
south.
Uses allstates3.dta & scheme vg_s2c

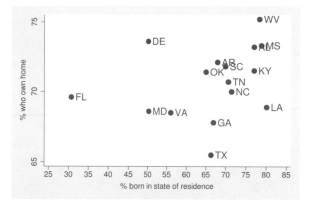

```
scatter ownhome borninstate, mlabel(stateab) xlabel(25(5)85)
```

Example 2 of 4. Let's modify the
markers to be large, black, hollow
circles. Using the Pointer ⬈ tool, click
on any marker. In the Contextual
Toolbar, change the **Color** to Black,
the **Size** to Large, and the **Symbol** to
Hollow Circle. The label for AL
overlaps with the marker for MS, and
the label for AR overlaps with the label
for SC. Let's try to fix that next.
Uses allstates3.dta & scheme vg_s2c

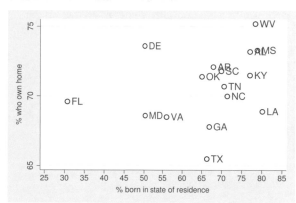

```
scatter ownhome borninstate, mlabel(stateab) xlabel(25(5)85)
```

Example 3 of 4. Let's try changing the
position of the marker labels to avoid
any overlaps. Click on any marker label
(e.g., MS) and, in the Contextual
Toolbar, change the **position** to 12
o'clock. This change does not work
because WV is now off the graph. You
can undo this change by clicking on the
Undo ↰ icon in the Standard Toolbar.
However, it is actually faster to simply
change the **position** back to
3 o'clock.
Uses allstates3.dta & scheme vg_s2c

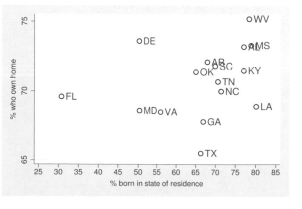

Introduction   Editor   Twoway   Matrix   Bar   Box   Dot   Pie   Options   Standard options   Styles   Appendix

Overview   Browser   Modifying   Adding   Moving   Hiding/Showing   Locking/Unlocking   Graph Recorder   Editor vs. commands

```
scatter ownhome borninstate, mlabel(stateab) xlabel(25(5)85)
```

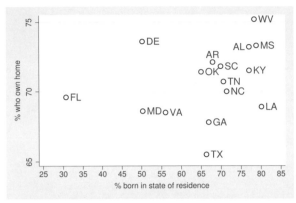

Example 4 of 4. Let's change the position of the label just for AL. Right-click on AL (the label, not the marker) and choose **Observation Properties**. Change the **Position** to 9 o'clock in the resulting dialog box. (Any further changes made to the AL label will only affect the AL label.) Let's also fix AR by right-clicking on AR, choosing **Observation Properties**, and changing the **Position** to 12 o'clock. See Options : Markers (307) and Options : Marker labels (320) for details.
Uses allstates3.dta & scheme vg_s2c

```
graph twoway (scatter propval100 urban) (lfit propval100 urban)
    (qfit propval100 urban)
```

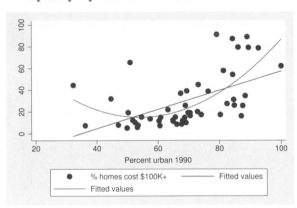

Example 1 of 3. This graph has a legend in it, giving us a chance to see how we can modify the legend. We have returned to the *allstates* data file.
Uses allstates.dta & scheme vg_s2c

```
graph twoway (scatter propval100 urban) (lfit propval100 urban)
    (qfit propval100 urban)
```

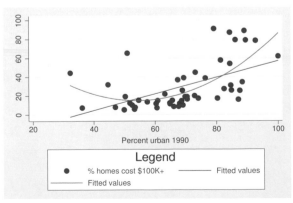

Example 2 of 3. Let's add the word Legend as a title to the legend. We can do this by going to the Object Browser and clicking on the plus sign next to **legend** to expand it. Below **legend** is **title** (as well as **subtitle**, **note**, and **caption**). Double-click on **title** and change the **Text** to Legend.
Uses allstates.dta & scheme vg_s2c

```
graph twoway (scatter propval100 urban) (lfit propval100 urban)
    (qfit propval100 urban)
```

Example 3 of 3. Click on the words *Fitted Values* for the linear fit and, in the Contextual Toolbar, change the text to `Linear Fit`. Likewise, click on the words *Fitted Values* for the quadratic fit and, in the Contextual Toolbar, change the text to `Quadratic Fit`. See Options : Legend (361) for more examples of modifying legends.
Uses allstates.dta & scheme vg_s2c

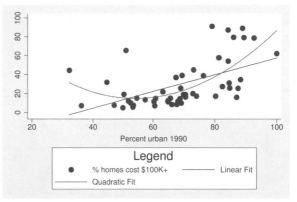

```
twoway rconnected high low tradeday, sort
```

Example 1 of 3. Let's use the *spjanfeb2001* data file and make a graph of the high and low price for the first two months of 2001.
Uses spjanfeb2001.dta & scheme vg_s2c

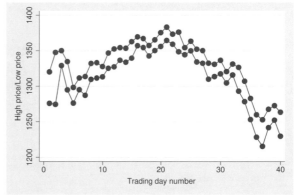

```
twoway rconnected high low tradeday, sort
```

Example 2 of 3. Let's see what these data would look like as a range cap plot. Click on any of the markers and, in the Contextual Toolbar, change the **plottype** to `Rcap`. The graph immediately changes to a range plot with capped spikes.
Uses spjanfeb2001.dta & scheme vg_s2c

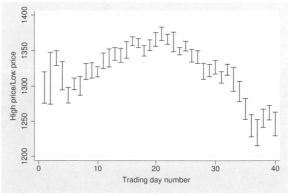

Introduction   Overview   Browser   Editor   Modifying   Twoway   Adding   Matrix   Moving   Bar   Hiding/Showing   Box   Dot   Locking/Unlocking   Pie   Options   Graph Recorder   Standard options   Editor vs. commands   Styles   Appendix `twoway rconnected high low tradeday, sort`

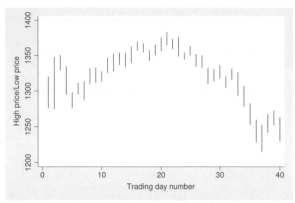

Example 3 of 3. Let's see what these data would look like as a spike plot. Click on any of the markers and, in the Contextual Toolbar, change the **plottype** to `Rspike`. The graph immediately changes to a range plot with spikes.

Uses spjanfeb2001.dta & scheme vg_s2c

To summarize, there are a variety of ways that you can modify an object using the Graph Editor. They include

- Clicking on an object and using the Contextual Toolbar to change one of a subset of properties.
- Selecting **More...** from the Contextual Toolbar to access dialog boxes of all properties.
- Double-clicking on an object to access the dialog boxes of all properties.
- Double-clicking on the object name in the Object Browser to access the dialog boxes of all properties.
- Right-clicking on an object and selecting the option to modify the properties of that specific object.

2.4   Adding objects

This section discusses how you can add objects to your graphs, namely, how you can add text, lines (including arrows), and markers to your graphs. Of course, once you add such objects, it is likely that you will want to move them to more finely adjust their position in the graph. The topic of moving objects is covered in Editor : Moving (62); refer to that section for more details once you have added objects. Let's start with examples you might use if you wanted to teach students about residuals, leverage, and influence. Let's start by using the `allstates` data file and running this `regress` command:

```
. vguse allstates
. regress married pov
```

Then we can create a leverage-versus-squared-residual plot.

```
lvr2plot, mlabel(stateab) mlabpos(12)
```

Example 1 of 6. This graph shows the leverage by the residual squared. We can see that DC has a high residual but has low leverage and that MS and LA are high in leverage but have small residual values. Let's add labels to describe these observations.

Uses allstates.dta & scheme vg_s2c

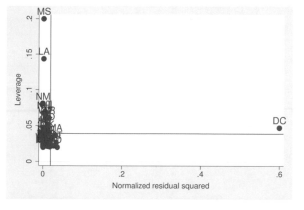

```
lvr2plot, mlabel(stateab) mlabpos(12)
```

Example 2 of 6. To add text to the graph, click on the Add Text **T** tool. Click in the lower right portion of the graph (near DC) and for the **Text** enter "High Residual" "Low Leverage". Click on **Apply**. To make the text larger, change the **Size** to Large and then click on **OK**.

Uses allstates.dta & scheme vg_s2c

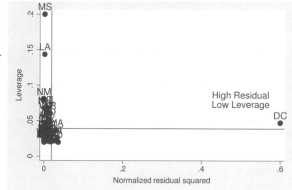

```
lvr2plot, mlabel(stateab) mlabpos(12)
```

Example 3 of 6. Let's place a label near MS and LA to indicate that these points are high in leverage but have low values of their residuals. Check that the Add Text **T** tool is still selected. Click in the upper left portion of the graph and for the **Text** enter "High Leverage" "Low Residual". Let's change the **Size** to Large. For fun, click on the *Box* tab and check **Place box around text**.

Uses allstates.dta & scheme vg_s2c

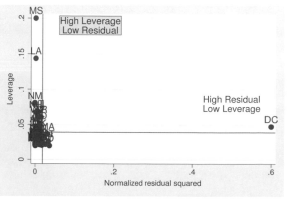

Introduction   Editor   Twoway   Matrix   Bar   Box   Dot   Pie   Options   Standard options   Styles   Appendix

Overview   Browser   Modifying   Adding   Moving   Hiding/Showing   Locking/Unlocking   Graph Recorder   Editor vs. commands

`lvr2plot, mlabel(stateab) mlabpos(12)`

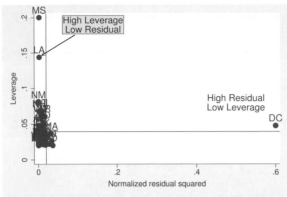

Example 4 of 6. Now let's add an arrow pointing from the textbox at the upper left to the marker for LA. Click on the Add Line ⬊ tool and click and hold at the bottom of the textbox. Stretch the line to point to the marker for LA. Then select the Pointer ▸ tool and click on the line. In the Contextual Toolbar, change the **Width** to Medium thick, **Arrowhead** to Head, and the **Arrow size** to Medium-large.

Uses allstates.dta & scheme vg_s2c

`lvr2plot, mlabel(stateab) mlabpos(12)`

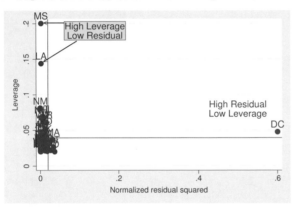

Example 5 of 6. Let's add a horizontal arrow pointing from the textbox at the upper left to the marker for MS. Click on the Add Line ⬊ tool. While holding the *Shift* key, click and hold at the top of the textbox and stretch the line to point to the marker for MS. (Holding the *Shift* key forces the line to be perfectly horizontal, vertical, or diagonal.) Change the attributes of the line as in the previous example.

Uses allstates.dta & scheme vg_s2c

`lvr2plot, mlabel(stateab) mlabpos(12)`

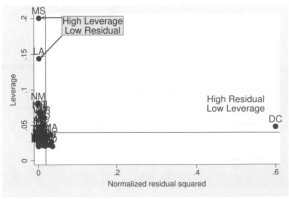

Example 6 of 6. Let's make the diagonal line be perfectly diagonal. Using the Pointer tool, right-click on the diagonal line and select **Delete**. Then choose the Add Line ⬊ tool. Starting from the marker for LA, press the *Shift* key, and click, hold, and stretch the line to the textbox. In the Contextual Toolbar, change the **Width** to Medium thick, **Arrowhead** to Tail, and the **Arrow size** to Medium-large.

Uses allstates.dta & scheme vg_s2c

The Graph Editor allows you to set the default characteristics (e.g., size and color) of text, lines, and markers. If you click on the Add Text **T** tool, the Contextual Toolbar appears, giving you the opportunity to change the default characteristics of added text. If you change the **Color** to Red and the **Size** to Large, added text would, by default, be red and large for all subsequent graphs (even after you close and restart Stata). You can even click on the **More...** button from the Contextual Toolbar to specify other default characteristics for added text. If you want to return to the Stata factory default settings, you can click on the **More...** button from the Contextual Toolbar, select the *Advanced* tab, and click on **Reset Defaults**. Likewise, you can click on the Add Line $\diagdown$ tool and then use the Contextual Toolbar to specify defaults for added lines, or click on the Add Marker $\bullet_+$ tool and then use the Contextual Toolbar to specify defaults for added markers. For any given tool, the settings can be reset to the factory defaults by first selecting the tool, then clicking on the **More...** button from the Contextual Toolbar, then selecting the *Advanced* tab, and clicking on **Reset Defaults**.

```
graph twoway (scatter married pov, mlabpos(12) mlabel(stateab))
   (lfit married pov)
```

Example 1 of 6. Here is another way to illustrate the leverage and residuals for these data, using a scatterplot and fit line. As before, let's annotate this graph. Click the Add Text **T** tool and, in the Contextual Toolbar, change the **Size** to Large, permanently changing the default size of added text. Click next to DC and for the **Text** enter High Residual. The added text is Large.
Uses allstates.dta & scheme vg_s2c

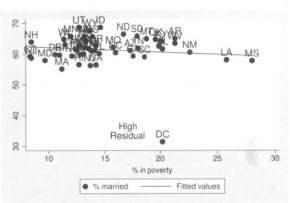

```
graph twoway (scatter married pov, mlabpos(12) mlabel(stateab))
   (lfit married pov)
```

Example 2 of 6. With the Add Text **T** tool still selected, click below MS and for the **Text** enter "High" "Leverage". By default, the text displays as large. With the Add Text **T** tool selected, you can change other defaults with the Contextual Toolbar and the **More...** button. You can restore the factory defaults by clicking on **More...** and then within the *Advanced* tab click on **Reset Defaults**.
Uses allstates.dta & scheme vg_s2c

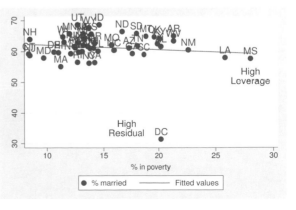

```
graph twoway (scatter married pov, mlabpos(12) mlabel(stateab))
   (lfit married pov)
```

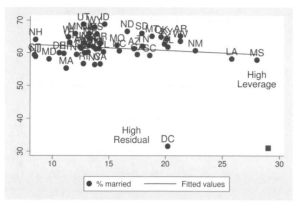

Example 3 of 6. Let's add a marker for an observation that would have both a high residual and high leverage. Select the Add Marker ⊕₊ tool and then click on a point in the lower right corner. Select the Pointer ↖ tool and click on the marker. In the Contextual Toolbar, change the **Color** to Red and the **Symbol** to Square.

Uses allstates.dta & scheme vg_s2c

```
graph twoway (scatter married pov, mlabpos(12) mlabel(stateab))
   (lfit married pov)
```

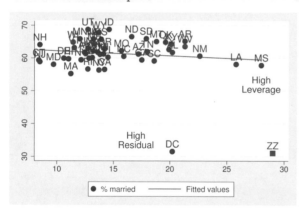

Example 4 of 6. Let's label the observation for this fictitious state as "zz". With the Pointer ↖ tool selected, double-click on the new observation. Click the *Label* tab and change the **Label** to ZZ, the **Size** to Large, the **Color** to Red, and the **Position** to 12 o'clock.

Uses allstates.dta & scheme vg_s2c

```
graph twoway (scatter married pov, mlabpos(12) mlabel(stateab))
   (lfit married pov)
```

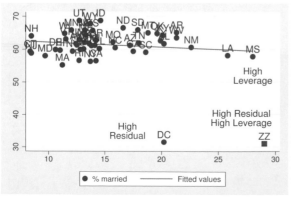

Example 5 of 6. Now let's label this lower right corner. Select the Add Text **T** tool and click in the lower right corner. For the text enter "High Residual" "High Leverage".

Uses allstates.dta & scheme vg_s2c

```
graph twoway (scatter married pov, mlabpos(12) mlabel(stateab))
   (lfit married pov)
```

Example 6 of 6. Let's now add a hypothetical regression line representing the exaggerated influence of this new observation. Select the Add Line ╲ tool and point to the top left corner of the graph. Click and hold and stretch the line to the bottom right corner of the graph. Then select the Pointer ↖ tool, click on the line, and change the **Width** to Thick.

Uses allstates.dta & scheme vg_s2c

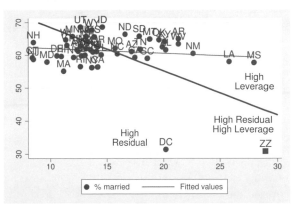

```
graph twoway scatter married pov age, mlabel(stateab)
   title(Married and Poverty by Age) ytitle(Married/Poverty)
```

Example 1 of 2. Consider this graph with four lines added. The black line goes from the FL marker to the marker below it. The red line goes from the AK marker to the bottom of the graph, and the green line goes from the bottom of the graph to UT. The pink line goes from the *x* title to the title.

Uses allstates.dta & scheme vg_s2c

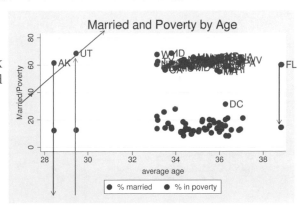

```
graph twoway scatter married pov age, mlabel(stateab)
   title(Married and Poverty by Age) ytitle(Married/Poverty)
```

Example 2 of 2. Now we changed the *y* axis to range from −20 to 120 and the *x* axis to range from 20 to 50. The black line still points from FL to the marker below it. The pink line still points from the *x* title to the title. The green line remains unchanged, but it no longer points to UT. The red line still originates from AK but no longer reaches the bottom of the graph.

Uses allstates.dta & scheme vg_s2c

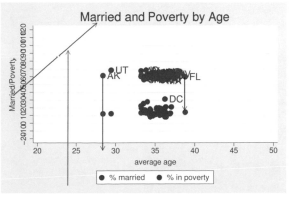

You can further understand this by using the Object Browser. Click on the plus sign to expand all the objects in the Object Browser. Click on the black line; this is within **added lines** under the **plotregion1**. When the axes are resized, the line is resized along with the plot region, and the starting and finishing positions are properly adjusted. The red line also belongs to **plotregion1**, so it is properly adjusted in the x axis and shrunk along with the y axis. The green line would appear to be the same as the red line, but remember it originated from outside the plot region. In the Object Browser, this line is located in **added lines** belonging to the overall graph. When the axes are modified, the green line stayed put, but the object it pointed to moved. The pink line also belongs to the entire graph and stayed put. This explains why these different lines behaved differently in response to rescaling the y axis.

Let's conclude with a series of examples that combine graphs, returning to the scatterplot and regression line along with the leverage-versus-squared-residual plot. Using the commands below, let's create these graphs.

```
. clear all
. vguse allstates
. regress married pov
. graph twoway (scatter married pov, mlabpos(12) mlabel(stateab))
      (lfit married pov), name(g1)
. lvr2plot, mlabel(stateab) mlabpos(12) name(g2)
```

Now let's combine these graphs and see how we can add objects to a graph that is actually composed of two graphs.

```
graph combine g1 g2
```

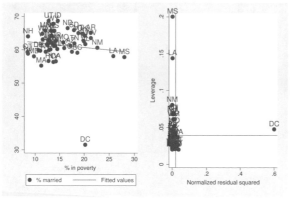

Example 1 of 4. We combine the two graphs so we can see both the scatterplot and the leverage-versus-squared-residual plot at the same time.
Uses allstates.dta & scheme vg_s2c

graph combine g1 g2

Example 2 of 4. Let's add an arrow pointing from the DC marker in the scatterplot to the DC marker in the leverage-versus-squared-residual plot. Select the Add Line ╲ tool. Click and hold near DC in the left graph and stretch the line to point to the marker for DC in the right graph. The line stops at the edge of the left graph. *Uses allstates.dta & scheme vg_s2c*

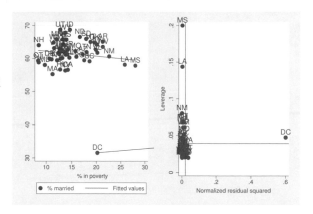

The Object Browser can help you understand why this happened. In the Object Browser, look under **Graph**, under **plotregion1**, under **graph1** (which represents the left graph), and then under **plotregion1** (which represents the plot region for the left graph). There you will find **added lines**. Select the Pointer ↖ tool and click on **added lines**; you will see the entire line displayed in red. Because this line belongs to the plot region of the left graph, the line terminates at the edge of that region. Let's try this again. Right-click on this line and click on **Delete**.

graph combine g1 g2

Example 3 of 4. Let's try this again by adding a line that belongs to the entire graph. Select the Add Line ╲ tool. Click and hold below the legend of the left graph and stretch the line to point to the DC marker in the right graph. In the Object Browser, there should be an entry for **added lines** at the bottom associated with this line that now belongs to the entire graph. *Uses allstates.dta & scheme vg_s2c*

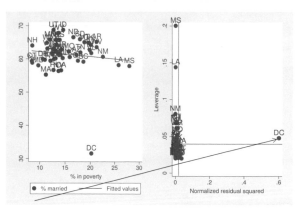

`graph combine g1 g2`

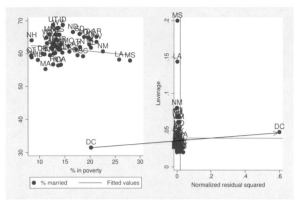

Example 4 of 4. Now select the line
(which might be tricky, so use the
Object Browser and click on the object
named **added lines** at the bottom).
Then click and hold the starting point
at the left and move the starting point
to the marker for DC in the left graph.
This is a great segue into the next
section, which elaborates on how to
move objects within the Graph Editor.
Uses allstates.dta & scheme vg_s2c

2.5   Moving objects

This chapter is about moving objects with the Graph Editor. The Graph Editor offers
you three ways to move objects: 1) using the Pointer ➹ tool to select and move the object,
2) changing the properties of the object such as the position or justification that changes
where the object is displayed, and 3) using the Grid Edit ⊞ tool to select and move the
object. This section covers each of these methods, starting with the Pointer tool.

Moving with the Pointer tool

The Pointer ➹ tool is the default tool selected in the Graph Editor. It allows you to
select objects by clicking on them once, move them by clicking and holding and dragging
them, and modify their properties by double-clicking on them. This portion of the section
focuses on moving objects by selecting them with the Pointer ➹ tool and dragging them
to a new location.

```
lvr2plot, mlabel(stateab) mlabpos(12)
```

Example 1 of 8. Consider this graph that we made in Editor : Adding (54). Suppose that we were unhappy with the locations of the added text and wanted to move them both toward the top right corner of the graph. The following examples illustrate how we can move these objects using the Pointer ↖ tool.
Uses allstates.dta & scheme vg_s2c

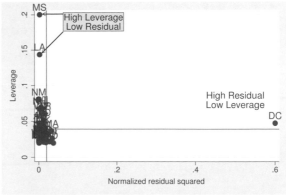

```
lvr2plot, mlabel(stateab) mlabpos(12)
```

Example 2 of 8. Select the Pointer ↖ tool and click on *High Residual Low Leverage*. A red box will appear around it, and the icon will turn into four arrowheads (ready to move). Click and hold and drag the text toward the upper right corner. You can repeat this process to make fine adjustments in the position.
Uses allstates.dta & scheme vg_s2c

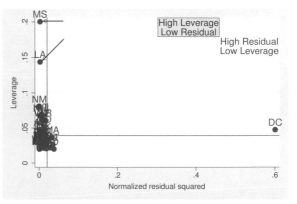

```
lvr2plot, mlabel(stateab) mlabpos(12)
```

Example 3 of 8. As before, select the Pointer ↖ tool and click on *High Leverage Low Residual*. The black box surrounding the text is converted to a red box, and the icon turns into four arrowheads (ready to move). Click and hold and drag the text toward the upper right corner. You can repeat this process to make fine adjustments in the position.
Uses allstates.dta & scheme vg_s2c

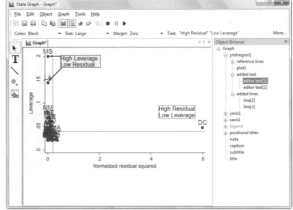

Introduction   Editor   Twoway   Matrix   Bar   Box   Dot   Pie   Options   Standard options   Styles   Appendix

Overview   Browser   Modifying   Adding   Moving   Hiding/Showing   Locking/Unlocking   Graph Recorder   Editor vs. commands

`lvr2plot, mlabel(stateab) mlabpos(12)`

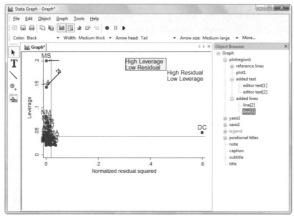

Example 4 of 8. We can move lines/arrows in the same way as we moved the added text, by selecting and moving them (this is not illustrated). We can also select either the start or the end point of an arrow and stretch it to a new position. Let's click on the start point for the arrow pointing to LA. With the four-arrow icon showing, click on the start point and stretch it to extend to the textbox.

Uses allstates.dta & scheme vg_s2c

`lvr2plot, mlabel(stateab) mlabpos(12)`

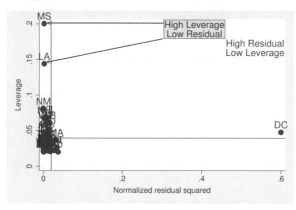

Example 5 of 8. We can also move lines/arrows in horizontal, vertical, and diagonal angles by holding the *Shift* key during the select-and-move process. While holding the *Shift* key, select the start point for the line pointing to MS and then stretch the start point to extend to the textbox.

Uses allstates.dta & scheme vg_s2c

`lvr2plot, mlabel(stateab) mlabpos(12)`

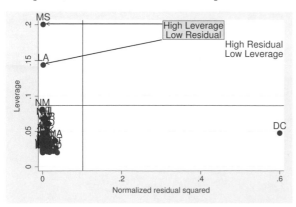

Example 6 of 8. We can even move the added lines (the red lines). Click on the horizontal line and drag it up. When dragging this line, the icon is a double-headed arrow pointing up and down, indicating that this line can only be shifted up and down. Likewise, select the vertical line and drag it to the right.

Uses allstates.dta & scheme vg_s2c

`lvr2plot, mlabel(stateab) mlabpos(12)`

Example 7 of 8. We have added a line
pointing to the marker for DC, but the
marker label was in the way. We could
try to move it with the Pointer ⟍ tool,
but such labels are not movable that
way. Instead, we must modify the
properties by double-clicking directly
on DC (the label, not the marker) and
changing the **Position** to 9 o'clock.
This is an example of moving an object
with the properties of the object,
discussed later in this section.
Uses allstates.dta & scheme vg_s2c

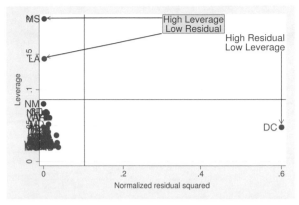

`lvr2plot, mlabel(stateab) mlabpos(12)`

Example 8 of 8. We could try to move
the *y* axis to the right side of the graph
with the Pointer ⟍ tool. Instead, we
must select the Grid Edit ⊞ tool and
then click on the *y* axis (it will turn
pale red). Then drag the *y* axis to the
far right of the graph (when the right
border turns dark red). Then, with the
Pointer ⟍ tool, double-click on that
axis and click on **Advanced**. Change
the **Position** to `Right`. Grid editing is
further discussed later in this section.
Uses allstates.dta & scheme vg_s2c

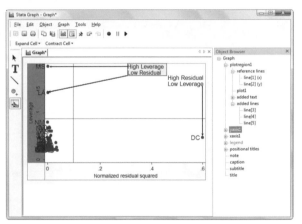

   As you have seen, using the Pointer tool is a simple way to move objects, but it is
not always the best tool, and sometimes it is not available at all. Sometimes you need to
(or want to) move an object by changing its properties (as illustrated when we moved the
marker label for DC above). Likewise, sometimes you need to (or want to) move an object
with the Grid Edit ⊞ tool (as was illustrated in moving the *y* axis from the left to the
right). Let's now look at how to move objects with their properties.

Moving by object properties

   Let's dive right into an example that shows how to move objects by changing their
properties.

```
twoway (scatter propval100 urban, mlabel(stateab)) (lfit propval100 urban)
  (qfit propval100 urban), title("This is a Very Long" "Title for a Graph")
```

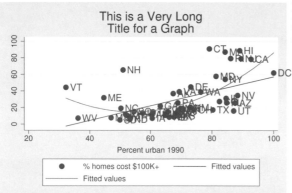

Example 1 of 6. This graph provides lots of opportunities to move objects by changing their properties.
Uses allstates.dta & scheme vg_s2c

```
twoway (scatter propval100 urban, mlabel(stateab)) (lfit propval100 urban)
  (qfit propval100 urban), title("This is a Very Long" "Title for a Graph")
```

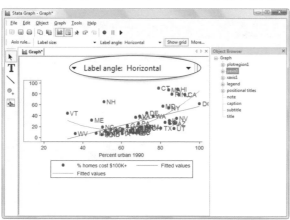

Example 2 of 6. Although it is more rotating than moving, we might want the labels for the y axis to appear horizontal. We can click on the y axis and, in the Contextual Toolbar, change the **Label angle** to Horizontal.
Uses allstates.dta & scheme vg_s2c

```
twoway (scatter propval100 urban, mlabel(stateab)) (lfit propval100 urban)
  (qfit propval100 urban), title("This is a Very Long" "Title for a Graph")
```

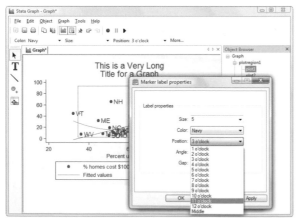

Example 3 of 6. The marker label for DC is off the right edge of the graph. We can reposition the marker labels by double-clicking on a marker label (e.g., NH) and then change the **Position** to 11 o'clock. (Double-clicking on DC does not work because that label is outside the plot region.)
Uses allstates.dta & scheme vg_s2c

```
twoway (scatter propval100 urban, mlabel(stateab)) (lfit propval100 urban)
   (qfit propval100 urban), title("This is a Very Long" "Title for a Graph")
```

Example 4 of 6. Let's reposition the title. Double-click on the title. From the *Format* tab, change the **Position** to West and change the **Justification** to Left. Click on **OK** and the title is left-justified on the left-hand side of the graph.

Uses allstates.dta & scheme vg_s2c

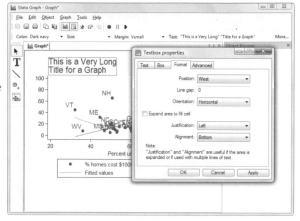

```
twoway (scatter propval100 urban, mlabel(stateab)) (lfit propval100 urban)
   (qfit propval100 urban), title("This is a Very Long" "Title for a Graph")
```

Example 5 of 6. Although we could manually move the objects within the legend to create one row of items, it is much simpler to click on **legend** in the Object Browser and, in the Contextual Toolbar, change the **Columns** to 3 and press *Enter*.

Uses allstates.dta & scheme vg_s2c

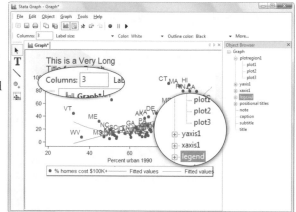

```
twoway (scatter propval100 urban, mlabel(stateab)) (lfit propval100 urban)
   (qfit propval100 urban), title("This is a Very Long" "Title for a Graph")
```

Example 6 of 6. Because the legend is a bit too wide, we can stack the symbols and text. Although we can do so by manually moving the objects, it is much simpler to double-click on **legend** in the Object Browser and change **Stack symbols and text** to Yes.

Uses allstates.dta & scheme vg_s2c

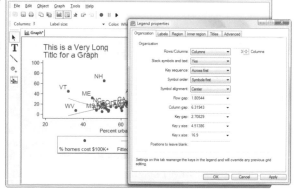

Moving with the Grid Edit tool

The main purpose of this section is to show you how to use the Grid Edit ⬚ tool. This section illustrates how objects within a graph are hierarchically organized and displayed in a nested grid structure.

```
scatter married pov, title(title) subtitle(subtitle) note(note)
    caption(caption) legend(on)
```

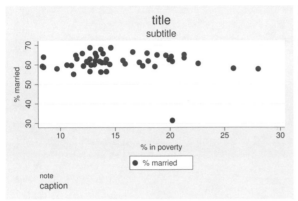

Example 1 of 5. This graph might be boring, but it is useful for understanding the nature of the grid system in which objects are organized and displayed in graphs, as further explained below.
Uses allstates.dta & scheme vg_s2c

Select the Grid Edit ⬚ tool and then click on the title of the graph (see figure 2.2).

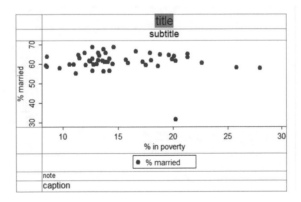

Figure 2.2: Grid Edit selection

Notice the red gridlines that are displayed. These gridlines, which are generally invisible to you, reflect a structure in which the objects are displayed. This structure is much like a spreadsheet, composed of cells that contain objects. We can see two columns: the left column contains the y axis, and the right column contains the rest of the objects. There are seven rows. The first row contains the title, the second row contains the subtitle, the third row contains the y axis and the plot region, all the way down to the seventh row that contains the caption. The objects are positioned with respect to the cell in which they live. For example, the note is left-justified, but because it lives in the second column, the note is justified against the line that divides the first and second columns. Objects can be moved from one cell to another and, unlike a spreadsheet, they can be moved into a cell that is occupied by another object so the two objects are displayed together. As objects are moved, if a cell (or row or column) is no longer needed, the Graph Editor may collapse it.

```
scatter married pov, title(title) subtitle(subtitle) note(note)
    caption(caption) legend(on)
```

Example 2 of 5. Let's explore this more. With the Grid Edit 🔲 tool selected, click and hold *note* and drag it to the cell to the left of it. All four lines surrounding the cell will turn dark red indicating that we are dragging the object to that cell (pictured at the right). Release the mouse and the note now appears in column 1 (but still in row 6), and it is still left-justified.
Uses allstates.dta & scheme vg_s2c

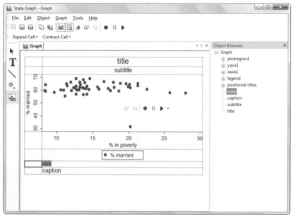

```
scatter married pov, title(title) subtitle(subtitle) note(note)
    caption(caption) legend(on)
```

Example 3 of 5. Now click and hold *title* and drag it into the plot region (all four lines surrounding the plot region will turn red when we have the title in the proper position; pictured at the right). Then release the mouse. The first row was eliminated because it is no longer needed. The title now inhabits the same space as the scatterplot, and it retains its same positioning there (top and center).
Uses allstates.dta & scheme vg_s2c

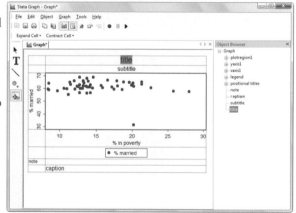

```
scatter married pov, title(title) subtitle(subtitle) note(note)
    caption(caption) legend(on)
```

Example 4 of 5. As with objects in a spreadsheet, we can change the vertical and horizontal alignment of the title within the "cell" of the plot region. Select the Pointer ↖ tool. Double-click on *title* and select the *Format* tab. Change the **Position** to Northeast and click on **Apply**. The title is now justified in the top right corner of the graph. Play with other positions (followed by **Apply**) to see how you can move the title within the plot region.
Uses allstates.dta & scheme vg_s2c

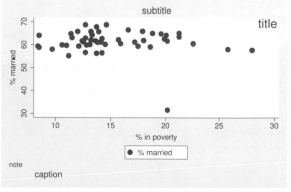

Introduction · Editor · Twoway · Matrix · Bar · Box · Dot · Pie · Options · Standard options · Styles · Appendix

Overview · Browser · Modifying · Adding · Moving · Hiding/Showing · Locking/Unlocking · Graph Recorder · Editor vs. commands

```
scatter married pov, title(title) subtitle(subtitle) note(note)
    caption(caption) legend(on)
```

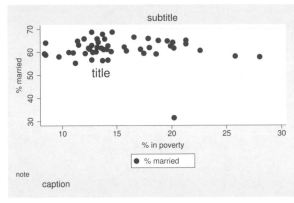

Example 5 of 5. We can move the title within the plot region with the Pointer ➤ tool. If you had simply used the Pointer ➤ tool to move the title from its original position to the plot region, an empty space (row) would have remained where the title had originally been. The benefit of using the Grid Edit ⊞ tool to move the title is that the original row is no longer needed and is thus collapsed.
Uses allstates.dta & scheme vg_s2c

```
scatter married pov, title(title) subtitle(subtitle) note(note)
    caption(caption) legend(on title(leg title) subtitle(leg subtitle)
    note(leg note) caption(leg caption))
```

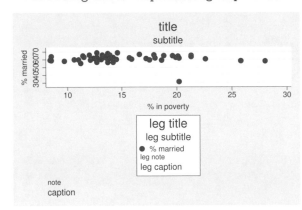

Example 1 of 7. Certainly, this is not the kind of graph that we would ever make, but it is a useful example for learning more about the Grid Edit ⊞ tool.
Uses allstates.dta & scheme vg_s2c

```
scatter married pov, title(title) subtitle(subtitle) note(note)
    caption(caption) legend(on title(leg title) subtitle(leg subtitle)
    note(leg note) caption(leg caption))
```

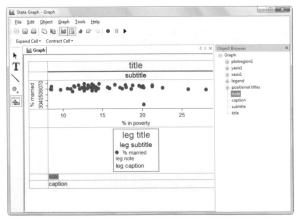

Example 2 of 7. Let's move the note between the title and the subtitle. Using the Grid Edit ⊞ tool, click and hold *note* and drag it between *title* and *subtitle*. We will know that we are on the right spot when the line between the title and subtitle turns red (pictured at left). When we release the mouse, the note appears between the title and the subtitle, creating a new row in the grid. The row formerly occupied by the note is collapsed.
Uses allstates.dta & scheme vg_s2c

```
scatter married pov, title(title) subtitle(subtitle) note(note)
    caption(caption) legend(on title(leg title) subtitle(leg subtitle)
    note(leg note) caption(leg caption))
```

Example 3 of 7. Let's move the title to the left of the y title. Using the Grid Edit 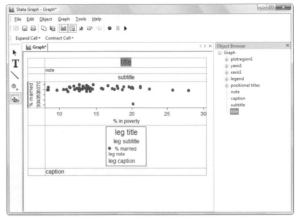 tool, click and hold *title* and drag it to the left of the y title (pictured at the right). When the grid line to the left of the y title turns red, release the mouse. The row formerly occupied by the title is collapsed, and a new column is created for the title.
Uses allstates.dta & scheme vg_s2c

Before we proceed to the next example, let's explore the grid structure of the legend for this graph. With the Grid Edit tool selected, go to the Object Browser and click on **legend**. The entire legend is now highlighted in red. From the items nested within **legend**, click on **note** and a new grid appears within the legend that has one column and five rows (the note representing the fourth row). Click on **key region** and the key region is now highlighted. Click on the plus sign next to **key region** and then click on **key**. The **key region** is actually composed of one row with four columns. Thus the overall graph contains the legend object that is composed of 5 rows, but the third row (the **key region**) is itself further composed of 4 columns. Let's explore this further.

```
scatter married pov, title(title) subtitle(subtitle) note(note)
    caption(caption) legend(on title(leg title) subtitle(leg subtitle)
    note(leg note) caption(leg caption))
```

Example 4 of 7. Using the Grid Edit 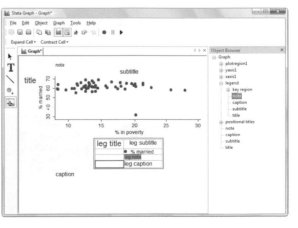 tool, select the legend title and drag it to the left of the legend subtitle, creating a new column for the legend title and removing the row for the legend title. Now select the legend note and move it to the cell to the left of the legend caption (pictured at the right).
Uses allstates.dta & scheme vg_s2c

```
scatter married pov, title(title) subtitle(subtitle) note(note)
    caption(caption) legend(on title(leg title) subtitle(leg subtitle)
    note(leg note) caption(leg caption))
```

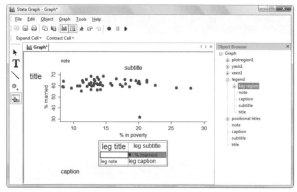

Example 5 of 7. Now let's move the **key region** to the left column. The important part is to select the entire **key region** and not just one of the elements within it. To do this, go to the Object Browser and click on **key region** (pictured at the left). Using the Grid Edit 🔲 tool, move the key region to the left column.
Uses allstates.dta & scheme vg_s2c

```
scatter married pov, title(title) subtitle(subtitle) note(note)
    caption(caption) legend(on title(leg title) subtitle(leg subtitle)
    note(leg note) caption(leg caption))
```

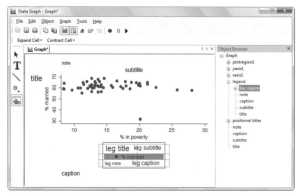

Example 6 of 7. In the Object Browser, select the **key region** and, in the Contextual Toolbar, select **Expand Cell** and **Right 1 cell**. The key region that formerly occupied the left column has now been expanded to subsume both columns. The first and third rows have two columns, whereas the second row spans both columns (pictured at the left).
Uses allstates.dta & scheme vg_s2c

```
scatter married pov, title(title) subtitle(subtitle) note(note)
    caption(caption) legend(on title(leg title) subtitle(leg subtitle)
    note(leg note) caption(leg caption))
```

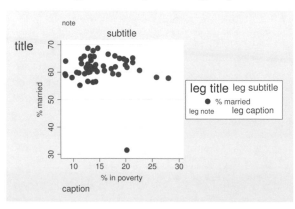

Example 7 of 7. Let's move the entire legend to the right side of the plot region. It is important (and tricky) to select the entire legend, so go to the Object Browser and click on **legend**. With the Grid Edit 🔲 tool selected, click and hold the legend and drag it to the right side of the graph. Then center it by double-clicking on **legend** in the Object Browser and, in the *Advanced* tab, changing the **Alignment** to `Center`.
Uses allstates.dta & scheme vg_s2c

2.6   Hiding and showing objects

One of the ways to modify graphs is by hiding objects that are currently displayed or by showing objects that are currently hidden. You can hide an object in one of two ways: 1) you can right-click directly on an object and select **Hide**, or 2) you can right-click on an object within the Object Browser and select **Hide**. Either way, the object is then hidden and becomes dimmed in the Object Browser to indicate that it is hidden. Hidden objects can be shown only in one way: by locating the name of the hidden (dimmed) object in the Object Browser, right-clicking on it, and selecting **Show**. The following examples will illustrate this process with an emphasis on using the Object Browser (because once an object is hidden, you need to know how to find it again in the Object Browser to reverse the process and show it again). Let's first create the following graph.

```
graph twoway (scatter propval100 urban) (lfit propval100 urban)
```

Example 1 of 4. Let's start by showing this graph of a scatterplot with a fitted regression line. The legend is not useful, so we might want to hide it.
Uses allstates.dta & scheme vg_s2c

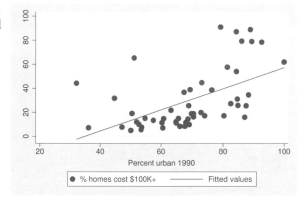

```
graph twoway (scatter propval100 urban) (lfit propval100 urban)
```

Example 2 of 4. Because the legend is not useful, let's hide it. In the Object Browser, right-click on **legend** and select **Hide** (pictured at the right). In the Object Browser, **legend** and all the objects within it are now dimmed, indicating that they are hidden. Now the *y* axis is no longer well labeled. Let's fix that in the next example.
Uses allstates.dta & scheme vg_s2c

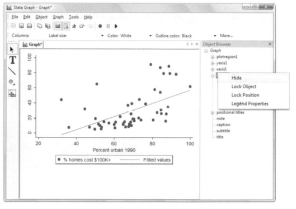

`graph twoway (scatter propval100 urban) (lfit propval100 urban)`

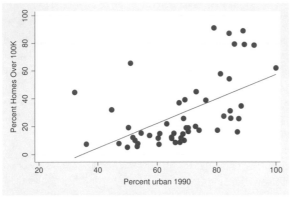

Example 3 of 4. You might think that the title for the y axis is hidden, but if we go to **yaxis1** and then **title**, we see this object is not dimmed (not hidden). If you double-click on it, the title is simply empty (which is different from hidden). I raise this as a way of distinguishing objects that are hidden from those that are simply blank. Let's make the title read `Percent Homes Over 100K`.

Uses allstates.dta & scheme vg_s2c

`graph twoway (scatter propval100 urban) (lfit propval100 urban)`

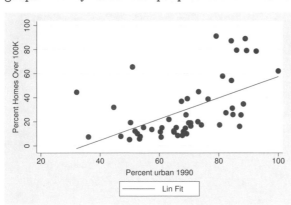

Example 4 of 4. Let's bring back the legend, but only for labeling the fit line. In the Object Browser, right-click on **legend** and select **Show**. Then expand the **key region** under the **legend**, right-click on **key[1]**, and select **hide**. Right-click on **label[1]** and select **hide**.

Uses allstates.dta & scheme vg_s2c

`graph twoway (scatter propval100 rent700 urban), by(nsw)`

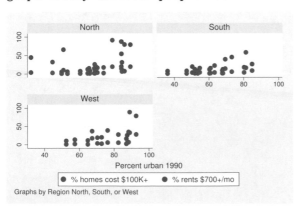

Example 1 of 6. Graphs that incorporate the `by()` option raise lots of possibilities for hiding and showing objects.

Uses allstates.dta & scheme vg_s2c

```
graph twoway (scatter propval100 rent700 urban), by(nsw)
```

Example 2 of 6. Within **plotregion1** note how the legends named **legend[1]**, **legend[2]**, and **legend[3]** are all dimmed, indicating that they are hidden. Let's right-click on each one and click on **Show** to show them.

Uses allstates.dta & scheme vg_s2c

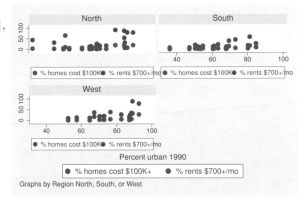

```
graph twoway (scatter propval100 rent700 urban), by(nsw)
```

Example 3 of 6. First, let's hide the legends that we showed in the last example by right-clicking on each and selecting **Hide**. Now the *x* axis for *North* is hidden. This corresponds to **xaxis1[1]**. Right-click on this and select **Show** to display it.

Uses allstates.dta & scheme vg_s2c

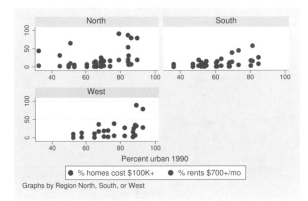

```
graph twoway (scatter propval100 rent700 urban), by(nsw)
```

Example 4 of 6. The *y* axis for *South* is also hidden. This corresponds to **yaxis1[2]**. Right-click on this and select **Show** to display it.

Uses allstates.dta & scheme vg_s2c

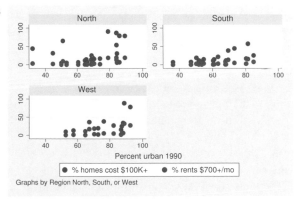

```
graph twoway (scatter propval100 rent700 urban), by(nsw)
```

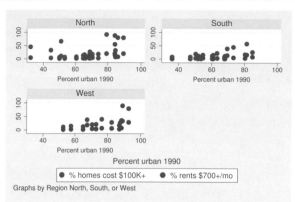

Example 5 of 6. **xaxis1[1]** contains an object named **title** that is dimmed. Right-click on this and select **Show** to display it. Repeat this for **xaxis1[2]** and **xaxis1[3]**.
Uses allstates.dta & scheme vg_s2c

```
graph twoway (scatter propval100 rent700 urban), by(nsw)
```

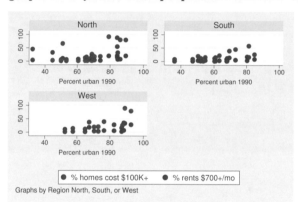

Example 6 of 6. Now the bottom title is no longer needed, so let's hide it. First, click on it once, and in the Object Browser we see that this is **bottom 1** of the **positional titles**. Right-click on the title and select **Hide**.
Uses allstates.dta & scheme vg_s2c

2.7   Locking and unlocking objects

When editing a graph, it is possible that you might make inadvertent changes to an object. To help avoid this, the Graph Editor offers the ability to lock the object—it locks the position and properties of the object until you unlock it. Locked objects have a little padlock on them in the Object Browser. You can lock an object by right-clicking on the object in the graph (or in the Object Browser) and selecting **Lock Object**. Once locked, you can no longer use the Pointer ⬉ tool to either select, move, or modify the object. To unlock an object, you must right-click on it in the Object Browser and select **Unlock Object**.

You can also lock the position of an object such that you are permitted to select and modify an object with the Pointer ⌖ tool, but you are prevented from moving the object with the Pointer ⌖ tool. You can do this by right-clicking the object in the graph (or in the Object Browser) and selecting **Lock Position**. There is no symbol that indicates that the position of an Object is locked. When only the position of an object is locked, you can unlock the position by right-clicking the object in the graph (or the corresponding name in the Object Browser) and selecting **Unlock Position**.

The position of some objects is permanently locked, such as the x and y axis, and cannot be moved with the Pointer ⌖ tool (but can be moved with the Grid Edit ⊞ tool); see Editor : Moving (62). Let's have a quick look at examples illustrating locking and unlocking.

```
graph twoway scatter propval100 urban
```

Example 1 of 4. Let's right-click on the x title of this graph and select **Lock Object**. We are no longer able to select the object with the Pointer ⌖ tool, so we can no longer select, move, or modify it. Note the padlock next to **title** under the **xaxis1** in the Object Browser (pictured at the right).
Uses allstates.dta & scheme vg_s2c

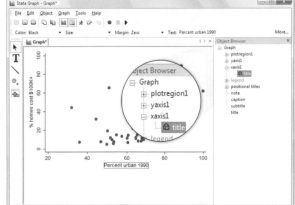

```
graph twoway scatter propval100 urban
```

Example 2 of 4. To unlock the x title, in the Object Browser right-click on **title** under the **xaxis1** and select **Unlock Object**. The padlock is removed (pictured at the right), and we can use the Pointer ⌖ tool to select, move, or modify it.
Uses allstates.dta & scheme vg_s2c

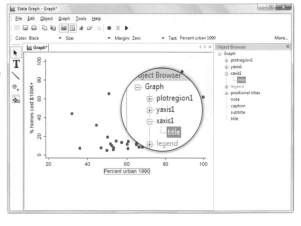

Introduction   Editor   Twoway   Matrix   Bar   Box   Dot   Pie   Options   Standard options   Styles   Appendix

Overview   Browser   Modifying   Adding   Moving   Hiding/Showing   Locking/Unlocking   Graph Recorder   Editor vs. commands

```
graph twoway scatter propval100 urban
```

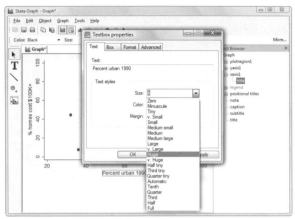

Example 3 of 4. Let's now lock the position of the x title by right-clicking on it and selecting **Lock Position**. Although we cannot move the object, we can still double-click on it and modify the properties. Here we double-click on the x title and change the **Size** to Huge.
Uses allstates.dta & scheme vg_s2c

```
graph twoway scatter propval100 urban
```

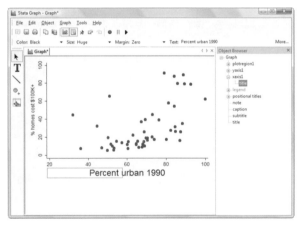

Example 4 of 4. Now right-click on the title and select **Unlock Position**. To prove that we have unlocked the position, select the x-axis title with the Pointer ➤ tool and move it (as illustrated here, the text can be moved to the left).
Uses allstates.dta & scheme vg_s2c

2.8   Using the Graph Recorder

One of the most powerful and useful features of the Stata Graph Editor is the Graph Recorder. With the Graph Recorder, you can record customizations for a graph for later playback. This is useful when you create a graph and would like to add some customizations that you want to repeat (replay) in the future without laboring through the process of repeating all the points and clicks. Further, you can create customizations that are more general and that you might want to apply to a wide variety of graphs. Both types of uses are illustrated in this section.

```
graph twoway (scatter married pov, mlabpos(12) mlabel(stateab))
    (lfit married pov)
```

Consider this graph from earlier in this chapter. Previously, we added text to this graph using the Add Text **T** tool to note that the observation for DC has a high residual. But, if we re-created this graph, the annotation would be lost. In the next example, we will *record* the customizations so we can play them back later.

Uses allstates.dta & scheme vg_s2c

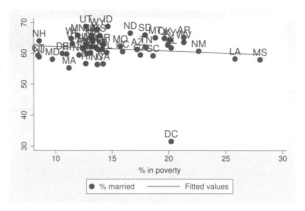

```
graph twoway (scatter married pov, mlabpos(12) mlabel(stateab))
    (lfit married pov)
```

Click on the Start/Stop Recording ● icon to begin the process of recording your customizations. Click on the Add Text **T** tool. Then click next to DC and for the **Text** enter "High" "Residual". Select the Pointer ![pointer] tool, click on the added text, and, in the Contextual Toolbar, change the **Size** to Large. Stop the recording by clicking on the Start/Stop Recording ● icon. Name the recording highres.

Uses allstates.dta & scheme vg_s2c

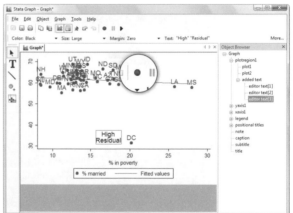

```
graph twoway (scatter married pov, mlabpos(12) mlabel(stateab))
    (lfit married pov)
```

Now enter this graph command again. The graph no longer has the annotation in it. Click on the Play Recording ▶ icon to see a list of your recordings (pictured at the right). Select highres, and the customizations saved in highres are played back, adding the text to the graph in a large font. When a recording is played, it moves to the top of your list of recordings.

Uses allstates.dta & scheme vg_s2c

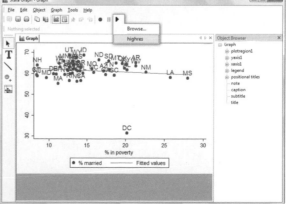

Introduction Editor Twoway Matrix Bar Box Dot Pie Options Standard options Styles Appendix

Overview Browser Modifying Adding Moving Hiding/Showing Locking/Unlocking Graph Recorder Editor vs. commands

The Play Recording ▶ icon selects and plays recordings. Suppose that you set up your graphs to display in the same window (as described in the paragraph about Graph Preferences on page 39). You can then quickly play the same recording for multiple graphs by tabbing to the next graph, pressing the Play Recording ▶ icon, selecting your recording, tabbing to the next graph, pressing the Play Recording ▶ icon, and so forth.

If you prefer, you can play back recordings with graph commands. One method is to add the `play()` option to a graph to play back a graph recording. For example, you could create the above graph with this command:

```
. graph twoway (scatter married pov, mlabpos(12) mlabel(stateab))
     (lfit married pov), play(highres)
```

The graph is first created, and then the customizations from `highres` are played back, creating the annotated graph. This is an excellent way to automate the process of playing back graph customizations, especially if you have a series of commands stored in a do-file.

You can also use the `graph play` command to play back a graph recording for the graph that is currently displayed. You could, for example, create a graph with the following command:

```
. graph twoway (scatter married pov, mlabpos(12) mlabel(stateab))
     (lfit married pov)
```

You can then issue the following command to play back the customizations we previously recorded in `highres`:

```
. graph play highres
```

The `graph play highres` command plays back the customizations recorded in `highres`, creating the annotated graph. This feature can be particularly useful if you wanted to play back more than one recording because you can repeat the `graph play` command as many times as you wish.

In the next example, let's make a recording that we could apply to a variety of graphs.

```
graph twoway scatter propval100 urban
```

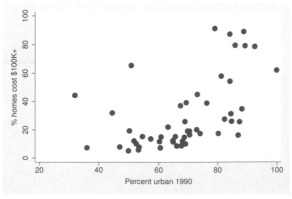

Start the Graph Recorder by clicking on the Start/Stop Recording ● icon. Click on the y axis. From the Contextual Toolbar change the **Label Size** to Large and the **Label Angle** to Horizontal. Click on **Show Grid** and **Axis Rule...**. Then click on **Suggest # of ticks** and enter 10. Click on the x axis and from the Contextual Toolbar change the **Label Size** to Large. Click on **Axis Rule...** and **Suggest # of ticks** and enter 10. Stop the Graph Recorder by clicking on the Start/Stop Recording ● icon. Name the recording `fancy`.

Uses allstates.dta & scheme vg_s2c

`graph twoway histogram tenure,` `play(fancy)`

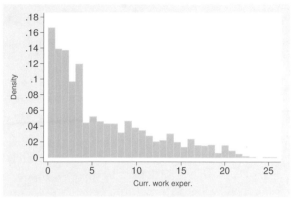

Adding the play(fancy) option to the command plays the customizations recorded in `fancy`. The recorded settings are successfully played for this graph, even though it is a different graph (and even a different data file).
Uses nlsw.dta & scheme vg_s2c

`graph bar hours, over(occ7) play(fancy)`

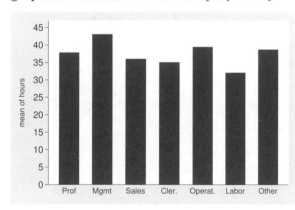

Recall that `fancy` had customizations with respect to both the x and y axes. For this graph, the x axis is a categorical variable, and the customizations from `fancy` are not appropriate; however, the recording does not fail. The appropriate customizations are played back, and any that are not appropriate for the graph are ignored, with a note written to the Results window for each edit that is not appropriate.
Uses nlsw.dta & scheme vg_s2c

One final feature that I did not address is the Pause Recording ▌▌ icon. If you are in the middle of a recording and want to pause the recording, you can click the Pause Recording ▌▌ button and the recording will be suspended. You can make changes to the graph and they will not be reflected in the recording. When you are ready, you can press the Pause Recording ▌▌ button again and the graph recording will resume.

As you can see, the Graph Recorder is a tool for bridging the gap between the Graph Editor and graph commands. With the `play()` option or the `graph play` command, you can enjoy the reproducibility and automation that you get with graph commands combined with the ease of use afforded by the Graph Editor. Now let's take a look at the Graph Editor versus Stata commands.

Introduction Editor Twoway Matrix Bar Box Dot Pie Options Standard options Styles Appendix

Overview Browser Modifying Adding Moving Hiding/Showing Locking/Unlocking Graph Recorder Editor vs. commands

2.9   Graph Editor versus Stata commands

The title of this section is a bit misleading. It sounds like a battle between the rookie Graph Editor versus the veteran Stata commands. Instead, the purpose of this section is to discuss the situations where you might find that graph commands are more useful than the Graph Editor, and vice versa. But the ultimate goal, despite the competitive-sounding title of this section, is to help you to use Stata commands and the Graph Editor together to create the best graphs with the least effort on your part.

I approach this section with the perspective (or bias) that it is most desirable to use graph commands to create graphs because such graphs can be easily, and automatically, reproduced. By contrast, graphs created using the Graph Editor require manual labor to reproduce. Because of this, my general recommendation is to create graphs with graph commands as often as possible. But, if there remain any customizations that are especially difficult with graph commands and simple with the Graph Editor, then do such customizations in the Graph Editor. You can use the Graph Recorder to record those customizations, and you can use either the `play()` option or the `graph play` command to automatically incorporate those customizations into future graphs; see Editor : Graph Recorder (78) for more details. Then, just before you are ready to "publish" your graph (e.g., print it, put it in your slide presentation, paste it into your document), you can use the Graph Editor to do any last minute aesthetic touch-ups.

This section begins by describing situations where the graph commands offer a big edge over the Graph Editor, followed by a discussion of situations where the Graph Editor offers a big edge over graph commands. It concludes with suggestions for how to use these two tools cooperatively.

Operations where graph commands offer an edge

Aside from the general recommendation to use graph commands whenever possible, here are some situations where I feel the graph commands are especially better than their Graph Editor counterparts.

Legend customizations. Although the Graph Editor offers a nice suite of tools allowing you to customize legends, using these tools has a hidden peril. Suppose that you make customizations to the legend such as changing the names of the key labels or rearranging the order of keys. Then, with the legend selected, you innocently use the Contextual Toolbar to change the number of columns in the legend. Suddenly, all the customizations that you previously made are lost. (If you double-clicked on the legend and changed the number of columns in the dialog box, you would be warned of this loss of previous changes.) This is one reason I prefer to use the `legend()` option for modifying legends, in addition to the fact that the `legend()` option is both powerful and logical for controlling the display of the legend. See Options : Legend (361) for more details.

Moving scale to alternate positions. The x scale, by default, is located at the bottom of the plot region, and the y scale, by default, is located at the left of the plot region. Alternatively, you can relocate the x scale to the top of the plot region with the `xscale(alt)` option, and you can relocate the y scale to the right with the `yscale(alt)` option. Although it is possible to use the Grid Edit ⬚ tool to move the x and y scales to their alternate positions, the graph command offers a much simpler way of doing this.

Sorting bars, boxes, and dots. One incredible feature of the `graph bar`, `graph box`, and `graph dot` commands is the detailed options they offer for sorting the bars, boxes, and dots. In fact, a section for each of these commands is devoted to the sorting options available; see Bar : Sorting (188), Box : Sorting (235), and Dot : Sorting (271). By contrast, for simplicity the Graph Editor offers a more limited set of sorting options.

Labeling axes with text. The graph commands offer a simple and powerful way to label an axis with descriptive text. For example, in Options : Axis labels (330) there is an example where the x axis is `region` and the axis is labeled with the region names by using the option `xlabel(1 "NorthEast" 2 "NorthCentral" 3 "South" 4 "West")`. Although this same labeling can be done with the Graph Editor, the nature of point-and-click interfaces requires opening a dialog box for each value to be labeled (i.e., opening the dialog box four times). The `xlabel()` option permits you to do this labeling in one fell swoop.

Unequal spacing of labels. I would say that 99% of the time you want the labels of your axes to be equally spaced, for example, from 0 to 100 in increments of 10. Both the graph commands and the Graph Editor offer you simple means for such labeling. But, 1% of the time you may want unequal spacing of the labels, such as when a scale is labeled using a log scale. The Graph Editor does not offer unequal scaling options, but you can use the `xlabel()` option for such scaling, for example, `xlabel(1 10 100 1000 10000)`.

Operations where the Graph Editor offers an edge

It might seem that the Graph Editor is taking it on the chin. But wait until you see the nice things that you can do with the Graph Editor, which can be cumbersome or impossible with graph commands.

Adding objects. Below we see an example of the commands needed when using the graph command to add text, with an arrow pointing to one observation to indicate that it is a possible outlier.

Introduction   Editor   Twoway   Matrix   Bar   Box   Dot   Pie   Options   Standard options   Styles   Appendix

Overview   Browser   Modifying   Adding   Moving   Hiding/Showing   Locking/Unlocking   Graph Recorder   Editor vs. commands

```
graph twoway (scatter ownhome propval100)
   (pcarrowi 42.5 26 42.5 61.3, lwidth(medthick) lcolor(black)),
   text(42.5 12 "Possible Outlier", size(large)) legend(off)
```

I retried this command 20 times to
adjust the coordinates for the
placement of the text and arrow.
Uses allstatesdc.dta & scheme vg_s2c

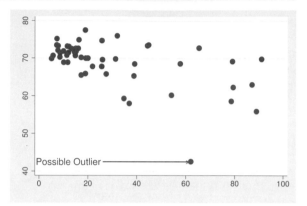

Using the Graph Editor, it would have been a snap to add the text and arrow to the
graph and simple to make the fine adjustments until the size and positions were exactly to
your liking; see Editor : Adding (54) for examples.

Modifying one object. As a variation to the previous example, let's call attention to the
outlying observation by making it look different, say, displayed as a red square. Consider
the additional syntax needed in the following graph to do this with the graph command.
By contrast, if you were to do this in the Graph Editor, you would simply right-click on the
outlying observation, select **Observation Properties**, and change the **Symbol** to Square
and the **Color** to Red; see Editor : Modifying (47) for more details. You can also use this
method in the Graph Editor to control the display of individual bars created by graph bar,
individual boxes created by graph box, and individual dots created by graph dot.

```
graph twoway (scatter ownhome propval100)
   (scatteri 42.6 62.1, msymbol(S) color(red)),
   legend(off) ytitle(Percent Own Home) xtitle(Property Value)
```

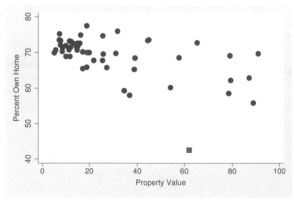

With syntax, we need to obtain the
coordinates for the outlying
observation, and then add the
scatteri command to plot a point
exactly over the outlying observation.
We then need to suppress the legend,
which would appear for a plot created
with the scatteri() command, and we
need to supply a title for the *x* and *y*
titles, which disappear with the
scatteri() command.
Uses allstatesdc.dta & scheme vg_s2c

A related capability is controlling the position of individual marker labels. With the graph command, you need to create a separate variable that contains the marker positions for each observation and then use the `mlabvpos()` option; see Options: Marker labels (320). By contrast, in the Graph Editor, you can right-click on a marker label and select **Observation Properties**. Within the dialog box, you can, not only change the position of the individual marker label, but you can also change the size, color, angle, and even the label itself.

Axes and titles with by().

The section on using the `by()` option [Options: By (346)] illustrates the use of options like `iyaxes`, `ixaxes`, `iytitle`, and `ixtitle` to control whether individual axes and titles are displayed for each *bygroup*. The Graph Editor gives you finer (and simpler) control over the display of these titles and axes. Consider the following example.

```
twoway scatter ownhome borninstate,
    by(north, total ixaxes ixtitle iyaxes iytitle)
```

Here we use the `ixaxes ixtitle iyaxes iytitle` options to individually display the *x* and *y* axes and titles for all the graphs created from the `by(north)` option.
Uses allstates.dta & scheme vg_s2c

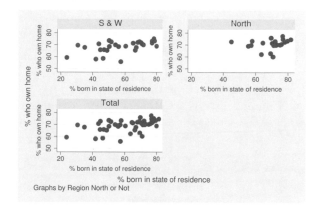

Let's explore the ways to control axes and titles using the Graph Editor. Create the graph from the previous example and start the Graph Editor. In the Object Browser, expand the object named **plotregion1**; you will see objects named **yaxis1[1]**, **yaxis1[2]**, and **yaxis1[3]**. These correspond to the *y* axes for each of the three graphs (S & W, North, and Total). You can expand each of these objects to find an object named **title**, which corresponds to the *y*-axis title for the three graphs. Likewise, you will find objects named **xaxis1[1]**, **xaxis1[2]**, and **xaxis1[3]**, which correspond to the *x* axis for each of the three graphs, each of which has an object named **title** below it, which corresponds to the title of the corresponding *x* axis. Let's use the Graph Editor to hide the entire *y* axis for the second graph (North) and the *x*-axis title for the first graph (S & W).

Introduction   Editor   Twoway   Matrix   Bar   Box   Dot   Pie   Options   Standard options   Styles   Appendix

Overview   Browser   Modifying   Adding   Moving   Hiding/Showing   Locking/Unlocking   Graph Recorder   Editor vs. commands

```
twoway scatter ownhome borninstate,
   by(north, total ixaxes ixtitle iyaxes iytitle)
```

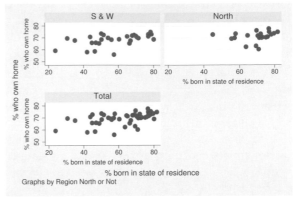

To hide the entire y axis for the second graph (North), right-click on **yaxis2[1]** and select **Hide**. To hide the x-axis title for the first graph (S & W), find **title** below **xaxis1[1]**, right-click on it, and select **Hide**.

Uses allstatesdc.dta & scheme vg_s2c

Using the Object Browser in the Graph Editor, you can select any of the axes and titles, right-click on them, and **Show** or **Hide** whatever combination of axes and titles that you want, illustrating the degree of control you have with the Graph Editor.

Rotating axes for bar, box, and dot plots. One of the most simple and powerful features in the Graph Editor is the **Rotate Categories** button that is available when you create bar, box, or dot plots. For example, consider the following graph.

```
graph hbar prev_exp tenure ttl_exp, over(occ5) over(union)
```

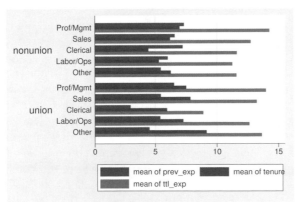

Here we see that previous experience, current experience, and total experience are broken down by occupation and whether one is in a union. It can be cumbersome with graph commands to rotate the display of the categories, but this is easily done with the Graph Editor. Within the Object Browser, simply double-click on **bar region** and then click on the **Rotate Categories** button.

Uses nlsw.dta & scheme vg_s2c

In fact, you can click on **Rotate Categories** as described in the previous example to get six different views of your data. This is a quick and easy way to get the view that expresses your data most effectively. Figure 2.3 shows the six different views. The text has been shrunk in size to be able to show all six graphs on one page.

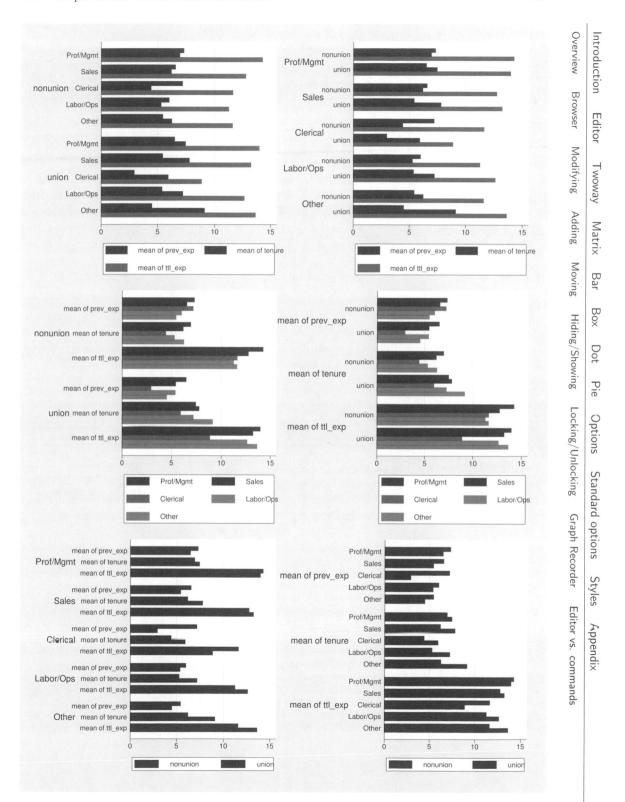

Figure 2.3: Bar graphs resulting from Rotate Categories button

Immediate feedback. This not a specific feature but just the general nature of the Graph Editor. If you are wondering how the graph might look if you made a particular change, you see the impact immediately with the Graph Editor, whereas you need to redraw the entire graph to see the impact by changing and reissuing a graph command. This is a useful segue into the final topic of this section, using the Graph Editor and graph commands collaboratively.

Using the Graph Editor and graph commands collaboratively

Get quick feedback with the Graph Editor. Even if you are building a graph command, there is no reason that you cannot use the Graph Editor to get quick feedback as to how the changes will look. You can change the color of objects, resize them, change the thicknesses of lines, change the scaling of axes, and so forth. You can quickly get the graph looking the way that you like because of the immediate feedback you get from the Graph Editor. Once you have the graph looking exactly the way that you like, go back and modify your graph command to include the customizations that you made with the Graph Editor.

Use graph commands to help you learn the Graph Editor. Maybe you are familiar with certain graph commands, but you are not sure how they are implemented with the Graph Editor. You can issue a command with the option specified, and then see if you can locate within the Graph Editor how this change was made.

Automate customizations with the Graph Recorder. You might want to leave some final customizations for the Graph Editor. The best part is that with the Graph Recorder, you can record those customizations and automate the process of incorporating those customizations by adding the `play()` option or by using the `graph play` command. These commands are discussed in more detail in Editor : Graph Recorder (78). This way you do not need to worry that you might forget what these final touch-ups are or that you might forget to do them. By using the `play()` option or by using the `graph play` command, you can be sure that the customizations will be automatically included when you re-create the graphs.

I do not feel that using the Graph Editor or the graph commands is an either/or situation. I hope this section has helped and encouraged you to use the two tools together for creating your graphs.

3   Twoway graphs

The graph twoway command represents not just one kind of graph but actually over thirty different kinds of graphs. Many of these graphs are similar in appearance and function, so I have grouped them into nine families, which form the first nine sections of this chapter. These first nine sections, which discuss scatterplots to contour plots, cover the general features of these graphs and briefly mention some important options. The next section gives an overview of the options you can use with twoway graphs. [For further details about the options you can use with twoway graphs, see Options (307).] The chapter concludes with a section illustrating how you can overlay twoway graphs. For more details about twoway graphs, see [G-2] **graph twoway**.

3.1   Scatterplots

This section covers the use of scatterplots. Because scatterplots are so commonly used, this section will cover more details about the use of these graphs than subsequent sections. Also, this section introduces some of the options that we can use with many twoway plots, with cross-references to Options (307).

```
graph twoway scatter ownhome propval100
```

Here is a basic scatterplot. This command starts with graph twoway, which indicates that this is a twoway graph. scatter indicates that we are creating a twoway scatterplot. We next list the variables to be placed on the y and x axis, respectively.
Uses allstates.dta & scheme vg_s2c

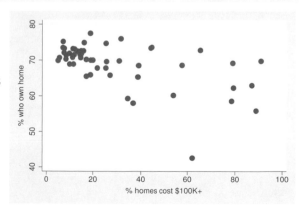

`twoway scatter ownhome propval100`

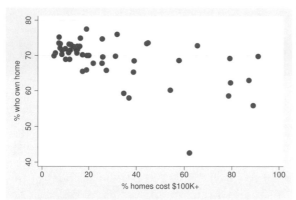

Because it can be cumbersome to type
`graph twoway scatter`, Stata allows
us to shorten this command to `twoway`
`scatter`.
Uses allstates.dta & scheme vg_s2c

`scatter ownhome propval100`

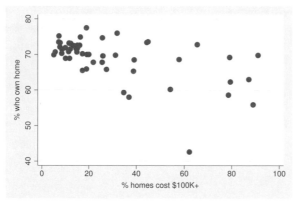

In fact, some `graph twoway` commands
are so frequently used that Stata
permits us to omit `graph twoway`. Here
we start the command with `scatter`.
Although this omission can save some
typing, it can sometimes conceal the
fact that the command is really a
twoway graph and that these are a
special class of graphs. For clarity, I
will generally present these graphs
starting with `twoway`.
Uses allstates.dta & scheme vg_s2c

`twoway scatter ownhome propval100, msymbol(Sh)`

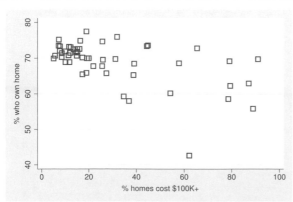

We can control the marker symbol with
the `msymbol()` option. Here we make
the symbols large, hollow squares. See
Options : **Markers** (307) for more details
about controlling the marker symbol,
size, and color. See **Styles** : **Symbols**
(427) for the available symbols.

Click on a marker and then, in the
Contextual Toolbar, change the
Symbol to **Hollow square**.
Uses allstates.dta & scheme vg_s2c

`twoway scatter ownhome propval100, mcolor(maroon)`

We can control the marker color with the `mcolor()` option. Here we change the marker color to maroon. See Styles : Colors (412) for other colors, and also see Options : Markers (307) for more details about markers.

Click on a marker and then, in the Contextual Toolbar, change the **Color** to maroon.

Uses allstates.dta & scheme vg_s2c

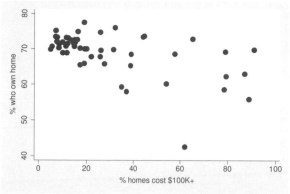

`twoway scatter ownhome propval100, msize(vlarge)`

We can control the marker size with the `msize()` option. By using `msize(vlarge)`, we make the markers very large. We switched to the `vg_outc` scheme, showing white-filled markers, which can be useful when the markers are large. See Styles : Markersize (425) for other available sizes and also see Options : Markers (307) for more details about markers.

Click on a marker and then, in the Contextual Toolbar, change the **Size** to v Large.

Uses allstates.dta & scheme vg_outc

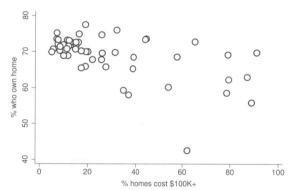

`twoway scatter ownhome propval100 [aweight=rent700], msize(small)`

We can also use a weight variable to determine the size of the symbols. By using `[aweight=rent700]`, the symbols are sized according to the proportion of rents that exceed $700 per month, allowing us to graph three variables at once. We add the `msize(small)` option to shrink the size of all the markers so they do not get too large. See Options : Markers (307) for more details.

Uses allstates.dta & scheme vg_outc

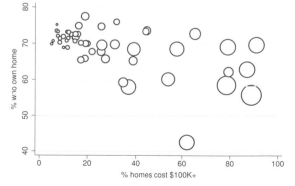

Introduction Editor Twoway Matrix Bar Box Dot Pie Options Standard options Styles Appendix

Scatter Fit CI fit Line Area Bar Range Distribution Contour Options Overlaying

`twoway scatter ownhome propval100, mlabel(stateab)`

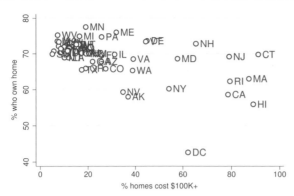

The `mlabel(stateab)` option adds a marker label with the state abbreviation. See Options : Marker labels (320) for more details about controlling the size, position, color, and angle of marker labels.

Uses allstates.dta & scheme vg_outc

`twoway scatter ownhome propval100, mlabel(stateab) mlabsize(vlarge)`

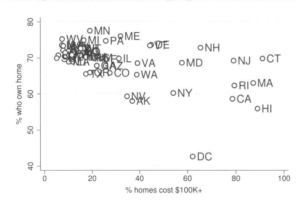

The `mlabsize()` option controls the marker label size. Here we make the marker label very large.

Click on a marker label (e.g., NY) and then, in the Contextual Toolbar, change the **Size** to v Large.

Uses allstates.dta & scheme vg_outc

`twoway scatter ownhome propval100, mlabel(stateab) mlabposition(12)`

The `mlabposition()` option controls the marker label position with respect to the marker. Here we place the marker labels at the twelve o'clock position, directly above the markers. See Options : Marker labels (320) for more examples.

Click on a marker label (e.g., NY) and then, in the Contextual Toolbar, change the **Position** to 12 o'clock.

Uses allstates.dta & scheme vg_outc

```
twoway scatter ownhome propval100, mlabel(stateab)
    mlabposition(0) msymbol(i)
```

The `mlabposition(0)` option places
the marker label in the center. The
`msymbol(i)` option makes the marker
symbol invisible. This replaces the
marker symbols with the marker labels.

Click on any marker and then, in
the Contextual Toolbar, change the
Symbol to `None`. Then select any
marker label (e.g., NY) and, in the
Contextual Toolbar, change the
Position to `Middle`.

Uses allstates.dta & scheme vg_outc

```
twoway scatter fv propval100
```

Say that we ran the following
commands:

```
. regress ownhome propval100
. predict fv
```

The variable `fv` represents the fit
values, and here we graph `fv` against
`propval100`. As we expect, all the
points fall along a line, but they are not
connected. The next few examples will
discuss options we can use to connect
points; see Options : Connecting (323)
for more details. For variety, we have
switched to the `vg_past` scheme.

Uses allstates.dta & scheme vg_past

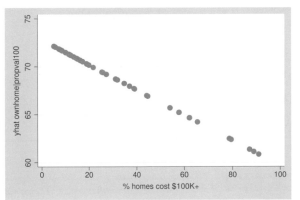

```
twoway scatter fv propval100, connect(l) sort
```

We add the `connect(l)` option to
indicate that the points should be
connected with a line. We also add the
`sort` option, which is generally
recommended when we connect
observations and the data arc not
already sorted on the *x* variable.

Uses allstates.dta & scheme vg_past

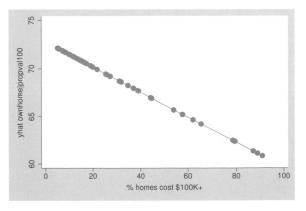

Scatter   Fit   CI fit   Line   Area   Bar   Range   Distribution   Contour   Options   Overlaying

Introduction   Editor   Twoway   Matrix   Bar   Box   Dot   Pie   Options   Standard options   Styles   Appendix

```
twoway scatter fv ownhome propval100, connect(l i) sort
```

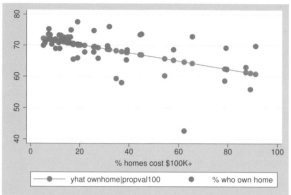

We can show both the observations and the fit values in one graph. The `connect(l i)` option specifies that the first *y* variable should be connected with straight lines (`l` for line) and the second *y* variable should not be connected (`i` for invisible connection).
Uses allstates.dta & scheme vg_past

```
twoway scatter fv ownhome propval100, msymbol(i .) connect(l i)
    sort
```

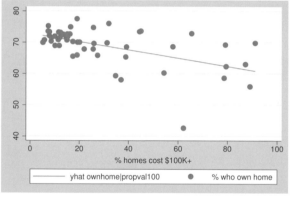

The `msymbol(i .)` option specifies that the first *y* variable should not have symbols displayed (`i` for invisible symbol) and that the second *y* variable should have the default symbols displayed.
Uses allstates.dta & scheme vg_past

```
twoway scatter fv ownhome propval100, msymbol(i .) connect(l i)
    sort legend(label(1 Pred. Perc. Own))
```

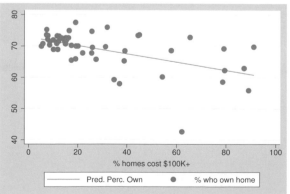

The `legend()` option controls the legend. We use `label()` within the `legend()` option to specify the contents of the first item in the legend. See Options: Legend (361) for more details on legends.

Click on a label in the legend and then, in the Contextual Toolbar, type in the **Text** for the label.
Uses allstates.dta & scheme vg_past

```
twoway scatter fv ownhome propval100, msymbol(i .) connect(l i)
    sort legend(label(1 Pred. Perc. Own) order(2 1))
```

The order() option within the
legend() option specifies the order in
which the items in the legend are
displayed.
Uses allstates.dta & scheme vg_past

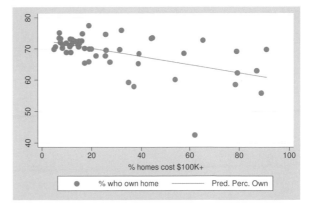

```
twoway scatter fv ownhome propval100, msymbol(i .) connect(l i)
    sort legend(label(1 Pred. Perc. Own) order(2 1) cols(1))
```

The cols(1) option makes the items in
the legend display in one column.
Select the legend by going to the
Object Browser and clicking on
legend. Then, in the Contextual
Toolbar, change the **Columns** to 1.
Uses allstates.dta & scheme vg_past

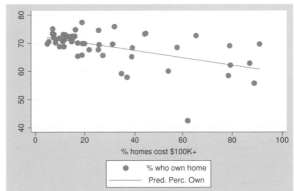

```
twoway scatter ownhome propval100,
    xtitle("Percent homes over $100K") ytitle("Percent who own home")
```

The xtitle() and ytitle() options
specify the titles for the x and y axes.
See Options: Axis titles (327) for more
details about how to control the display
of axes. We are now using the vg_s2m
scheme, one that you might favor for
graphs that will be printed in black and
white.
Click on the x-axis title and then, in
the Contextual Toolbar, change the
title. We can do the same for the y-axis
title.
Uses allstates.dta & scheme vg_s2m

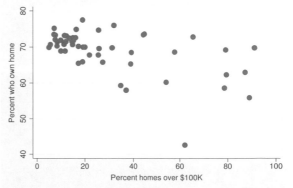

Introduction   Editor   Twoway   Matrix   Bar   Box   Dot   Pie   Options   Standard options   Styles   Appendix

Scatter   Fit   CI fit   Line   Area   Bar   Range   Distribution   Contour   Options   Overlaying

```
twoway scatter ownhome propval100,
    ytitle("Percent who own home", size(huge))
```

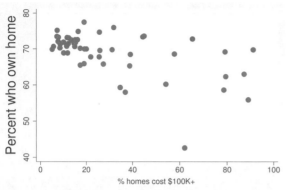

Here we use the `size(huge)` option to make the title on the y axis huge. For other text sizes, see **Styles : Textsize** (428).

🖱 Click on the y-axis title and then, in the Contextual Toolbar, change the **Size** to Huge.

Uses allstates.dta & scheme vg_s2m

```
twoway scatter ownhome propval100, xlabel(#10) ylabel(#5)
```

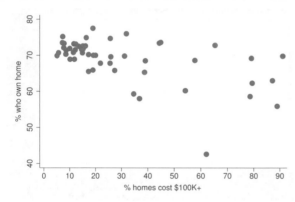

In this example, we use the `xlabel(#10)` option to ask Stata to use approximately 10 nice labels and the `ylabel(#5)` option to use approximately 5 nice labels. Here our gentle request was observed exactly, but sometimes Stata will choose somewhat different values to create axis labels it believes are logical. See **Options : Axis labels (330)** for more details on labeling axes.

🖱 See the next graph.

Uses allstates.dta & scheme vg_s2m

```
twoway scatter ownhome propval100, xlabel(#10) ylabel(#5)
```

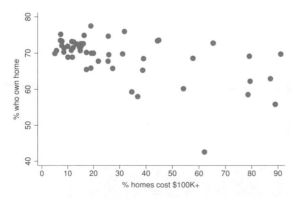

🖱 Using the Graph Editor, click on any number labeling the x axis and, in the Contextual Toolbar, click on **Axis Rule...**. Select **Suggest # of Ticks** and choose 10. Then click on any number labeling the y axis and, in the Contextual Toolbar, click on **Axis Rule...**. Select **Suggest # of Ticks** and choose 5.

Uses allstates.dta & scheme vg_s2m

```
twoway scatter ownhome propval100, xlabel(#10) ylabel(#5, nogrid)
```

We use the `nogrid` option within the `ylabel()` option to suppress the grid for the y axis (and we could show the grid by adding the `grid` option). You can also specify `grid` or `nogrid` within the `xlabel()` option to control grids for the x axis. For more details, see Options : Axis labels (330).

📊 Click on the y axis and then, in the Contextual Toolbar, click on **Show Grid** to toggle the display of the grid.
Uses allstates.dta & scheme vg_s2m

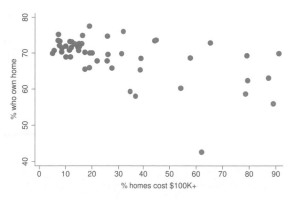

```
twoway scatter ownhome propval100, xlabel(#10) ylabel(#5, nogrid)
    yline(55 75)
```

The `yline()` option adds a horizontal reference line where y equals 55 and 75. You can add a vertical reference line with the `xline()` option.

📊 Double-click on the y axis, click on **Reference Line**, and then enter 55 75 for the **Y axis value**. You can add a vertical reference line in the same way after double-clicking on the x axis.
Uses allstates.dta & scheme vg_s2m

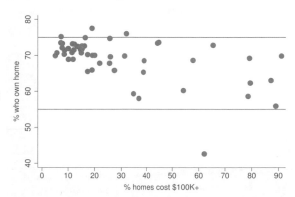

```
twoway scatter ownhome propval100, xscale(alt)
```

Here we use the `xscale()` option to request that the x axis be placed in its alternate position, which is at the top instead of at the bottom. To learn more about axis scales, including suppressing, extending, or relocating them, see Options : Axis scales (339).
Uses allstates.dta & scheme vg_s2m

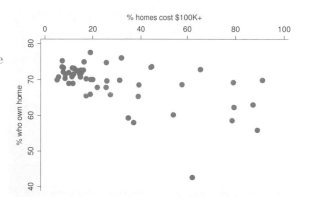

Introduction   Editor   Twoway   Matrix   Bar   Box   Dot   Pie   Options   Standard options   Styles   Appendix

Scatter   Fit   CI fit   Line   Area   Bar   Range   Distribution   Contour   Options   Overlaying

`twoway scatter ownhome propval100, by(nsw)`

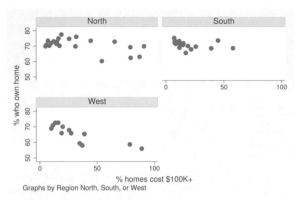

We use the `by(nsw)` option here to make separate graphs for states in the North, South, and West. At the bottom left corner, there is a note that describes how the graphs are separated, which is based on the variable label for `nsw`. If this variable had not been labeled, the variable name would have used, i.e., *Graphs by nsw*. See Options: By (346) for more details about using the `by()` option.
Uses allstates.dta & scheme vg_s2m

`twoway scatter ownhome propval100, by(nsw, total)`

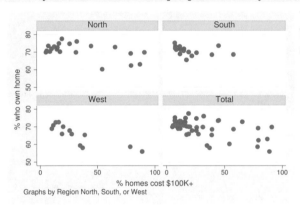

We can use the `total` option within the `by()` option to add another graph showing all the observations.
Uses allstates.dta & scheme vg_s2m

`twoway scatter ownhome propval100, by(nsw, total compact)`

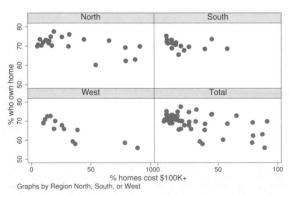

The `compact` option within the `by()` option makes the graphs display more compactly.
Uses allstates.dta & scheme vg_s2m

`twoway scatter ownhome propval100, text(47 62 "Washington, DC")`

The `text()` option adds text to the
graph. Here we add text to label the
observation belonging to Washington,
DC. See Options : Adding text (374) for
more information about adding text.

Click on the Add Text **T** tool, click
where to add the text, and then enter
`Washington, DC`.

Uses allstates.dta & scheme vg_s2m

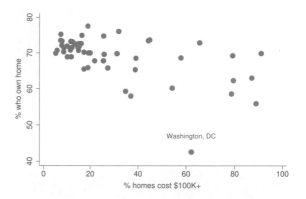

`twoway scatter ownhome propval100, text(47 62 "Washington, DC", size(large)`
`   lwidth(vthick) box)`

Here we make the text large and
surround it with a thick-lined box. See
Options : Textboxes (379) for more
details.

Continuing from the previous
example, double-click on *Washington,
DC*, and change the **Size** to **Large**. In
the *Box* tab, check **Place box around
text** and change the **Outline width** to
v Thick.

Uses allstates.dta & scheme vg_s2m

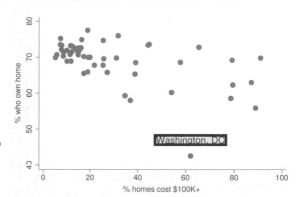

`twoway (scatter ownhome propval100) (scatteri 42.6 62.1 "DC")`

This graph uses the `scatteri` (scatter
immediate) command to plot and label
a point for Washington, DC. The values
`42.6` and `62.1` are the values for
`ownhome` and `propval100` for
Washington, DC, and are followed by
`"DC"`, which acts as a marker label for
that point.

See the previous two examples.
Uses allstates.dta & scheme vg_s2m

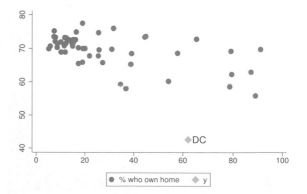

Introduction   Editor   Twoway   Matrix   Bar   Box   Dot   Pie   Options   Standard options   Styles   Appendix

Scatter   Fit   CI fit   Line   Area   Bar   Range   Distribution   Contour   Options   Overlaying

```
twoway (scatter ownhome propval100)
   (scatteri 42.6 62.1 "DC" 55.9 89 (8) "HI"), legend(off)
```

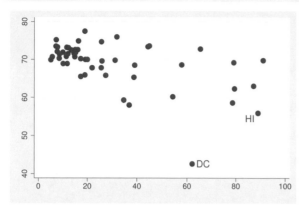

This graph also labels Hawaii at the eight o'clock position. The legend(off) option also suppresses the legend. Finally, this graph uses the vg_samec scheme so the markers created with scatteri look identical to the other markers.

Label Hawaii as shown in the previous examples. In the Object Browser, right-click on **legend** and then **hide** to hide the legend.

Uses allstates.dta & scheme vg_samec

Next let's turn to more graph commands that make graphs similar to twoway scatter, namely, twoway spike, twoway dropline, and twoway dot. Most of the options that have been illustrated apply to these graphs as well, so they will not be repeated here. The rest of the graphs in this section use the vg_blue scheme.

```
twoway scatter r yhat
```

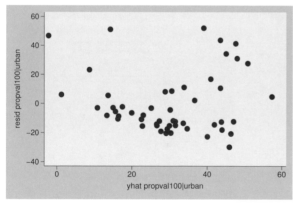

Imagine that we ran a regression predicting propval100 from urban and generated the residual, calling it r, and the predicted value, calling it yhat. Consider this graph using the scatter command to display the residual by the predicted value.

Uses allstates.dta & scheme vg_blue

twoway spike r yhat

This same graph could be shown using
the spike command. This command
produces a spike plot, with each spike,
by default, originating from 0.
Uses allstates.dta & scheme vg_blue

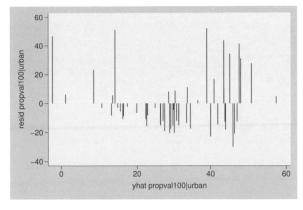

twoway spike r yhat, lcolor(red) lwidth(thick)

We can use the lcolor() (line color)
option to set the color of the spikes and
the lwidth() (line width) option to set
the width of the spikes. Here we make
the spikes thick and red. See
Styles : Colors (412) for more details
about specifying colors, and see
Styles : Linewidth (422) for more details
about specifying line widths.

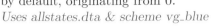 Click on a spike and, in the
Contextual Toolbar, change the **Color**
to Red and the **Width** to Thick.
Uses allstates.dta & scheme vg_blue

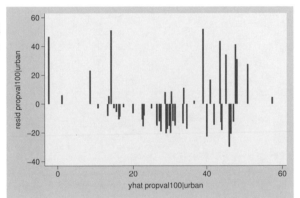

twoway spike r yhat, base(10)

By default, the base is placed at 0,
which is a logical choice when
displaying residuals because our interest
is in deviations from 0. For illustration,
we use the base(10) option to set the
base of the y axis to 10, and the spikes
are displayed with respect to 10.
Uses allstates.dta & scheme vg_blue

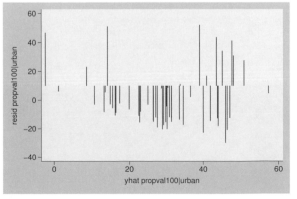

Introduction   Editor   Twoway   Matrix   Bar   Box   Dot   Pie   Options   Standard options   Styles   Appendix

Scatter   Fit   CI fit   Line   Area   Bar   Range   Distribution   Contour   Options   Overlaying

```
twoway spike r yhat, horizontal xtitle(Title for x axis)
    ytitle(Title for y axis)
```

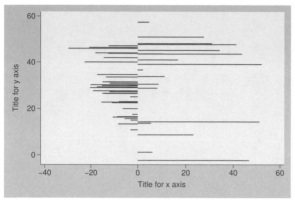

The `horizontal` option swaps the position of the `r` and `yhat` variables. The x axis remains at the bottom, and the y axis remains at the left.
Uses allstates.dta & scheme vg_blue

```
twoway dropline r yhat, msymbol(D)
```

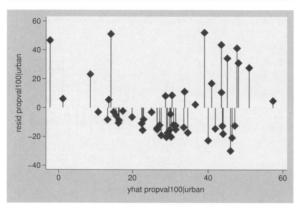

A `twoway` `dropline` plot is much like a spike plot but permits a symbol, as well. It supports the `horizontal`, `base()`, `lcolor()`, and `lwidth()` options just like `twoway spike`. Here we add the `msymbol(D)` option to obtain diamonds as the symbols; see Options : Markers (307) for more details.

📊 Click on a marker and use the Contextual Toolbar to change the **Symbol** to Diamond.
Uses allstates.dta & scheme vg_blue

```
twoway dropline r yhat, msymbol(D) msize(large) mcolor(purple)
    mlwidth(thick) lcolor(red)
```

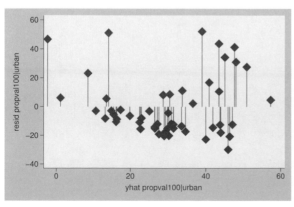

Here we make the symbols large and purple and the lines thick and red. For more information, see Options : Markers (307).

📊 Double-click on a marker and, in the *Spikes* tab, change the **Color** to Red and the **Width** to Thick. In the *Markers* tab, change the **Size** to Large and the **Color** to Purple.
Uses allstates.dta & scheme vg_blue

```
twoway dot close tradeday, msize(large) msymbol(O)
   mfcolor(eltgreen) mlcolor(emerald) mlwidth(thick)
```

A dot plot is similar to a scatterplot
but shows dotted lines for each x
variable value, making it more useful
when the x values are equally spaced.
Here we look at the closing price of the
S&P 500 by trading day and make the
markers filled with eltgreen with thick
emerald outlines.
Uses spjanfeb2001.dta & scheme
vg_blue

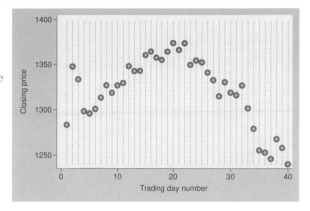

This section concludes with a special kind of scatterplot, a paired coordinate plot. As
its name implies, this plot takes as its input a pair of coordinates, for example, (y_1, x_1) and
(y_2, x_2), and then plots a line between each coordinate pair. The line may be plain (if you
use `pcspike`), capped with symbols (if you use `pccapsym`), terminated with an arrow (if you
use `pcarrow`), capped with bidirectional arrows (if you use `pcbarrow`), or a pair of marker
symbols (if you use `pcscatter`). To illustrate these graphs, we will begin by using one of
the built-in data files from Stata named `nlswide1`, which contains nine observations with
aggregate data on nine different occupation classes for 1968 and for 1988. We will focus
on hourly wages (`wage68` and `wage88`) and hours worked per week (`hours68` and `hours88`).
These paired coordinate graphs will help us see how these two variables have changed over
this 20-year span for these nine occupations.

```
twoway (pcspike wage68 hours68 wage88 hours88)
```

The command's syntax is to provide
the (y_1, x_1) variables for time 1 and
then the (y_2, x_2) for time 2. This graph
is interesting, but without any labels it
is hard to understand the trends being
conveyed.
Uses nlswide1.dta & scheme vg_blue

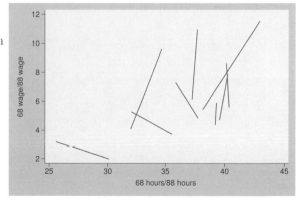

```
twoway (pcspike wage68 hours68 wage88 hours88)
   (scatter wage88 hours88, msymbol(i) mlabel(occ) mlabsize(small))
```

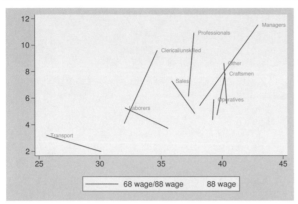

We can combine the previous command with a `scatter` command to label the data in 1988 with the names of the occupations. This graph is easier to interpret.

Uses nlswide1.dta & scheme vg_blue

```
twoway pccapsym wage68 hours68 wage88 hours88,
   mlabel(occ) mlabsize(small) headlabel
```

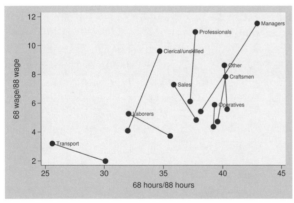

A simpler solution is to use the `pccapsym` command with the `mlabel()` option to label the lines with the occupation names. The `headlabel` option labels the head of the line (i.e., 1988) rather than the default of labeling the tail (i.e., 1968). A related command is `pcscatter` (not illustrated), which would display the symbols without the lines.

Uses nlswide1.dta & scheme vg_blue

```
twoway pcarrow wage68 hours68 wage88 hours88,
   mlabel(occ) mlabsize(small)
```

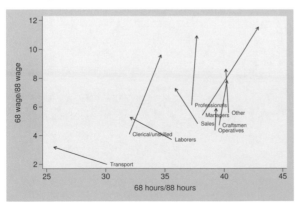

Because these data represent changes in time, the graph may be more compelling by displaying the lines as arrows. We now omit the `headlabel` option so the occupation names are displayed at the tail of the arrow. A related command is `pbcarrow` (not illustrated), which would display bidirectional arrows.

Uses nlswide1.dta & scheme vg_blue

In addition to the traditional forms of these commands, there are two immediate forms of these commands that permit us to supply coordinates as part of the command. The `twoway pci` command displays lines, whereas the `twoway pcarrowi` command displays arrows. We now switch to the `allstates` data file. We use an outlier in a scatterplot, with lines and arrows to call attention to the outlier.

```
graph twoway (scatter ownhome propval100)
    (pci 42.5 26 42.5 61.3, lwidth(medthick) lcolor(black))
```

Here we add a line from (42.5,26) to (42.5,61.3) to help call attention to this outlying observation. However, an arrow might be more effective.

Uses allstates.dta & scheme vg_blue

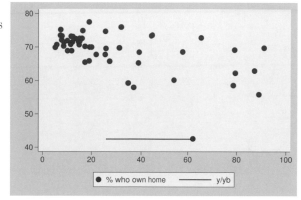

```
graph twoway (scatter ownhome propval100)
    (pcarrowi 42.5 26 42.5 61.3, lwidth(medthick) lcolor(black)
    msize(5) barbsize(3) mcolor(black))
```

By using `pcarrowi`, we create an arrow with an arrowhead of size 5 and barb of size 3 that points to this outlying observation.

Uses allstates.dta & scheme vg_blue

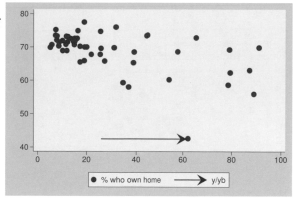

Introduction Editor Twoway Matrix Bar Box Pie Options Standard options Styles Appendix

Scatter Fit CI fit Line Area Bar Range Distribution Contour Options Overlaying

`no command`

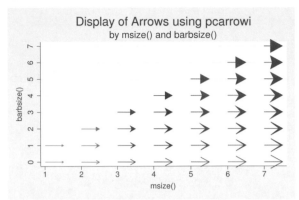

Here is a display of arrows using `pcarrowi` to illustrate the effect of `msize()` and `barbsize()`. The values of `msize()` range from 1 to 7, and the values of `barbsize()` range from 0 to 7. As you move to the right, the overall arrowhead size gets larger (due to increasing `msize()` from 1 to 7). As you move up, the amount of the arrowhead that is filled up increases (due to increasing `barbsize` from 0 to 7).

Uses none.dta & scheme vg_s2c

3.2   Regression fits and splines

This section focuses on the twoway commands you can use for displaying fit values: `lfit`, `qfit`, `fpfit`, `mband`, `mspline`, and `lowess`. For more information, see [G-2] **graph twoway lfit**, [G-2] **graph twoway qfit**, [G-2] **graph twoway fpfit**, [G-2] **graph twoway mband**, [G-2] **graph twoway mspline**, and [G-2] **graph twoway lowess**. We use the `allstates` data file, omitting Washington, DC, and show the graphs using the `vg_s2c` scheme.

`twoway (scatter ownhome pcturban80) (lfit ownhome pcturban80)`

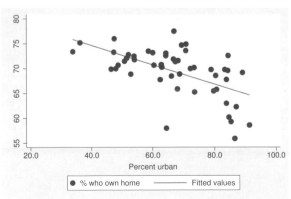

Here we show a scatterplot of `ownhome` by `pcturban80`. We also overlay a linear fit (`lfit`) predicting `ownhome` from `pcturban80`. See Twoway : Overlaying (152) for more information about overlaying twoway graphs.

Uses allstatesdc.dta & scheme vg_s2c

```
twoway (scatter ownhome pcturban80) (lfit ownhome pcturban80),
    pcycle(1)
```

In the previous graph, the line color
was different from the marker color. By
adding the `pcycle(1)` option, we
indicate that the pen characteristics
should repeat after one iteration. So
the scatterplot shows as blue markers,
and then the line is shown as a
matching blue line.
Uses allstatesdc.dta & scheme vg_s2c

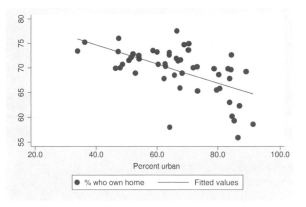

```
twoway (scatter ownhome pcturban80) (lfit ownhome pcturban80)
    (qfit ownhome pcturban80)
```

It is sometimes useful to overlay fit
plots to compare the fit values. Here we
overlay a linear fit (`lfit`) and quadratic
fit (`qfit`) and can see some
discrepancies between them.
Uses allstatesdc.dta & scheme vg_s2c

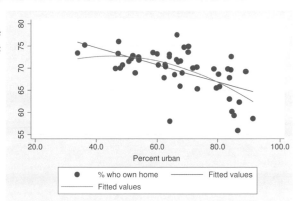

```
twoway (scatter ownhome pcturban80) (mspline ownhome pcturban80)
    (fpfit ownhome pcturban80) (lowess ownhome pcturban80)
```

Stata supports several other fit
methods. Here we show a median spline
(`mspline`) overlaid with a fractional
polynomial fit (`fpfit`) and a locally
weighted scatterplot smoothing
(`lowess`).
Uses allstatesdc.dta & scheme vg_s2c

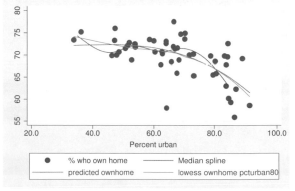

```
twoway (scatter ownhome pcturban80) (mband ownhome pcturban80)
    (lpoly ownhome pcturban80)
```

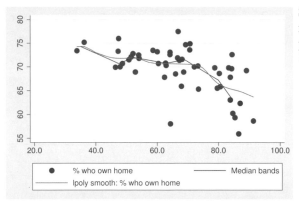

Stata supports two other fit methods: median band (`mband`) and local polynomial smooth (`lpoly`).
Uses allstatesdc.dta & scheme vg_s2c

```
lpoly ownhome pcturban80, degree(2)
```

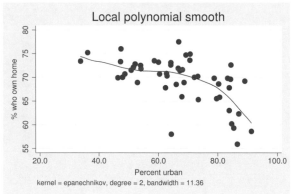

We can also use the `lpoly` command on its own. Here we add the `degree(2)` option to specify a second-degree polynomial for the smoothing. See `help lpoly` for more options.
Uses allstatesdc.dta & scheme vg_s2c

3.3   Regression confidence interval fits

This section focuses on the twoway commands that are used for displaying confidence intervals (CI) around fit values: `lfitci`, `qfitci`, and `fpfitci`. The options permitted by these three commands are virtually identical, so I have selected `lfitci` to illustrate these options. (However, `fpfitci` does not permit the options `stdp`, `stdf`, and `stdr`.) For more information, see [G-2] **graph twoway lfitci**, [G-2] **graph twoway qfitci**, and [G-2] **graph twoway fpfitci**.

`twoway (lfitci ownhome pcturban80) (scatter ownhome pcturban80)`

This graph uses the `lfitci` command to produce a linear fit with CI. The CI, by default, is computed using the standard error of prediction. We overlay this fit with a scatterplot.
Uses allstatesdc.dta & scheme vg_rose

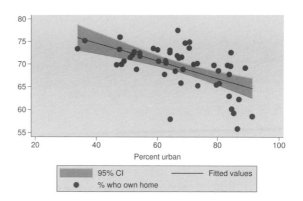

`twoway (scatter ownhome pcturban80) (lfitci ownhome pcturban80)`

This example is the same as the previous example; however, the order of the `scatter` and `lfitci` commands is reversed. The order matters because the points that fell within the CI are not displayed as they are masked by the shading of the CI.
Uses allstatesdc.dta & scheme vg_rose

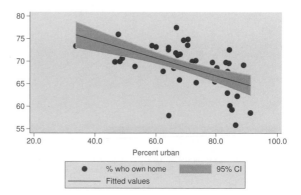

`twoway (lfitci ownhome pcturban80, stdf)`
`  (scatter ownhome pcturban80)`

Here we add the `stdf` option, which computes the CIs using the standard error of forecast. If samples were drawn repeatedly, this CI would capture 95% of the observations. With 50 observations, we would expect 2 or 3 observations to fall outside the CI. This expectation corresponds to the data shown here.
Uses allstatesdc.dta & scheme vg_rose

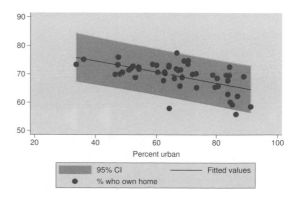

Introduction Editor Twoway Matrix Bar Box Dot Pie Options Standard options Styles Appendix

Scatter Fit CI fit Line Area Bar Range Distribution Contour Options Overlaying

```
twoway (lfitci ownhome pcturban80, stdf level(90))
    (scatter ownhome pcturban80)
```

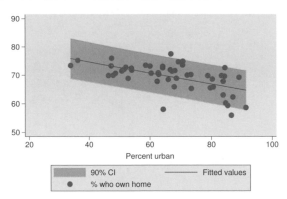

We can use the `level()` option to set the confidence level for the CI. Here we make the confidence level 90%.
Uses allstatesdc.dta & scheme vg_rose

```
twoway (lfitci ownhome pcturban80, nofit)
    (scatter ownhome pcturban80)
```

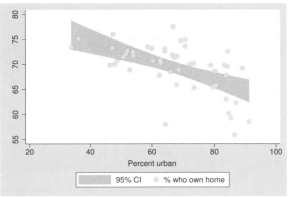

We now look at how to control the display of the fit line. We can use the `nofit` option to suppress the display of the fit line. We have switched to the `vg_brite` scheme for a different look of the graphs.
Uses allstatesdc.dta & scheme vg_brite

```
twoway (lfitci ownhome pcturban80, lpattern(dash) lwidth(thick))
    (scatter ownhome pcturban80)
```

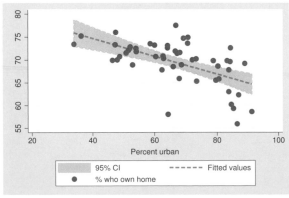

We can supply options like `connect()`, `lpattern()` (line pattern), `lwidth()` (line width), and `lcolor()` (line color) to control how the fit line will be displayed. Here we use the `lpattern(dash)` and `lwidth(thick)` options to make the fit line dashed and thick. See **Options : Connecting** (323) for more details.

▨ Click on the fit line and, in the Contextual Toolbar, change the **Width** to Thick and the **Pattern** to Dash.
Uses allstatesdc.dta & scheme vg_brite

```
twoway (lfitci ownhome pcturban80, bcolor(stone))
        (scatter ownhome pcturban80)
```

The bcolor(stone) option changes the color of the area and outline of the CI. You can use the options illustrated with twoway rarea to control the display of the area encompassing the CI, namely, bcolor(), fcolor(), lcolor(), lwidth(), and lpattern(). See Twoway : Range (123) and [G-2] **graph twoway rarea** for more details.

📖 Click the confidence band and, in the Contextual Toolbar, change the **Color** to Stone.

Uses allstatesdc.dta & scheme vg_brite

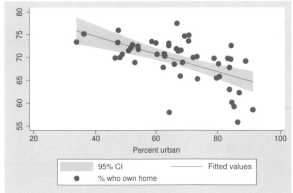

```
twoway (lfitci ownhome pcturban80, ciplot(rline))
        (scatter ownhome pcturban80)
```

The ciplot() option selects a different command for displaying the CI; the default is ciplot(rarea). Here we use the ciplot(rline) option to display the CI as two lines without a filled area. The valid options include rarea, rbar, rspike, rcap, rcapsym, rscatter, rline, and rconnected.

📖 Click on the confidence bands and, in the Contextual Toolbar, change the **Plottype** to Rline.

Uses allstatesdc.dta & scheme vg_brite

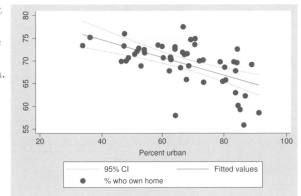

```
twoway (lfitci ownhome pcturban80, ciplot(rline) lcolor(green)
        lpattern(dash) lwidth(thick)) (scatter ownhome pcturban80)
```

Here we make the lines green, dashed, and thick. See Styles : Colors (412), Styles : Linepatterns (420), and Styles : Linewidth (422) for more details about colors, line patterns, and line widths.

📖 Click on the confidence bands and, in the Contextual Toolbar, change the **Color** to Green, the **Width** to Thick, and the **Pattern** to Dash. Then select the fit line and make the same changes.

Uses allstatesdc.dta & scheme vg_brite

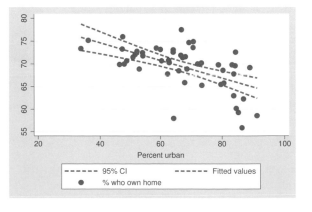

Introduction Editor Twoway Matrix Bar Box Dot Pie Options Standard options Styles Appendix

Scatter Fit CI fit Line Area Bar Range Distribution Contour Options Overlaying

`twoway (qfitci ownhome pcturban80) (scatter ownhome pcturban80)`

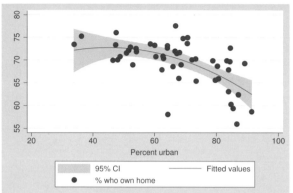

In addition to the `lfitci` command, the `qfitci` command produces a quadratic fit with a CI. We overlay this fit with a scatterplot.
Uses allstatesdc.dta & scheme vg_brite

`twoway (lpolyci ownhome pcturban80) (scatter ownhome pcturban80)`

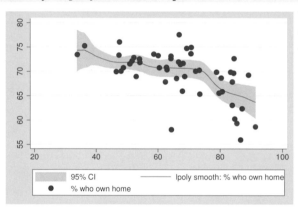

We can also use the `lpolyci` command to produce a local polynomial smooth with a CI.
Uses allstatesdc.dta & scheme vg_brite

3.4   Line plots

This section focuses on the twoway commands for creating line plots, including the `twoway line` and `twoway connected` commands. The `line` command is the same as `scatter`, except that the points are connected by default and marker symbols are not permitted, whereas the `twoway connected` command permits marker symbols. This section also illustrates `twoway tsline` and `twoway tsrline`, which are useful for drawing line plots when the x variable is a date variable. Because all these commands are related to the `twoway scatter` command, they support most of the options you would use with `twoway scatter`. For more information, see [G-2] **graph twoway line**, [G-2] **graph twoway connected**, and [G-2] **graph twoway tsline**.

twoway line close tradeday, sort

In this example, we use `twoway line` to
show the closing price across trading
days. The inclusion of the `sort` option
is recommended when we have points
connected in a Stata graph.
Uses spjanfeb2001.dta & scheme vg_s2c

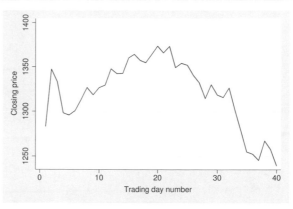

twoway line close tradeday, sort lwidth(vthick) lcolor(maroon)

Here we show options controlling the
width and color of the lines. By using
`lwidth(vthick)` (line width) and
`lcolor(maroon)` (line color), we make
the line very thick and maroon. See
Options : Connecting (323) for more
examples. You cannot use options that
control marker symbols with `graph
twoway line`.

📊 Click on the line and, using the
Contextual Toolbar, change the **Color**
to **Maroon** and the **Width** to v Thick.
Uses spjanfeb2001.dta & scheme vg_s2c

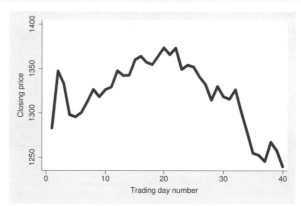

twoway connected close tradeday, sort

This `twoway connected` graph is
similar to the `twoway line` graphs
above, except that for `connected`, a
marker is shown for each data point.
Uses spjanfeb2001.dta & scheme vg_s2c

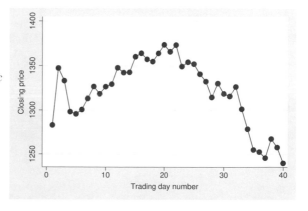

`twoway scatter close tradeday, connect(l) sort`

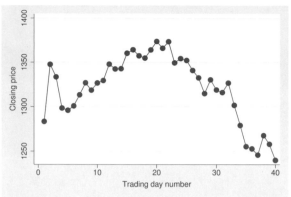

This graph is identical to the previous one, except this graph is made with the scatter command using the connect(l) option. This illustrates the convenience of using connected because we do not need to manually specify the connect() option.

Uses spjanfeb2001.dta & scheme vg_s2c

`twoway connected close tradeday, sort`
`    msymbol(Dh) mcolor(blue) msize(large)`

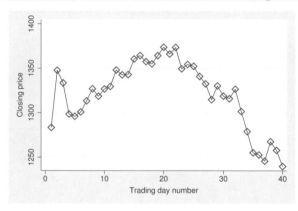

To control the marker symbols, we can use the options msymbol(), mcolor(), and msize(). Here we make the symbols large, blue, hollow diamonds. See **Options: Markers** (307) for more examples.

Click on a marker and, in the Contextual Toolbar, change the **Color** to Blue, the **Size** to Large, and the **Symbol** to Hollow Diamond.

Uses spjanfeb2001.dta & scheme vg_s2c

`twoway connected close tradeday, sort`
`    lcolor(cranberry) lpattern(dash) lwidth(thick)`

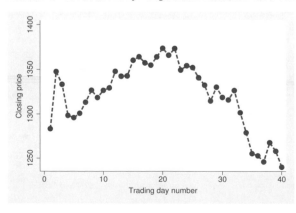

We can control the look of the lines with connect options such as lcolor(), lpattern() (line pattern), and lwidth(). Here we make the line cranberry, dashed, and thick. See **Options: Connecting** (323) for more details on connecting points.

Double-click on a marker and, in the *Connected properties* dialog box, choose the *Line* tab and change the **Color** to Cranberry, the **Width** to Thick, and the **Pattern** to Dash.

Uses spjanfeb2001.dta & scheme vg_s2c

```
twoway connected high low tradeday, sort
```

We can graph multiple variables on one
graph. Here we graph the high and low
prices across trading days.
Uses spjanfeb2001.dta & scheme vg_s2c

```
twoway connected high low tradeday, sort
   lwidth(thin thick) msymbol(Oh S)
```

When graphing multiple variables, we
can specify connect and marker symbol
options to control each line. In this
example, we use a thin line for the high
price and a thick line for the low price.
We also differentiate the two lines by
using different marker symbols: hollow
circles for the high price and squares for
the low price.
Uses spjanfeb2001.dta & scheme vg_s2c

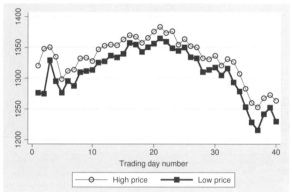

Stata has more commands for creating line plots where the x variable is a date variable,
namely, `twoway tsline` and `twoway tsrline`. The `tsline` command is similar to the `line`
command, and the `tsrline` is similar to the `rline` command. However, both of these `ts`
commands offer extra features, making it easier to reference the x variable for dates (see
[G-2] **graph twoway tsline**). To illustrate these commands, let's use the `sp2001ts` data
file, which has the prices for the S&P 500 index for 2001 with the trading date stored as a
date variable named `date`. Before saving the file `sp2001ts`, the `tsset date, daily` command was used to tell Stata that the variable `date` represents the time variable and that it
represents daily data.

twoway tsline close

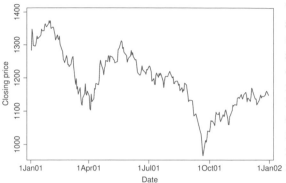

The `tsline` (time-series line) graph shows the closing price on the y axis and the date on the x axis. We did not specify the x variable in the graph command. Stata knew the variable representing time because we previously issued the `tsset date, daily` command before saving the `sp2001ts` file. If we save the data file, Stata remembers the time variable, and we do not need to set it again.

Uses sp2001ts.dta & scheme vg_s1c

twoway tsrline low high

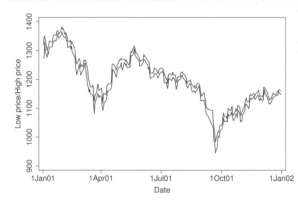

We can also use the `tsrline` (time-series range) graph to show the low price and high price for each day.

Uses sp2001ts.dta & scheme vg_s1c

twoway tsline close, lwidth(thick) lcolor(navy)

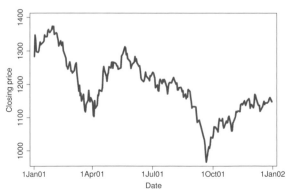

As with `twoway line`, we can use connect options to control the line. Here we make the line thick and navy.

Click on the line and, in the Contextual Toolbar, change the **Color** to Navy and the **Width** to Thick.

Uses sp2001ts.dta & scheme vg_s1c

```
twoway tsline close
    if (date >= mdy(1,1,2001)) & (date <= mdy(3,31,2001))
```

We can use `if` to subset cases to graph. Here we graph the closing prices between January 1, 2001, and March 31, 2001. See the next example for an easier way of doing this.

Uses sp2001ts.dta & scheme vg_s1c

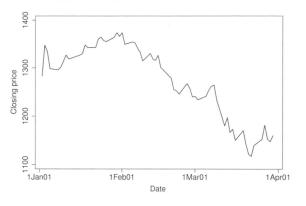

```
twoway tsline close if tin(01jan2001,31mar2001)
```

When using the `tsline` command, we can use `tin()` (time in between) to specify that we want to graph just the cases between January 1, 2001, and March 31, 2001, inclusively.

Uses sp2001ts.dta & scheme vg_s1c

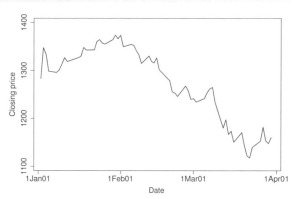

```
twoway tsline close, ttitle(Day of Year)
```

The `ttitle()` (time title) option gives a title to the time variable. We specify this as a `ttitle()` instead of `xtitle()` because this refers to the axis with the time variable.

📈 Click on the title *Date* and, in the Contextual Toolbar, change the **Text** to `Day of Year`.

Uses sp2001ts.dta & scheme vg_s1c

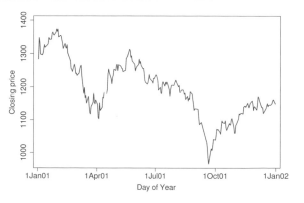

Introduction Editor Twoway Matrix Bar Box Dot Pie Options Standard options Styles Appendix

Scatter Fit CI fit Line Area Bar Range Distribution Contour Options Overlaying

```
twoway tsline close,
    tlabel(01jan2001 31mar2001 30jun2001 30sep2001 01jan2002)
```

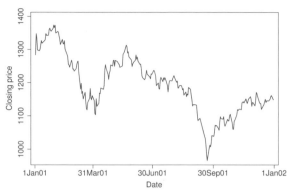

The tlabel() option labels the time points on the time axis. Here we specify these dates by using date literals; Stata knows how to interpret and appropriately label the graph with these values.

Uses sp2001ts.dta & scheme vg_s1c

```
twoway tsline close,
    tlabel(01jan2001 30jun2001 01jan2002) tmlabel(31mar2001 30sep2001)
```

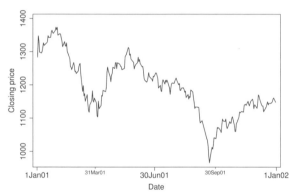

In this example, we use the tmlabel() option to include minor labels.

Uses sp2001ts.dta & scheme vg_s1c

```
twoway tsline close,
    tlabel(01jan2001 30jun2001 01jan2002) tmtick(31mar2001 30sep2001)
```

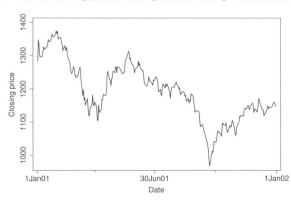

Instead of minor labels, we now use the tmtick() option to include minor ticks.

Uses sp2001ts.dta & scheme vg_s1c

```
twoway tsline close,
   tline(01apr2001 01jul2001 01oct2001)
```

We can use the `tline()` option to
include lines at certain time points.
Here we place lines at the start of the
second, third, and fourth quarters.
Uses sp2001ts.dta & scheme vg_s1c

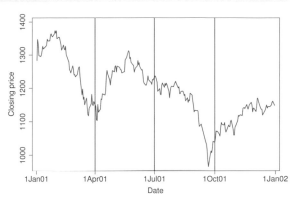

```
twoway tsline close,
   ttext(1035 01apr2001 "Start of Q2", orientation(vertical))
```

The `ttext()` option adds text to the
graph. The first coordinate refers to the
position on the *y* axis, and the second
coordinate is the position on the time
axis for the date.
Uses sp2001ts.dta & scheme vg_s1c

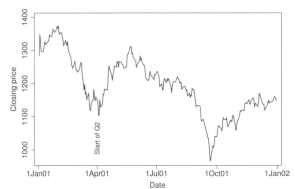

3.5   Area plots

This section illustrates the use of area graphs using `twoway area`. These graphs are
similar to `twoway line` graphs, except that the area under the line is shaded. As a result,
many of the options that you would use with `twoway line` are applicable; see Twoway : Line
(112) for more details. For even more details, see [G-2] **graph twoway area**. These examples use the `spjanfeb2001` data file, which has the prices for the S&P 500 index for
January and February 2001.

Introduction   Editor   Twoway   Matrix   Bar   Box   Dot   Pie   Options   Standard options   Styles   Appendix

Scatter   Fit   CI fit   Line   Area   Bar   Range   Distribution   Contour   Options   Overlaying

`twoway area close tradeday, sort`

This is an example of a `twoway area` graph. Because this graph is composed of connected points, the `sort` option is recommended in case the data are not already sorted by `tradeday`. If the data are not sorted and the `sort` option is not specified, the points are connected in the order they appear in the data file producing a graph that you do not desire.

Uses spjanfeb2001.dta & scheme vg_palec

`twoway area close tradeday, horizontal sort`
   `xtitle(Title for x axis) ytitle(Title for y axis)`

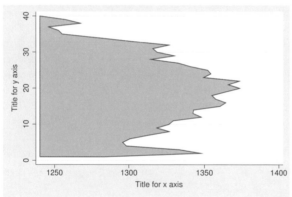

The `horizontal` option swaps the position of the `close` and `tradeday` variables. The *x* axis remains at the bottom, and the *y* axis remains at the left.

Uses spjanfeb2001.dta & scheme vg_palec

`twoway area close tradeday, sort base(1320.28)`

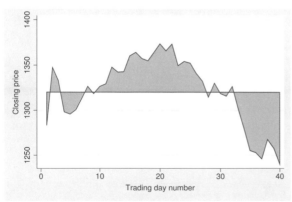

The `base()` option indicates a base from which the area is to be shaded. Here the base is the closing price on the first trading day, and thus all the subsequent points are a kind of deviation from the first day's closing price.

Uses spjanfeb2001.dta & scheme vg_palec

```
twoway area close tradeday, sort bcolor(khaki)
```

The `bcolor()` option sets the color of the shaded area and the line. Here we color the shaded area and the line khaki. Although they are not shown, you can also use the `fcolor()` and `lcolor()` options to control the fill color and line color and the `lwidth()` option to control the thickness of the outline.

🖳 Click on the shaded area and, in the Contextual Toolbar, change the **Color** to Khaki.

Uses spjanfeb2001.dta & scheme vg_palec

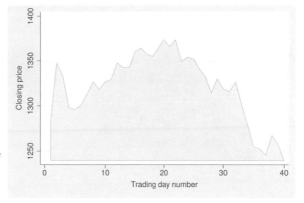

3.6 Bar plots

This section illustrates the use of twoway bar graphs using `twoway bar`. These graphs show a bar for each x value, where the height of the bar corresponds to the value of the y variable. For more details, see [G-2] **graph twoway bar**. We will continue to use the `spjanfeb2001` data file, which has the prices for the S&P 500 index for January and February 2001, but show the graphs using the `vg_s1m` scheme. `twoway bar` is useful for creating bar graphs with overlays of lines, points, or other plot types and can be used with evenly spaced x-variable data. `graph bar` is more useful for creating bar graphs with categorical data.

```
twoway bar close tradeday
```

Consider this bar chart, which shows the closing prices of the S&P 500 broken down by the trading day of the year.

Uses spjanfeb2001.dta & scheme vg_s1m

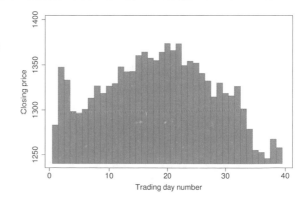

Introduction Editor Twoway Matrix Bar Box Dot Pie Options Standard options Styles Appendix

Scatter Fit CI fit Line Area Bar Range Distribution Contour Options Overlaying

```
twoway bar close tradeday, horizontal
   xtitle(Title for x axis) ytitle(Title for y axis)
```

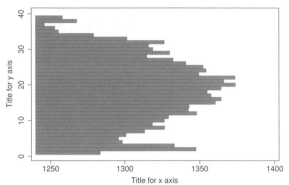

We can make the `close` and `tradeday` variables trade places by specifying the `horizontal` option. The x axis remains at the bottom, and the y axis remains at the left.

Uses spjanfeb2001.dta & scheme vg_s1m

```
twoway bar close tradeday, base(1200)
```

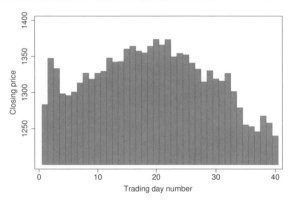

The base for the bar charts is the trading day with the lowest price, unless otherwise specified. The closing price on day 40 was 1,239.94, so if you had not specified the `base()` option, the base would have been 1,239.94. As a result, the bar for day 40 would have had a zero height. However, because we specified `base(1200)`, the base height is 1,200.

Uses spjanfeb2001.dta & scheme vg_s1m

```
twoway bar close tradeday, barwidth(.7)
```

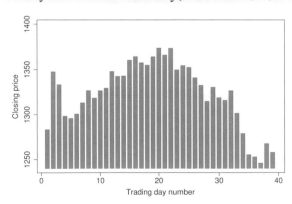

By default, the width of each bar is one x unit (here, one day). By making the width of the bars .7, we can obtain a small gap between the bars.

▨ Click on a bar and, in the Contextual Toolbar, change the **Width** to .7.

Uses spjanfeb2001.dta & scheme vg_s1m

```
twoway bar close tradeday, fcolor(gs15) lcolor(gs5)
```

The `fcolor()` (fill color) option sets
the color of the inside of the bars, and
the `lcolor()` (line color) option sets
the color of the bar outlines. Here we
make the bars light gray on the inside
and dark gray on the outside. See
Styles : Colors (412) for more colors.

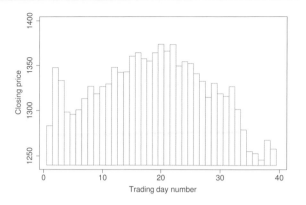

Double-click on a bar and change
the **Color** to Gray 15, and then check
Different outline color and change
the **Outline Color** to Gray 5.

*Uses spjanfeb2001.dta & scheme
vg_s1m*

3.7   Range plots

This section focuses on twoway commands that display range plots. The major characteristic these graphs share is that, for each x value, there are two corresponding y values. A common example is a confidence interval where, for each x value, there are upper
and lower confidence limits. We will first look at examples of all these types of graphs
and then consider the options we can use to customize them. For more information, see
[G-2] **graph twoway rarea**, [G-2] **graph twoway rbar**, [G-2] **graph twoway rspike**, [G-
2] **graph twoway rcap**, [G-2] **graph twoway rcapsym**, [G-2] **graph twoway rscatter**,
[G-2] **graph twoway rline**, and [G-2] **graph twoway rconnected**. Let's start by looking
at the `rconnected`, `rscatter`, `rline`, and `rarea` graphs, which use combinations of lines,
symbols, and shading to display range plots. These examples use the `spjanfeb2001` data
file.

```
twoway rconnected high low tradeday, sort
```

The `rconnected` (range connected)
graph shows the high and low prices by
`tradeday`, the number of days stocks
have been traded in the year. The
`rconnected` plot shows a separate line
for the high and low prices, and a
marker appears for each x value. The
`sort` option is recommended because
the points are connected by lines and is
needed if the data were not already
sorted on `tradeday`.

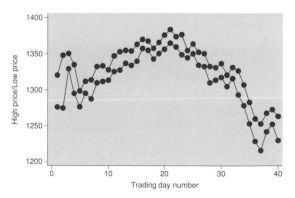

*Uses spjanfeb2001.dta & scheme
vg_rose*

`twoway rscatter high low tradeday`

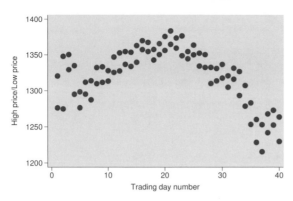

The `rscatter` graph is similar to the `rconnected` graph, except that lines connecting the symbols are not plotted.

📊 Continuing from the previous graph, click on any marker and, in the Contextual Toolbar, change the **Plottype** to `Rscatter`.

Uses spjanfeb2001.dta & scheme vg_rose

`twoway rline high low tradeday, sort`

The `rline` graph is similar to the `rconnected` graph, except that symbols are not plotted at each level of x. Note the inclusion of the `sort` option. This option is recommended because the points are connected by lines and is needed if the data were not already sorted on `tradeday`.

📊 Continuing from the previous graph, click on any marker and, in the Contextual Toolbar, change the **Plottype** to `Rline`.

Uses spjanfeb2001.dta & scheme vg_rose

`twoway rarea high low tradeday, sort`

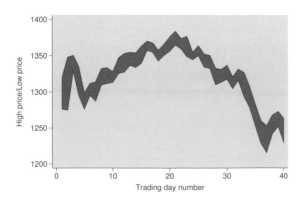

The `rarea` graph is similar to the `rline` graph, except that we can control the fill color of the area between the high and low values.

📊 Continuing from the previous graph, click on any line and, in the Contextual Toolbar, change the **Plottype** to `Rarea`.

Uses spjanfeb2001.dta & scheme vg_rose

Now let's look at `rcap`, `rspike`, and `rcapsym` graphs, which use combinations of spikes, caps, and symbols to display range plots. Then we will look at `rbar` graphs, which use bars to display range plots. The next examples are shown using the `vg_s2m` scheme.

`twoway rcap high low tradeday`

The `rcap` graph shows a spike ranging from the low to high values and puts a cap at the top and bottom of each spike.

📈 Continuing from the previous graph, click on any line and, in the Contextual Toolbar, change the **Plottype** to Rcap.

Uses spjanfeb2001.dta & scheme vg_s2m

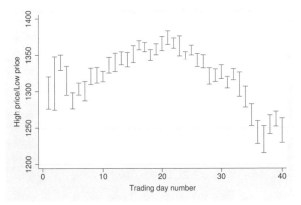

`twoway rspike high low tradeday`

The `rspike` graph is similar to the `rcap` graph, except that no caps are placed on the spikes.

📈 Continuing from the previous graph, click on any cap and, in the Contextual Toolbar, change the **Plottype** to Rspike.

Uses spjanfeb2001.dta & scheme vg_s2m

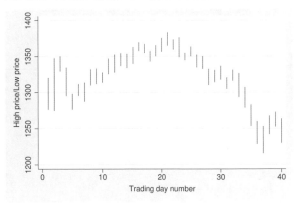

Introduction
Editor
Twoway
Matrix   Bar   Box   Dot   Pie   Options   Standard options   Styles   Appendix

Scatter   Fit   CI fit   Line   Area   Bar   Range   Distribution   Contour   Options   Overlaying

twoway rcapsym high low tradeday

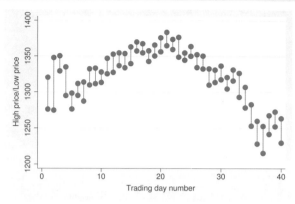

The rcapsym graph is similar to the rcap graph, except that instead of caps, symbols are placed at the top and bottom of each spike. We can choose among the symbols we would use for a scatterplot.

Continuing from the previous graph, click on any spike and, in the Contextual Toolbar, change the **Plottype** to Rcapsym.

Uses spjanfeb2001.dta & scheme vg_s2m

twoway rbar high low tradeday

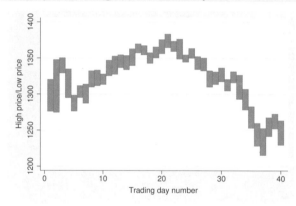

The rbar graph uses bars for each value of x to show the high and low values of y.

Continuing from the previous graph, you can click on any line and, in the Contextual Toolbar, change the **Plottype** to Rbar.

Uses spjanfeb2001.dta & scheme vg_s2m

Let's now consider options to use with the rconnected, rscatter, rline, and rarea graphs. We will start by looking at rconnected plots because many of the options used in that kind of graph also apply to rscatter, rline, and rarea graphs. These graphs use the vg_s1c scheme.

`twoway rconnected high low tradeday, sort`

Here is a general `rconnected` graph.
Uses spjanfeb2001.dta & scheme vg_s1c

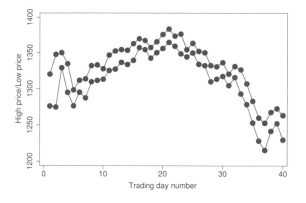

`twoway rconnected high low tradeday, sort horizontal`
    `xtitle(Title for x axis) ytitle(Title for y axis)`

With the `horizontal` option, we can
swap the axes where `high/low` and
`tradeday` appear. The *x* axis remains
at the bottom, and the *y* axis remains
at the left.
Uses spjanfeb2001.dta & scheme vg_s1c

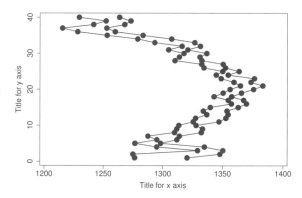

`twoway rconnected high low tradeday, sort`
    `msymbol(Oh) msize(large) mcolor(lavender)`

We can control the look of the marker
symbols with options such as
`msymbol()`, `msize()`, and `mcolor()`.
Here we make the marker symbols
large, lavender, hollow circles. For more
details about options related to
symbols, see Options: Markers (307).

Click on a marker and, in the
Contextual Toolbar, change the
Symbol to `Hollow circle`, the **Size**
to `Large`, and the **Color** to `Lavender`.
Uses spjanfeb2001.dta & scheme vg_s1c

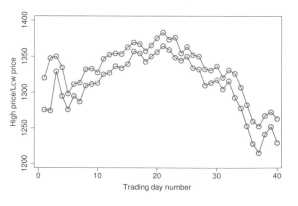

Introduction
Scatter · Fit · CI fit · Line · Area · Bar · Range · Distribution · Contour · Options · Overlaying
Editor
Twoway
Matrix · Bar · Box · Dot · Pie · Options · Standard options · Styles · Appendix

```
twoway rconnected high low tradeday, sort
    lwidth(thick) lcolor(dkgreen) lpattern(dash)
```

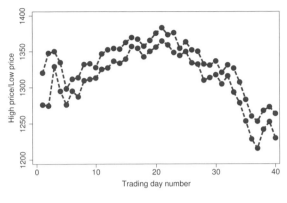

Here we use the `lwidth()`, `lcolor()`, and `lpattern()` options to make the lines thick, dark green, and dashed. See Options: Connecting (323) for more examples, and see Styles: Connect (416), Styles: Linewidth (422), Styles: Colors (412), and Styles: Linepatterns (420) for more details about the mentioned options.

📊 Double-click on a line and, in the *Lines* tab, change the **Color** to Dark green, the **Width** to Thick, and the **Pattern** to Dash.

Uses spjanfeb2001.dta & scheme vg_s1c

```
twoway rscatter high low tradeday, sort mcolor(red) msize(medium)
    msymbol(Sh)
```

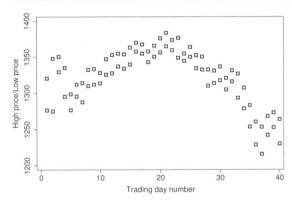

With `rscatter`, we can use the options that modify the markers. Here we change the color to red, the marker size to medium, and the markers to hollow squares. For more details about options related to marker symbols, see Options: Markers (307).

📊 Click on a marker and, in the Contextual Toolbar, change the **Color** to Red, the **Size** to Medium, and the **Symbol** to Hollow squares.

Uses spjanfeb2001.dta & scheme vg_s1c

```
twoway rline high low tradeday, sort
    lcolor(orange) lwidth(thick) lpattern(dash)
```

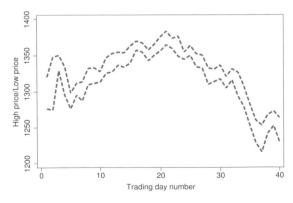

With `rline`, we can use the options that modify the lines to make them orange, thick, and dashed. For more details about connect options, see Options: Connecting (323).

📊 Click on a line and, in the Contextual Toolbar, change the **Color** to Orange, the **Width** to Thick, and the **Pattern** to Dash.

Uses spjanfeb2001.dta & scheme vg_s1c

`twoway rarea high low tradeday, sort bcolor(pink)`

The `rarea` graph is similar to the `rline` graph, but in addition to being able to control the characteristics of the line, we can control the color of the area between the low and high lines. Here we use the `bcolor()` option to make the color of the line and the area pink.

Click on the shaded area and, in the Contextual Toolbar, change the **Color** to Pink.

Uses spjanfeb2001.dta & scheme vg_s1c

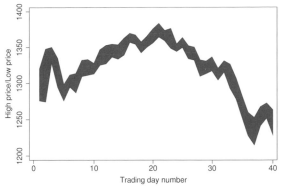

`twoway rarea high low tradeday, sort`
`    lcolor(gold) fcolor(teal) lwidth(thick)`

Here we make the color of the line gold with the `lcolor()` option, the fill color teal with the `fcolor()` option, and the line thick with the `lwidth()` option.

Double-click on the area and change the **Color** to Teal, change the **Outline Width** to Thick, check **Different outline color**, and change the **Outline color** to Gold.

Uses spjanfeb2001.dta & scheme vg_s1c

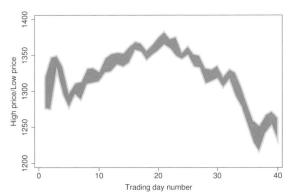

Now let's look at options for use with the `rcap`, `rspike`, and `rcapsym` graphs. The options permitted by the `rcap` option are similar to the options used with the `rspike` and `rcapsym` graphs. These examples use the `vg_s2c` scheme.

Introduction   Editor   Twoway   Matrix   Bar   Box   Dot   Pie   Options   Standard options   Styles   Appendix

Scatter   Fit   CI fit   Line   Area   Bar   Range   Distribution   Contour   Options   Overlaying

`twoway rcap high low tradeday`

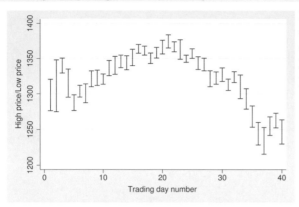

Here is a basic `rcap` graph. The `rcap` command supports the `horizontal` option, which would make the variables `high`/`low` and `tradeday` swap positions.

Uses spjanfeb2001.dta & scheme vg_s2c

`twoway rcap high low tradeday, msize(small)`

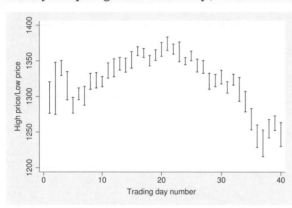

We usually use the `msize()` option to control the size of a marker, and it is adapted for this kind of graph to control the size of the cap. In this example, we make the cap small.

Double-click on a cap and change the **Cap size** to Small.

Uses spjanfeb2001.dta & scheme vg_s2c

`twoway rcap high low tradeday, lcolor(cranberry) lwidth(thick)`

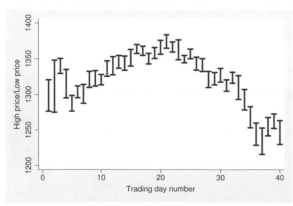

We use the `lcolor()` option to control the color of the line, here, cranberry. We use the `lwidth()` option to set the width of the line, here, thick. Although it is not shown here, you could also control the pattern of the line with the `lpattern()` option. See Options: Connecting (323) for more details.

Click on a cap and, in the Contextual Toolbar, change the **Color** to Cranberry and the **Width** to Thick.

Uses spjanfeb2001.dta & scheme vg_s2c

`twoway rspike high low tradeday, lcolor(red) lwidth(thin)`

The options used for `rspike` are
basically the same as those for `rcap`,
except that the `msize()` option is not
appropriate because there are no
markers to size. Here, for example, we
use `lcolor()` and `lwidth()` to make
the lines red and thin.

Click on a line and, in the
Contextual Toolbar, change the **Color**
to Red and the **Width** to Thin.

Uses spjanfeb2001.dta & scheme vg_s2c

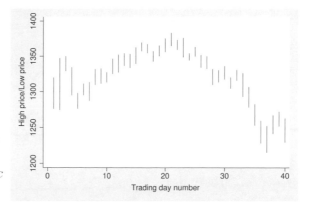

`twoway rcapsym high low tradeday, msymbol(Oh) lwidth(thick)`

The options used for `rcapsym` allow us
to modify both the markers and the
lines. Here we use the `msymbol()`
option to place hollow circles at the end
of the spikes and the `lwidth()` option
to make the lines thick.

Click on a marker and, in the
Contextual Toolbar, change the
Symbol to Hollow circle and the
Width to Thick.

Uses spjanfeb2001.dta & scheme vg_s2c

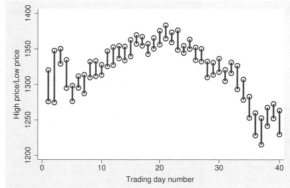

Let's now explore options to use with `twoway rbar`. These examples use the `vg_brite`
scheme.

`twoway rbar high low tradeday`

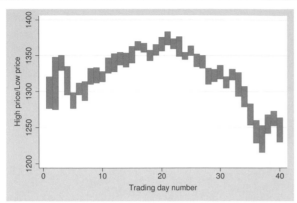

Here is a basic `rbar` graph. As with the other graphs in this family, you could have added the `horizontal` option to switch the position of the `high`/`low` and `tradeday` variables, but this is not shown.
Uses spjanfeb2001.dta & scheme vg_brite

`twoway rbar high low tradeday, barwidth(.4)`

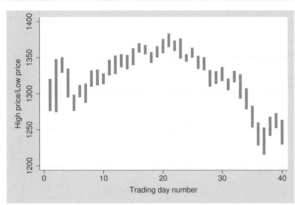

We can use the `barwidth()` option to set the width of the bar. This width is in units of the x variable. We set the bars to .4 units wide, so they no longer touch each other.
📈 Click on a bar and, in the Contextual Toolbar, change the **Width** to `.4`.
Uses spjanfeb2001.dta & scheme vg_brite

`twoway rbar high low tradeday, bcolor(sienna)`

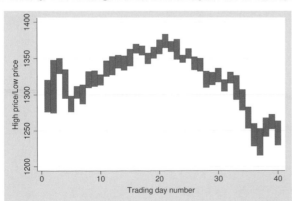

The `bcolor()` (bar color) option sets the color of the bar and the outline. Here we make the color sienna.
📈 Click on a bar and, in the Contextual Toolbar, change the **Color** to `Sienna`.
Uses spjanfeb2001.dta & scheme vg_brite

```
twoway rbar high low tradeday, fcolor(sienna)
    lcolor(cranberry) lwidth(thick)
```

With the `fcolor()` (fill color),
`lcolor()` (line color), and `lwidth()`
(line width) options, we make
sienna-filled bars, outlined with thick
cranberry lines.

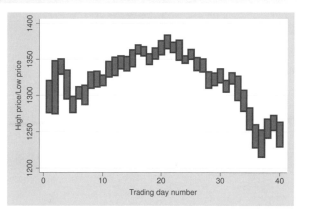

Double-click on a bar and change
the **Color** to Sienna, change the
Outline Width to Thick, check
Different outline color, and change
the **Outline color** to Cranberry.
*Uses spjanfeb2001.dta & scheme
vg_brite*

3.8   Distribution plots

This section describes the use of `twoway histogram` and `twoway kdensity` for showing
the distribution of one variable. This section also shows the use of `twoway function` for
showing the relationship between x and y using a function that you specify. See [G-2] **graph
twoway histogram**, [G-2] **graph twoway kdensity**, and [G-2] **graph twoway function**
for more information. This section starts by showing the `twoway histogram` command and
then illustrates options that allow you to control such things as the number of bins, the
width of the bins, and the starting point for the bins. The section concludes with options
that control the scaling of the y axis. The next few graphs use the `vg_past` scheme.

```
twoway histogram ttl_exp
```

We begin by showing a `histogram` of
the variable total work experience.
Unlike many other twoway plots, this
command takes only one variable that
is graphed on the x axis. The y axis
represents the density, such that the
sum of the areas of the bars equals 1. If
we are not going to combine this graph
with other twoway graphs, the
`histogram` command may be preferable
to `twoway histogram`.
Uses nlsw.dta & scheme vg_past

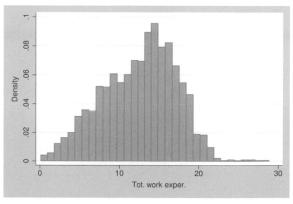

`twoway histogram ttl_exp, bin(10)`

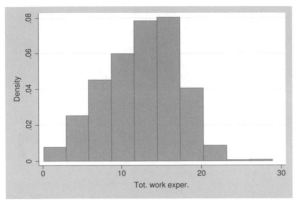

We can control the number of bins that are used to display the histogram by specifying the `bin()` option. In this example, we request that 10 bins be used.

Uses nlsw.dta & scheme vg_past

`twoway histogram ttl_exp, width(5)`

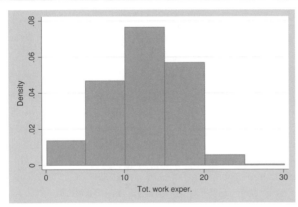

We can control the width of each bar using the `width()` option. Here we make each bar 5 units wide. As one might imagine, we can use either the `bin()` option or the `width()` option, but not both.

Uses nlsw.dta & scheme vg_past

`twoway histogram ttl_exp, start(-2.5) width(5)`

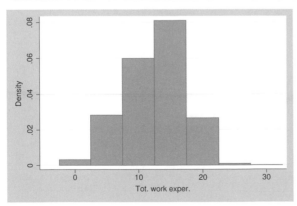

We add the `start()` option to indicate that we want the lower limit of the first bin to start at −2.5.

Uses nlsw.dta & scheme vg_past

`twoway histogram ttl_exp, fraction width(1)`

If we use the `fraction` option, the y
axis is scaled such that the height of
each bar is the probability of falling
within the range of x values represented
by the bar. Thus, if we specify the
width of bars to 1, the sum of the
heights of the bars is 1.
Uses nlsw.dta & scheme vg_past

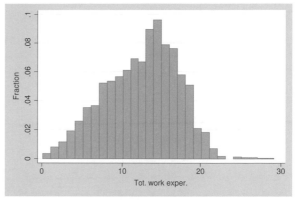

`twoway histogram ttl_exp, percent width(1)`

The `percent` option is similar to the
`fraction` option, except that the y axis
is represented as a percentage instead
of a proportion. If we also specify a bar
width of 1, the sum of the heights of
the bars is 100%.
Uses nlsw.dta & scheme vg_past

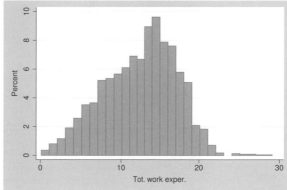

`twoway histogram ttl_exp, frequency width(1)`

The `frequency` option changes the
scaling of the y axis to represent the
number of cases that fall within the
range of x values represented by the
bar. If we specify a bar width of 1, the
sum of the heights of the bars equals
the number of nonmissing values for
`ttl_exp`.
Uses nlsw.dta & scheme vg_past

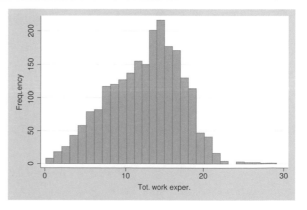

Let's now consider options that control the width of the bars and other characteristics of the bars, such as color. Then we will see how to display the graph as a horizontal histogram as well as options that allow you to treat *varname* as a discrete variable. These graphs use the `vg_blue` scheme.

`twoway histogram ttl_exp, gap(20)`

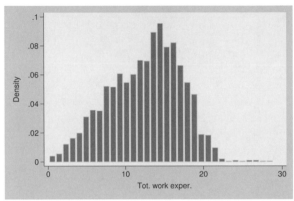

The `gap()` option specifies the gap between each of the bars. The gap is created by reducing the width of the bars. By default, the gap is 0, meaning that the bars touch exactly and the bars are reduced by 0%. Here we reduce the size of the bars by 20%, making a small gap between the bars. *Uses nlsw.dta & scheme vg_blue*

`twoway histogram ttl_exp, gap(99.99)`

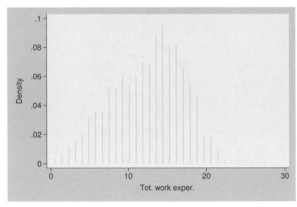

In this example, we reduce the size of the bars 99.99%, making the bars 0.01% of their normal size. *Uses nlsw.dta & scheme vg_blue*

`twoway histogram ttl_exp, barwidth(.5)`

Another way you can control the width
of the bars is with the `barwidth()`
option. Here we indicate that we want
each bar .5 *x*-units wide.

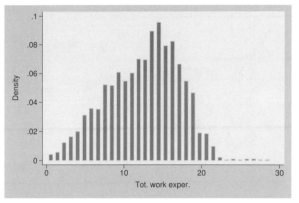

Click on a bar and, in the
Contextual Toolbar, change the **Width**
to `.5`.

Uses nlsw.dta & scheme vg_blue

`twoway histogram ttl_exp, fcolor(olive_teal) lcolor(teal)`
`    lwidth(thick)`

The `fcolor()` (fill color) option makes
the fill color of the bar olive-teal, and
the `lcolor()` (line color) option makes
the bar line color teal. The `lwidth()`
(line width) option makes the line
around the bar thick. The section
Styles : Colors (412) shows more about
colors, and Styles : Linewidth (422)
shows more about line widths.

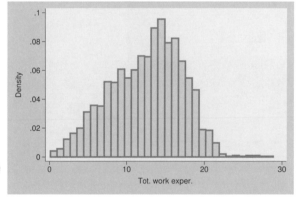

Double-click on a bar and change
the **Color** to `Olive teal`, the **Outline**
width to `Thick`, and the **Outline**
color to `Teal`.

Uses nlsw.dta & scheme vg_blue

   Let's now briefly consider other options that can be used with `twoway histogram`, seeing how to swap the position of the *x* and *y* axes, and the `discrete` option for use with
discrete variables. The next set of graphs uses the `vg_s1m` scheme.

`twoway histogram ttl_exp, horizontal`

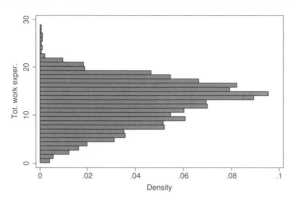

The `horizontal` option swaps the position of `ttl_exp` and its density, making a horizontal display of the histogram.

Uses nlsw.dta & scheme vg_s1m

`twoway histogram grade, discrete`

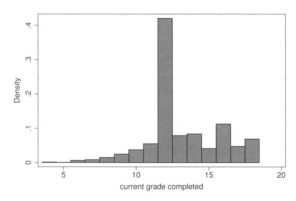

The `discrete` option tells Stata that the variable `grade` is a discrete variable and can take only integer values. Here each bin has a width of 1.

Uses nlsw.dta & scheme vg_s1m

`twoway histogram grade, discrete width(2)`

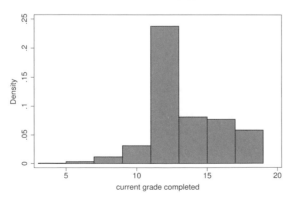

The `width()` option changes the bar's width. In this example, we change the width to 2.

Uses nlsw.dta & scheme vg_s1m

This section concludes with kernel-density plots created by using `twoway kdensity`. For more details, see [G-2] **graph twoway kdensity**. As with histograms, if you are not going to combine the kernel-density plot with other twoway plots, the `kdensity` command is more powerful (and hence preferable) to `twoway kdensity`. We will explore a handful of options that are useful for controlling the display of these graphs. These graphs use the `vg_s2c` scheme.

twoway kdensity ttl_exp

Here is a kernel-density plot of total work experience. You could have added the `horizontal` option to display the graph as a horizontal plot, but this option is not shown.
Uses nlsw.dta & scheme vg_s2c

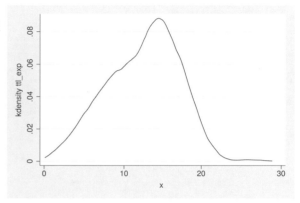

twoway kdensity ttl_exp, kernel(biweight)

By default, Stata uses an Epanechnikov kernel for computing the density estimates. Here we specify the `kernel(biweight)` option so that instead the biweight kernel for computing the densities is used. Methods include `biweight`, `cosine`, `gaussian`, `parzen`, `rectangle`, and `triangle`.
Uses nlsw.dta & scheme vg_s2c

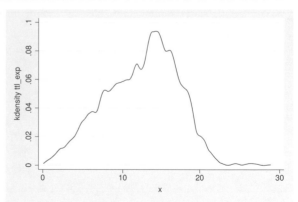

Introduction   Editor   Twoway   Matrix   Bar   Box   Dot   Pie   Options   Standard options   Styles   Appendix

Scatter   Fit   CI fit   Line   Area   Bar   Range   Distribution   Contour   Options   Overlaying

```
twoway (histogram ttl_exp, width(1) frequency)
   (kdensity ttl_exp, area(2246))
```

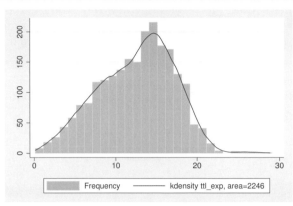

In this example, we overlay a histogram of `ttl_exp`, scaling the y axis as the frequency of values in each bin. We overlay this with a `kdensity` plot but want to scale the y axis in a commensurate manner. By using the `area()` option, we can specify that the sum of the area of the kernel density should sum to 2,246, the sample size.
Uses nlsw.dta & scheme vg_s2c

```
twoway kdensity ttl_exp, lwidth(thick) lpattern(dash)
```

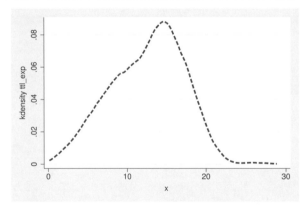

We can use the options `lcolor()`, `lwidth()`, and `lpattern()` to alter the characteristics of the line. Here we use the `lwidth()` and `lpattern()` options to make the line thick and dashed. See Styles : Linewidth (422), Styles : Linepatterns (420), and Styles : Colors (412) for more details.
Click the line and, in the Contextual Toolbar, change the **Width** to `Thick` and the **Pattern** to `Dash`.
Uses nlsw.dta & scheme vg_s2c

```
twoway function y=normalden(x), range(-4 4)
```

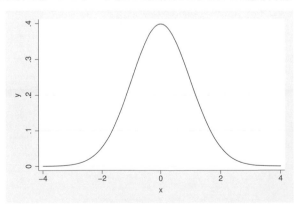

Here we show how a `twoway function` can be used to graph an arbitrary function. We graph the function `y=normalden(x)` to show a normal curve. We add the option `range(-4 4)` to specify that we want the x values to range from -4 to 4. By default, the graph would show the x values ranging from 0 to 1.
Uses nlsw.dta & scheme vg_s2c

3.9   Contour plots

This section focuses on the `twoway` command for creating contour plots, illustrating the `twoway contour` and `twoway contourline` commands. (For more information, see [G-2] **graph twoway contour** and [G-2] **graph twoway contourline**.) Contour plots are used in many disciplines as a means of representing three-dimensional data in two dimensions. In addition to the standard x and y axes, contour plots display a z dimension with contours that are represented using colors or lines. For example, consider the `sandstone.dta` dataset (that comes with Stata). This dataset contains geographic data regarding the depth of sandstone formations for a certain geographic region. Two dimensions are used to represent the location in the north or south direction and the east or west direction, and the third dimension indicates the depth of the sandstone layer. The following `twoway contour` command shows a contour plot for these variables using this dataset.

`twoway contour depth northing easting`

This `contour` graph shows the depth of the sandstone layers as a function of the location. The variable `depth` is graphed as the z axis, with `northing` as the y axis and `easting` as the x axis. Note how the first contour is blue, the second is cyan, the third is green, followed by yellow, and then red. The first contour color (i.e., blue) represents the smallest values for `depth`, and the last contour color (i.e., red) represents the highest values of `depth`.
Uses sandstone.dta & scheme vg_s2c

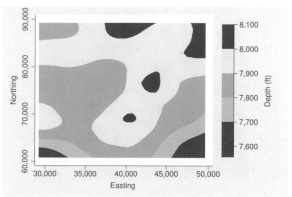

`twoway contour depth northing easting, levels(15)`

Note how the graph changes by adding the `levels(15)` option. Compared with the original graph that used 5 contours for `depth`, the depths are now represented using 15 contours. The key at the right shows the depth that corresponds to each of the 15 colors.
Uses sandstone.dta & scheme vg_s2c

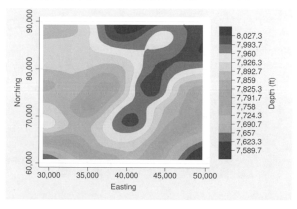

`twoway contour depth northing easting, ccuts(7500(100)8100)`

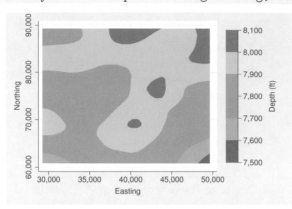

We can instead use the `ccuts()` option to specify the levels of the contours. In this example, we specify that the contours range from 7,500 to 8,100 in 100-foot increments. This yields a total of six contours, where blue is the smallest depth and orange is the greatest depth. Note how the colors are similar shades of green for the depths of 7,700–7,800 and 7,800–7,900.

Uses sandstone.dta & scheme vg_s2c

`twoway contour depth northing easting, ccuts(7500(100)8100)`
`        ccolors(red green blue pink orange purple yellow)`

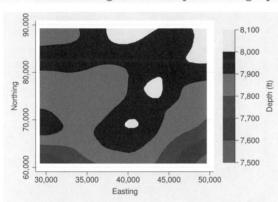

We can add the `ccolors()` option to specify the colors for each of the contours. Depths below 7,500 feet are colored red, 7,500–7,600 feet are green, 7,600–7,700 are blue, and so forth. The first color (red) for depths below 7,500 feet appears to be ignored, but this is because there are no observations where the depth is below 7,500 feet.

Uses sandstone.dta & scheme vg_s2c

`twoway contour depth northing easting, ccuts(7500(100)8100)`
`        ccolors(red green blue pink)`

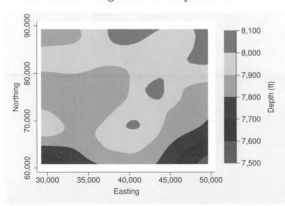

Actually, we do not need to specify the colors for all the contours. In this example, I specified just four colors. The contour depths below 7,500 feet is red, for 7,500–7,600 feet is green, for 7,600–7,700 feet is blue, and for 7,700–7,800 feet is pink. The remaining contours are colored using green, yellow, and orange, the colors used when we omitted the `ccolors()` option. As in the previous example, there are no observations where the depth is below 7,500 feet.

Uses sandstone.dta & scheme vg_s2c

If there is a large number of contours, it could be cumbersome to specify each color for all the contours. Instead, Stata provides you the option of specifying the starting contour color, the ending contour color, and a method for choosing the intermediate colors. This is illustrated in the following examples.

```
twoway contour depth northing easting, levels(15)
    crule(linear) scolor(red) ecolor(yellow)
```

This example uses the crule(linear) option combined with the scolor(red) and ecolor(yellow) options. The result is that the colors start with pure red at the lowest depths and end at pure yellow at the highest depths. As depth increases, the color is a mixture of red and yellow, with increasing amounts of yellow for higher depths. The scolor() option specifies the starting color, and the ecolor() option specifies the ending color.
Uses sandstone.dta & scheme vg_s2c

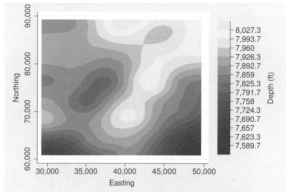

```
twoway contour depth northing easting, levels(15)
    crule(linear) scolor(red) ecolor(blue)
```

This example is similar to the previous example except that the ecolor(blue) option specifies that blue be the ending color. The smallest depths are red, the highest depths are blue, and the intermediate depths are mixtures of red and blue, including greater amounts of blue with increasing depth.
Uses sandstone.dta & scheme vg_s2c

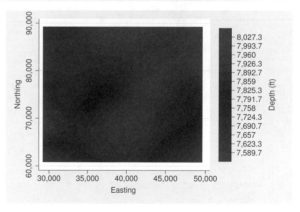

```
twoway contour depth northing easting, levels(15)
   crule(linear) scolor(white) ecolor(black)
```

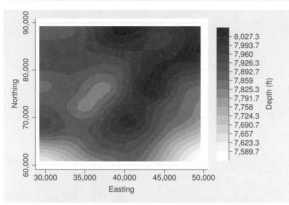

It might be simpler to use black and white in such graphs. In this example, the starting color is white and the ending color is black. The smallest value of depth is white, and the highest depth is black; darker colors correspond to greater values of depth.

Uses sandstone.dta & scheme vg_s2c

```
twoway contour depth northing easting, levels(15)
   crule(intensity)
```

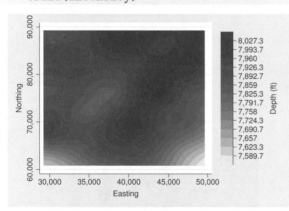

Rather than shifting from a starting and ending color, this example uses the crule(intensity) option. The result is that the contours are indicated by greater intensity of a single hue. The smallest levels of depth are displayed in very low intensity (nearly white). As depth increases, the intensity increases to light pink, then darker pink, and the greatest depth is shown in red.

Uses sandstone.dta & scheme vg_s2c

```
twoway contour depth northing easting, levels(15)
   crule(intensity) ecolor(blue)
```

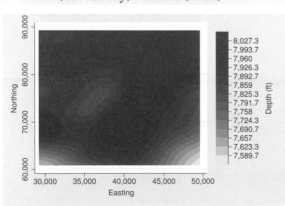

The ecolor() option can be used with the crule(intensity) option to specify the color used to represent the contours. In this example, contours are represented with different intensities of blue. Specifying the scolor() option has no effect when using the crule(intensity) option.

Uses sandstone.dta & scheme vg_s2c

no command

A third method for indicating the
colors of contours involves specifying a
starting and ending color, and the
intermediate colors are selected as the
hues between the starting and ending
colors. This pie chart illustrates a
360-degree color wheel that starts with
red (at 0 degrees), then runs through
orange, yellow, green, light blue, dark
blue, and purple, and finally comes full
circle back to red again. The following
examples will refer back to this
360-degree color wheel.

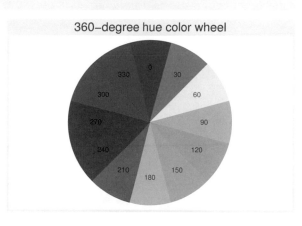

twoway contour depth northing easting, levels(15)
scolor(yellow) ecolor(purple) crule(hue)

In this example, the starting contour
color is yellow and the ending contour
is purple. Note how the intermediate
values correspond to a clockwise
rotation of hues from the color wheel,
progressing from yellow to green, cyan,
blue, and then ending at purple. I
included the crule(hue) option,
although not necessary, to emphasize
that the intermediate values are chosen
as a function of the hue.

Uses sandstone.dta & scheme vg_s2c

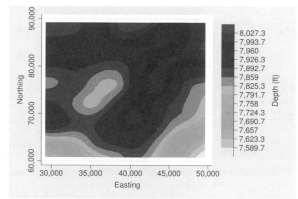

twoway contour depth northing easting, levels(15)
scolor(purple) ecolor(yellow) crule(hue)

In this example, the starting color is
purple and the ending color is yellow.
Because purple follows yellow on the
color wheel (300 vs. 60), the hues in
this example go in a counterclockwise
direction based on the color wheel. As
depth increases, the hues change from
purple to blue, cyan, green, and then
finally to yellow.

Uses sandstone.dta & scheme vg_s2c

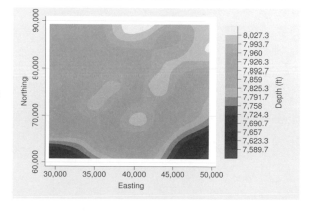

```
twoway contour depth northing easting, levels(15)
    scolor(purple) ecolor(yellow) crule(chue)
```

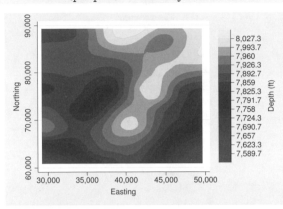

This example uses the crule(chue) option. The result is that the contour colors go from purple to yellow in a clockwise fashion, showing purple, red, orange, and then yellow. The crule(chue) option works like the crule(hue) option if the ending color follows the starting color on the 360-degree color wheel.

Uses sandstone.dta & scheme vg_s2c

```
twoway contour depth northing easting, levels(15)
    scolor(blue) ecolor(cyan) crule(chue)
```

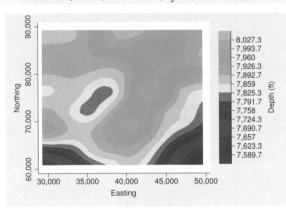

This command is similar to the previous command except that the contour color starts with blue and ends with cyan. Starting with blue, the contour colors increase to purple, and then red, and then circle around to orange, yellow, green, and then end at cyan.

Uses sandstone.dta & scheme vg_s2c

```
twoway contourline depth northing easting
```

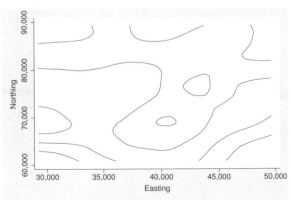

The twoway contourline command creates contour plots except that the contours are shown (by default) as contour lines.

Uses sandstone.dta & scheme vg_s2c

`twoway contourline depth northing easting, colorlines`

The `colorlines` option can be used to
display the contour lines in color.
Options previously illustrated with
`twoway contour` (such as `ccolors()`,
`scolor()`, and `ecolor()`) work
similarly with `twoway contourline`.
Uses sandstone.dta & scheme vg_s2c

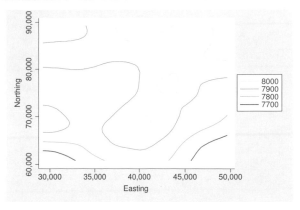

3.10 Options

This section discusses the use of options with `twoway`, showing the types of options we
can use. For more details, see Options (307). This section uses the `vg_outm` scheme for
displaying the graphs.

`twoway scatter ownhome propval100`

Consider this basic scatterplot. We will
use this for illustrating options.
Uses allstates.dta & scheme vg_outm

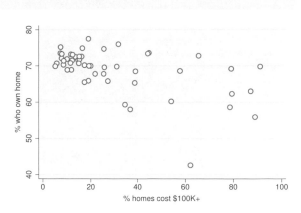

`twoway scatter ownhome propval100, msymbol(S)`

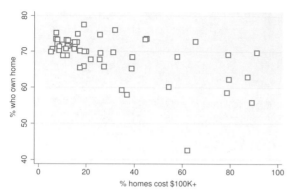

The `msymbol()` option controls the marker symbols. Here we use squares as symbols. See Options : Markers (307) for more details.

Click on a marker and, in the Contextual Toolbar, change the **Symbol** to `Square`.

Uses allstates.dta & scheme vg_outm

`twoway scatter ownhome propval100, msymbol(S) mlabel(stateab)`

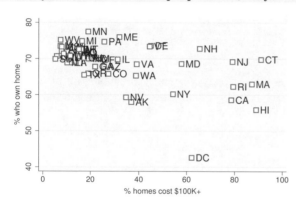

The `mlabel()` option controls the marker labels. Here we label each marker with the variable `stateab`, showing the two-letter abbreviation for each state. See Options : Marker labels (320) for more information about marker labels.

Uses allstates.dta & scheme vg_outm

`twoway scatter fv ownhome propval100, connect(l .) sort`

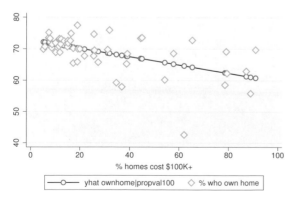

Say that we regressed `ownhome` on `propval100` and generated predicted values named `fv`. Here we make a scatterplot and a fit line in the same graph using the `connect(l .)` option, which connects the values of `fv` but not the values of `ownhome`. We also add the `sort` option, which is generally recommended when using the `connect()` option. See Options : Connecting (323) for more details.

Uses allstates.dta & scheme vg_outm

```
twoway scatter propval100 rent700 ownhome,
    xtitle(Percent of households that own their own home)
```

We add a title to the x axis using the
xtitle() option, as illustrated here.
See Options : Axis titles (327) for more
details about titles.
🖜 Click on the title for the x axis and,
in the Contextual Toolbar, change
Text to the title you wish.
Uses allstates.dta & scheme vg_outm

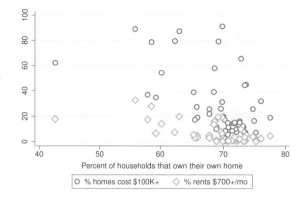

```
twoway scatter propval100 rent700 ownhome, ylabel(0(10)100)
```

We can label the y axis from 0 to 100,
incrementing by 10, using the
ylabel(0(10)100) option, as shown
here. See Options : Axis labels (330) for
more information about labeling axes.
🖜 Double-click on the y axis and select
Range/Delta and set the **Minimum
value** to 0, the **Maximum value** to
100, and the **Delta** to 10.
Uses allstates.dta & scheme vg_outm

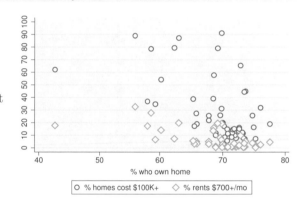

```
twoway scatter propval100 rent700 ownhome,
    ylabel(0(10)100) yscale(alt)
```

Stata provides several options you can
use to control the axis scale for both the
x and y axes. For example, here we use
yscale(alt) to move the y axis to its
alternate position, moving it from the
left to the right. See Options : Axis scales
(339) for more details about the options
for controlling the axis scales.
Uses allstates.dta & scheme vg_outm

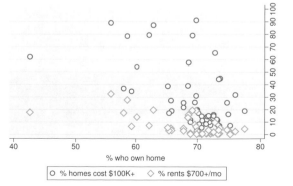

```
twoway (scatter propval100 ownhome)
   (scatter rent700 ownhome, yaxis(2))
```

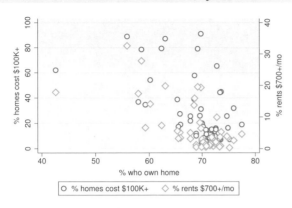

In this example, we show `propval100` by `ownhome` and also `rent700` by `ownhome`, but for this second plot, we put the *y* axis on the second *y* axis with the `yaxis(2)` option. See Options: Axis selection (343) for more information about using and controlling additional axes.

Uses allstates.dta & scheme vg_outm

```
twoway scatter propval100 rent700 ownhome,
   ylabel(0(10)100) yscale(alt) by(north)
```

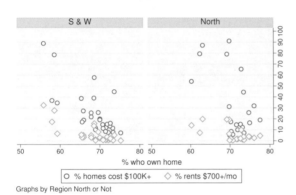

The `by()` option allows us to see a graph broken down by one or more `by()` variables. Here we show the previous graph further broken down by whether the state was part of the North, making two graphs that are combined into one graph. The section Options: By (346) shows more details and examples about the use of the `by()` option.

Uses allstates.dta & scheme vg_outm

```
twoway scatter propval100 rent700 ownhome, legend(cols(1))
```

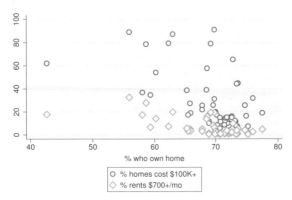

The `legend()` option allows us to control the contents and display of the legend. Here we use the `legend(cols(1))` option to indicate that we want the legend to display as one column. See Options: Legend (361) for more details about the `legend()` option.

⌨ In the Object Browser, double-click on **legend** and change the **Columns** to 1.

Uses allstates.dta & scheme vg_outm

`twoway scatter propval100 ownhome, text(62 45 "DC")`

In this graph, there is one observation that stands out from the rest. Rather than use the `mlabel()` option to label all the markers, we may want to label just the outlying point. Thus we use the `text()` option to add the text DC at the (y,x) coordinates of (62,45), in effect labeling that point; see Options: Adding text (374) for more details.

📊 See the next graph.

Uses allstates.dta & scheme vg_outm

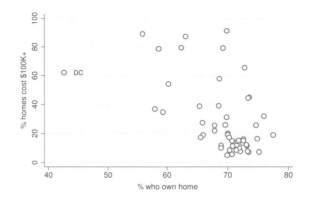

`twoway scatter propval100 ownhome`

📊 Using the Graph Editor, select the Add Text **T** tool. Click where the text should appear and then type DC as the text.

Uses allstates.dta & scheme vg_outm

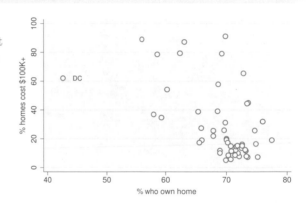

`twoway scatter propval100 ownhome,`
`    xtitle(Percent Home Ownership, box)`

Most items of text on a Stata graph actually display within a box. We illustrate this with the `xtitle()` option, showing how we can place a box around this text. These options are described in more detail in Options: Textboxes (379).

📊 Double-click on the title for the x axis and change the **Text** to the new title. Then select the *Box* tab and check **Place box around text**.

Uses allstates.dta & scheme vg_outm

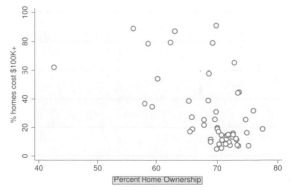

Introduction   Editor   Twoway   Matrix   Bar   Box   Dot   Pie   Options   Standard options   Styles   Appendix

Scatter   Fit   CI fit   Line   Area   Bar   Range   Distribution   Contour   Options   Overlaying

3.11   Overlaying plots

One of the features of twoway graphs is the ability to overlay them, giving you the flexibility to create more complex graphs. This section shows two strategies you can use. The first strategy is graphing multiple y variables against one x variable in a `twoway` command. The second strategy is specifying multiple commands within a `twoway` command, thus overlaying these graphs. It is also possible to create separate graphs and glue them together using the `graph combine` command, which is discussed in Appendix : Save/Redisplay/Combine (456). Let's start by looking at how you can specify multiple y variables against one x variable using a `twoway` command.

`twoway scatter propval100 rent700 urban`

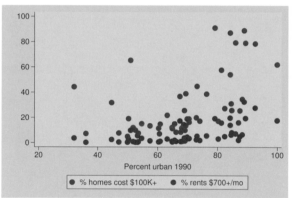

We can use `twoway scatter` to graph multiple y variables against one x variable in a plot. Here we show `propval100` and `rent700` against `urban`. We are now using the `vg_teal` scheme.
Uses allstates.dta & scheme vg_teal

`twoway scatter propval100 rent700 urban, msymbol(Oh t)`

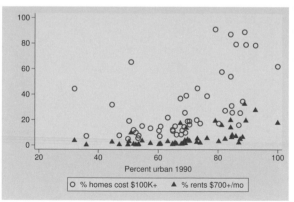

We can use the `msymbol()` option to select the marker symbols for the multiple y variables. Here we plot the variable `propval100` with hollow circles and `rent700` with triangles.
Uses allstates.dta & scheme vg_teal

```
twoway scatter propval100 rent700 urban, mstyle(p2 p8)
```

We can use the `mstyle()` (marker style) option to choose among marker styles. These composite styles set the symbol, size, fill, color, outline color, and outline width for the markers.
Uses allstates.dta & scheme vg_teal

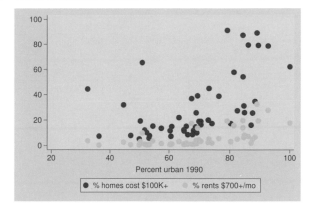

```
twoway (scatter propval100 urban) (scatter rent700 urban)
   (lfit propval100 urban, lwidth(thick))
   (lfit rent700 urban, lwidth(thick))
```

Here we use `twoway scatter` twice to show a scatterplot of `propval100` and `rent700` against `urban`, and then we use `twoway lfit` twice to show the linear fit amongst these same variables. The colors of the scatterplot, however, do not match the fit lines.
Uses allstates.dta & scheme vg_teal

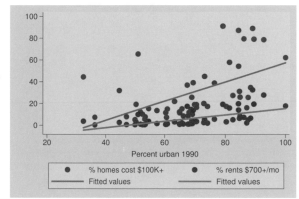

```
twoway (scatter propval100 urban) (scatter rent700 urban)
   (lfit propval100 urban, lwidth(thick))
   (lfit rent700 urban, lwidth(thick)), pcycle(2)
```

By adding `pcycle(2)`, we indicate that the pen characteristics should repeat after two iterations. The first scatterplot is in blue, the second is in red, the third plot (the fit plot) switches back to red, and then the next fit plot is blue. This is a simple solution to make the scatterplot and fit lines match in color.
Uses allstates.dta & scheme vg_teal

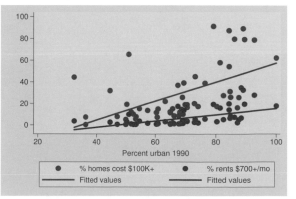

Introduction   Editor   Twoway   Matrix   Bar   Box   Dot   Pie   Options   Standard options   Styles   Appendix

Scatter   Fit   CI fit   Line   Area   Bar   Range   Distribution   Contour   Options   Overlaying

`twoway line high low close tradeday, sort`

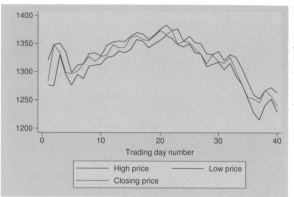

We will briefly switch to using the
`spjanfeb2001` data file. We can also
graph multiple *y* variables against one *x*
variable with a line graph. This works
with `twoway line`, as illustrated here,
as well as with `twoway connected` and
`twoway tsline`.
Uses spjanfeb2001.dta & scheme vg_teal

`twoway line high low close tradeday, sort lwidth(thick thick .)`

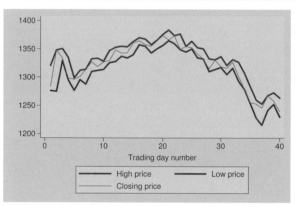

Here we use the `lwidth()` option to
change the width of the lines, making
the lines for the high and low prices
thick and leaving the line for the
closing price at the default width.
Uses spjanfeb2001.dta & scheme vg_teal

`twoway line high low close tradeday, sort lstyle(p1 p1 p2)`

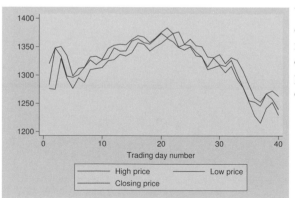

When we graph multiple *y* variables, we
can use `lstyle()` (line style) to control
many characteristics of the lines at
once. Here we plot the high and low
prices with the same style, `p1`, and the
closing price with a second style, `p2`.
Uses spjanfeb2001.dta & scheme vg_teal

```
twoway line high low close tradeday, sort lstyle(p1 p1 p2)
    lwidth(thick thick .)
```

In this example, we combine `lstyle()` and `lwidth()` to make the lines for the high and low prices the same style and both thick. The third line is drawn with the `p2` style, with the thickness left at its default value.

Uses spjanfeb2001.dta & scheme vg_teal

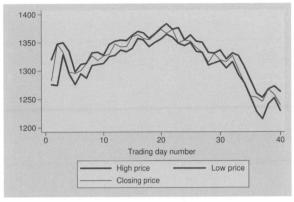

```
twoway (scatter propval100 urban) (lfit propval100 urban)
```

We return to the `allstates` data file. We can overlay multiple twoway graphs. Here we show a common kind of overlay: scatterplot overlaid with a linear fit between the two variables. Both the `scatter` command and the `lfit` command are surrounded by parentheses.

Uses allstates.dta & scheme vg_teal

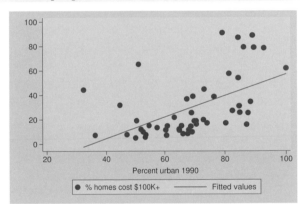

```
twoway (scatter propval100 urban) (lfit propval100 urban)
    (qfit propval100 urban)
```

We can add a quadratic fit to the previous graph by adding a `qfit` command, so we can compare a linear fit and quadratic fit to see if there are nonlinearities in the fit. The legend does not clearly differentiate between the linear and quadratic fit; we will show how the legend can be modified to label this more clearly below.

Uses allstates.dta & scheme vg_teal

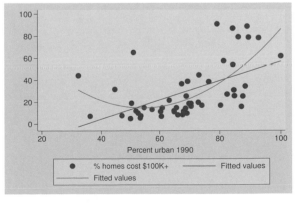

```
twoway (scatter propval100 urban, msymbol(Oh))
   (lfit propval100 urban, lpattern(dash))
   (qfit propval100 urban, lwidth(thick))
```

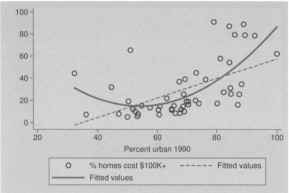

We add the `msymbol(Oh)` option to the `scatter` command, placing it after the comma, as it normally would be placed, but before the closing parenthesis, which indicates the end of the `scatter` command. We also add the `lpattern(dash)` option to the `lfit` command to make the line dashed and add the `lwidth(thick)` option to the `qfit` command to make the line thick.
Uses allstates.dta & scheme vg_teal

```
twoway (scatter propval100 urban) (lfit propval100 urban)
   (qfit propval100 urban), legend(label(2 Linear Fit) label(3 Quad Fit))
```

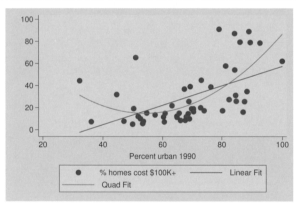

Although each graph subcommand can have its own options, some options can apply to the entire graph. As illustrated here, we add a legend to the graph to clarify the difference in the fit values. This option appears following a comma after the closing parenthesis of the `qfit` command. The `legend()` option appears at the end of the command because it applies to the entire graph.
Uses allstates.dta & scheme vg_teal

```
twoway (scatter propval100 urban) (lfit propval100 urban)
   (qfit propval100 urban, legend(label(2 Linear Fit) label(3 Quad Fit)))
```

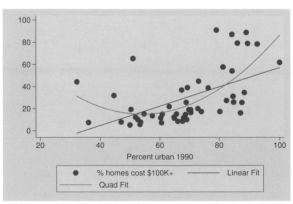

We can make the previous graph in a different, but less appropriate, way. The `legend()` option is given as an option of the `qfit()` command, not at the end as in the previous command. But Stata is forgiving of this, and even when such options are inappropriately given within a particular command, it treats them as though they were given at the end of the command.
Uses allstates.dta & scheme vg_teal

`twoway (qfitci propval100 urban) (scatter propval100 urban)`

Another common example of overlaying graphs is to overlay a fit line with a confidence interval and a scatterplot.
Uses allstates.dta & scheme vg_teal

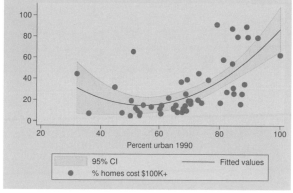

`twoway (scatter propval100 urban) (qfitci propval100 urban)`

However, note the order in which you overlay these two kinds of graphs. In this example, the `qfitci` was drawn after the `scatter`, and as a result, the points are obscured by the confidence interval.
Uses allstates.dta & scheme vg_teal

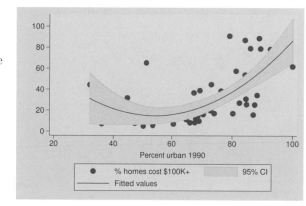

`twoway (rarea high low date) (spike volmil date)`

We now switch to the `sp2001ts` data file. Here we overlay the high and low closing prices with the volume of shares sold. However, because both are placed on the same y axis, it is difficult to see the spikes of `volmil` (volume in millions).
Uses sp2001ts.dta & scheme vg_teal

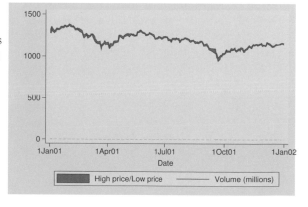

Scatter Fit CI fit Line Area Bar Range Distribution Contour Options Overlaying

Introduction Editor Twoway Matrix Bar Box Dot Pie Options Standard options Styles Appendix

```
twoway (rarea high low date) (spike volmil date, yaxis(2)),
    legend(span)
```

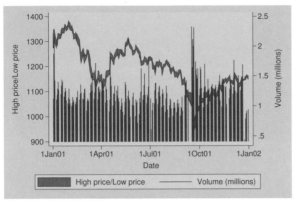

By placing `volmil` on the second y axis using the `yaxis(2)` option, we can now see the volume, but it obstructs the stock prices. We added the option `legend(span)` to allow the legend to be wider than the plot region of the graph.

Uses sp2001ts.dta & scheme vg_teal

```
twoway (rarea high low date) (spike volmil date, yaxis(2)),
    legend(span) yscale(range(500 1400) axis(1)) yscale(range(0 5) axis(2))
```

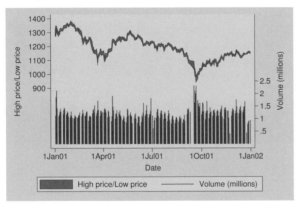

We use the `yscale()` option to modify the range for the first y axis to lift its range into the top third of the graph, and another `yscale()` option to modify the range for the second y axis, pushing the stock-market volume down to the bottom third.

Uses sp2001ts.dta & scheme vg_teal

Although the previous examples (and other examples in this book) have used the parenthetical notation for overlaid graphs, Stata also permits double vertical bars (||) for separating graphs. To illustrate this, let's repeat some of the graphs from above using this notation. These examples use the `vg_s2m` scheme.

```
twoway scatter propval100 urban || lfit propval100 urban
```

We switch back to the `allstates` data file. Here the || notation is used to separate the `scatter` command from the `lfit` command.

Uses allstates.dta & scheme vg_s2m

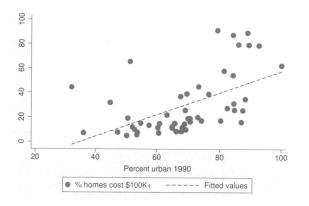

```
twoway scatter propval100 urban || lfit propval100 urban ||
    qfit propval100 urban
```

Here we create three overlaid graphs by using the || notation.

Uses allstates.dta & scheme vg_s2m

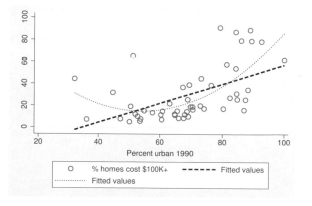

```
twoway scatter propval100 urban, msymbol(Oh) ||
    lfit propval100 urban, lwidth(thick) ||
    qfit propval100 urban, lwidth(medium)
```

This example shows how to use the || notation with options for each of the commands.

Uses allstates.dta & scheme vg_s2m

Introduction

Editor

Twoway

Matrix

Bar

Box

Dot

Pie

Options

Standard options

Styles

Appendix

Scatter Fit CI fit Line Area Bar Range Distribution Contour Options Overlaying

```
twoway scatter propval100 urban, msymbol(Oh) ||
   lfit propval100 urban, lwidth(thick) ||
   qfit propval100 urban, lwidth(medium) ||,
   legend(label(2 Linear Fit) label(3 Quad Fit))
```

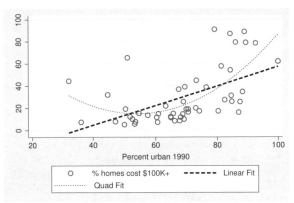

This is another example using the || notation, in this case illustrating how to have options on each of the commands, along with `legend()` as an overall option.

Uses allstates.dta & scheme vg_s2m

4   Scatterplot matrix graphs

This chapter explores the use of the `graph matrix` command for creating scatterplot matrices among two or more variables. Many of the options that you can use with `graph twoway scatter` apply to these kinds of graphs as well; see Twoway : Scatter (89) and Options (307) for related information. This chapter illustrates the use of marker options and marker labels, as well as options for controlling the display of axes. It also includes options specific to the `graph matrix` command, as well as how to use the `by()` option. For more details about scatterplot matrices, see [G-2] **graph matrix**.

4.1   Marker options

This section looks at controlling and labeling the markers in scatterplot matrices. This section will show how to change the marker symbol, size, and color (both fill and outline color) and how to label the markers. You can label markers by using the `graph matrix` command just as you could when using the `graph twoway scatter` command. See also Options : Markers (307) and Options : Marker labels (320) for more details. These examples will use the `vg_s1m` scheme.

`graph matrix propval100 ownhome borninstate, msymbol(Sh)`

We can control the marker symbol with the `msymbol()` option. Here we make the symbols hollow squares. Values that we could specify include D (diamond), O (circle), S (square), T (triangle), and X (x). Using a lowercase letter (d instead of D) makes the symbol smaller. For diamonds, circles, squares, and triangles, you can append an h (e.g., Oh) to indicate that the symbol should be hollow; see Styles : Symbols (427) for more examples.

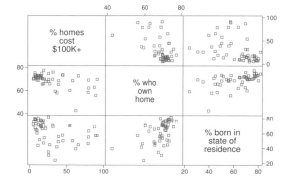

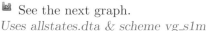 See the next graph.

Uses allstates.dta & scheme vg_s1m

`graph matrix propval100 ownhome borninstate`

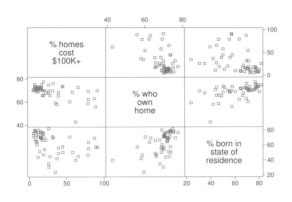

Using the Graph Editor, click on any marker and, in the Contextual Toolbar, change the **Symbol** to Hollow Square. We can also change the **Color** and **Size** of the markers in the Contextual Toolbar.

Uses allstates.dta & scheme vg_s1m

`graph matrix heatdd cooldd tempjan tempjuly, msymbol(p)`

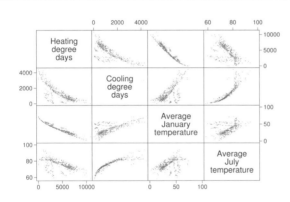

When there are many observations, the `msymbol(p)` option can be useful because it displays a small point for each observation and can help us to see the overall relationships among the variables. Here we switch to the `citytemp` data file to illustrate this option.

Click on a marker and, in the Contextual Toolbar, change the **Symbol** to Point.

Uses citytemp.dta & scheme vg_s1m

`graph matrix propval100 ownhome borninstate, msize(vlarge)`

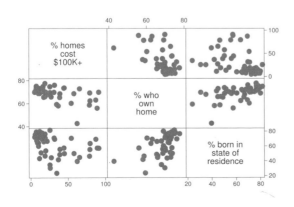

We can change the size of the markers by using the `msize()` (marker size) option. Here we make the markers very large. Values we could have chosen include vtiny, tiny, vsmall, small, medsmall, medium, medlarge, large, vlarge, huge, vhuge, and ehuge; see Styles : Markersize (425) for more details.

Click on a marker and, in the Contextual Toolbar, change the **Size** to v Large.

Uses allstates.dta & scheme vg_s1m

```
graph matrix propval100 ownhome borninstate, msize(*2)
```

If we prefer, we can specify the sizes as multiples of the marker's original size. Here we make the marker 2 times its original size by specifying `msize(*2)`. The asterisk is used to indicate that the value is # times the original.

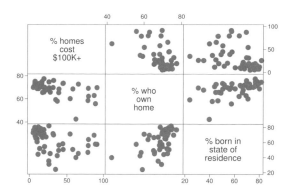

Double-click on a marker and, next to **Size**, type in *2 to double the size of the marker.

Uses allstates.dta & scheme vg_s1m

```
graph matrix propval100 ownhome borninstate, msize(vlarge) mcolor(gs13)
```

Stata offers gray-scale colors named `gs0` (black) to `gs16` (white). Here we use the `mcolor()` (marker color) option to display markers that are very light gray; see **Styles**: Colors (412) for more information about specifying colors.

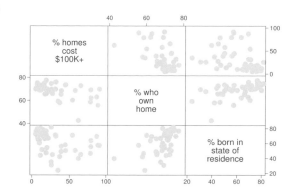

Click on a marker and, in the Contextual Toolbar, change the **Color** to Gray 13.

Uses allstates.dta & scheme vg_s1m

```
graph matrix propval100 ownhome borninstate, msize(vlarge)
    mfcolor(gs13) mlcolor(black)
```

Here we use the `mfcolor()` (marker fill color) option to make the inside color of the marker `gs13` (light gray) and the `mlcolor()` (marker line color) option to make the outline color `black`.

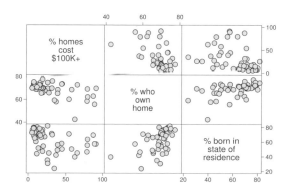

Double-click on a marker and change the **Color** to Gray 13, check **Different outline color**, and change the **Outline color** to Black.

Uses allstates.dta & scheme vg_s1m

Introduction   Editor   Twoway   Matrix   Bar   Box   Dot   Pie   Options   Standard options   Styles   Appendix

Marker options   Axes   Matrix options   By

`graph matrix propval100 ownhome borninstate, mlabel(stateab)`

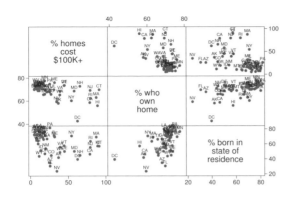

We can label the markers using the `mlabel()` (marker label) option. In this example, we label the markers with the two-letter state abbreviation by supplying the option `mlabel(stateab)`. Even though many of the labels overlap, the most interesting observations are those that stand out and have readable labels, such as DC and NV. For more details, see Options : Marker labels (320).

Uses allstates.dta & scheme vg_s1m

`graph matrix propval100 ownhome borninstate, mlabel(stateab)`
  `mlabsize(large)`

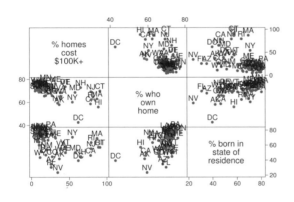

We can use the `mlabsize()` (marker label size) option to control the size of marker labels. Here we indicate that the marker labels should be large.

Click on a marker label (e.g., DC) and, in the Contextual Toolbar, change the **Size** to **Large**. The Contextual Toolbar also allows you to change the marker label **Color** and **Position**.

Uses allstates.dta & scheme vg_s1m

4.2   Controlling axes

This section looks at labeling axes in scatterplot matrices. It shows how to label axes of scatterplots, control the scale of axes, and insert titles along the diagonal. For more details, see Options : Axis labels (330), Options : Axis scales (339) and [G-3] *axis_options*. This section uses the `vg_s2c` scheme.

```
graph matrix urban propval100 borninstate
```

Let's look at a scatterplot matrix of
three variables: `urban`, `propval100`,
and `borninstate`.
Uses allstates.dta & scheme vg_s2c

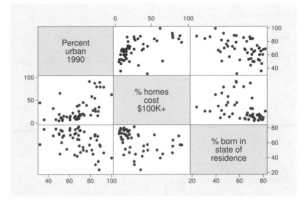

```
graph matrix urban propval100 borninstate,
    xlabel(20(10)100, axis(1))
```

The way we control the axis labels with
a scatterplot matrix is somewhat
different from other kinds of graphs.
Here we use the `xlabel()` option to
label the first variable, `urban`, using
labels going from 20 to 100 in
increments of 10. This option applies to
the first variable because we specified
`axis(1)` within the `xlabel()` option.
See the next graph.
Uses allstates.dta & scheme vg_s2c

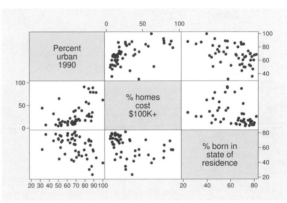

```
graph matrix urban propval100 borninstate
```

Using the Graph Editor,
double-click on the *x* axis for `urban`
and, within the **Axis rule** section,
select **Range/Delta** and enter 20 for
the **Minimum value**, 100 for the
Maximum value, and 10 for the
Delta.
Uses allstates.dta & scheme vg_s2c

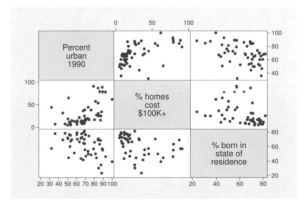

Introduction   Editor   Twoway   Matrix   Bar   Box   Dot   Pie   Options   Standard options   Styles   Appendix

Marker options   Axes   Matrix options   By

```
graph matrix urban propval100 borninstate,
   xlabel(20(10)100, axis(1)) ylabel(20(20)100, axis(1))
```

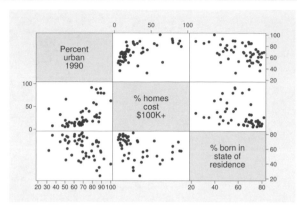

Here we alter the labels for the x axis and the y axis for the first variable, urban. The labels are changed on both axes because we specified axis(1) in both the xlabel() and ylabel() options. We labeled the x axis from 20 to 100 incrementing by 10, and the y axis from 20 to 100 incrementing by 20. See the previous example.

Uses allstates.dta & scheme vg_s2c

```
graph matrix urban propval100 borninstate,
   xlabel(0(20)100, axis(2)) ylabel(0(20)100, axis(2))
```

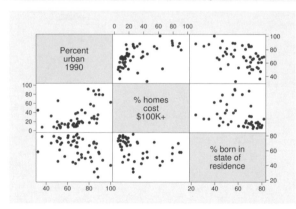

We can change the label for the second variable, propval100, in a similar manner, by specifying axis(2). Here we label the second variable ranging from 0 to 100 in increments of 20. Double-click on the x axis for propval100 and choose **Range/Delta** and enter 0 for the **Minimum value**, 100 for the **Maximum value**, and 20 for the **Delta**. Repeat the steps for the y axis, propval100.

Uses allstates.dta & scheme vg_s2c

```
graph matrix urban propval100 borninstate,
   xlabel(0(20)100, axis(1)) ylabel(0(20)100, axis(1))
   xlabel(0(20)100, axis(2)) ylabel(0(20)100, axis(2))
   xlabel(0(20)100, axis(3)) ylabel(0(20)100, axis(3))
```

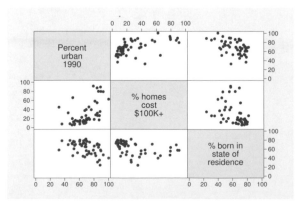

Let's label all these variables using the same scale, from 0 to 100 in increments of 20. As we can see, this involves quite a bit of typing, applying the xlabel() and ylabel() for axis(1), axis(2), and axis(3), which applies this to the first, second, and third variables. However, the next example shows a more efficient way to do this.

Uses allstates.dta & scheme vg_s2c

```
graph matrix urban propval100 borninstate,
    maxes(xlabel(0(20)100) ylabel(0(20)100))
```

Stata has a simpler way of applying the same labels to all the variables in the scatterplot matrix by using the maxes() (multiple axes) option. This example labels the *x* and *y* axes from 0 to 100 with increments of 20 for all variables.
Uses allstates.dta & scheme vg_s2c

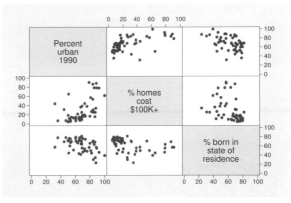

```
graph matrix urban propval100 borninstate,
    maxes(xlabel(0(20)100) ylabel(0(20)100))
    xlabel(20(20)100, axis(1)) ylabel(20(20)100, axis(1))
```

We might want to label most of the variables in the scatterplot matrix the same way but with one or more exceptions in a different way. In this example, we label all the variables from 0 to 100, incrementing by 20, but then override the labeling for **urban** to make it 20 to 100, incrementing by 20. We do this by adding more xlabel() and ylabel() options that apply just to axis(1).
Uses allstates.dta & scheme vg_s2c

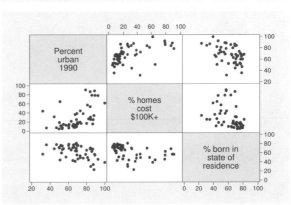

```
graph matrix urban propval100 borninstate,
    maxes(xlabel(0(20)100) ylabel(0(20)100) xtick(0(10)100) ytick(0(10)100))
```

Here we label all the variables from 0 to 100, incrementing by 20, and also add ticks from 0 to 100, incrementing by 10. The xtick() and ytick() options work the same way as the xlabel() and ylabel() options. We place these options within the maxes() option, and they apply to all the axes. See Options : Axis labels (330) and Options : Axis scales (339) for more details.
Uses allstates.dta & scheme vg_s2c

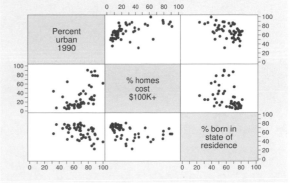

Introduction   Editor   Twoway   Matrix   Bar   Box   Dot   Pie   Options   Standard options   Styles   Appendix

Marker options   Axes   Matrix options   By

```
graph matrix urban propval100 borninstate,
   diagonal("% Urban" "% Homes Over $100K" "% Born in State")
```

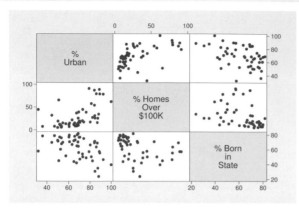

When we use `twoway scatter`, we control the titles by using the `xtitle()` and `ytitle()` options. When using `graph matrix`, we control the titles by using the `diagonal()` option. Here we use the `diagonal()` option to change the titles for all variables.

Double-click on any title and change **Text** to the new title.

Uses allstates.dta & scheme vg_s2c

```
graph matrix urban propval100 borninstate,
   diagonal("% Urban" . "% Born in State")
```

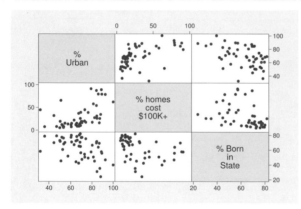

We do not have to change all the titles. If we want to change just some of the titles, we can place a period (.) for the labels that we want to stay the same. In this example, we change the titles for the first and third variables but leave the second as is.

Uses allstates.dta & scheme vg_s2c

```
graph matrix urban propval100 borninstate,
   diagonal("% Urban" . "% Born in State", fcolor(eggshell))
```

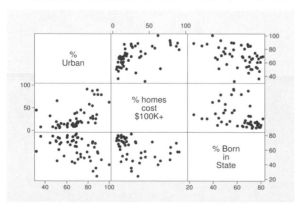

We can control the display of the titles by using textbox options. For example, we make the background color of the text area eggshell by using the `fcolor(eggshell)` option. See Options : Textboxes (379) for more examples of textbox options.

Double-click an empty part of the diagonal and, within the *Grid Properties* dialog, change the **Color** to Eggshell.

Uses allstates.dta & scheme vg_s2c

4.3 Matrix options

This section discusses options that we can use to control the look of the scatterplot matrix, including showing just the lower half of the matrix, jittering markers, and scaling the size of marker text. For more details, see [G-2] **graph matrix**. These graphs use the vg_s2m scheme.

`graph matrix propval100 ownhome region, half`

The `half` option displays only the lower diagonal of the scatterplot matrix.
Uses allstates.dta & scheme vg_s2m

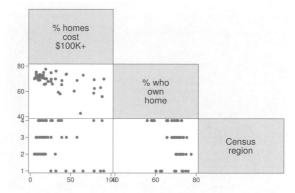

`graph matrix propval100 ownhome region, jitter(3)`

The `jitter()` option adds random noise to the points; the higher the value given, the more random noise is added. This option is especially useful when many observations have the same (x,y) values, so several observations can appear as one point.
Uses allstates.dta & scheme vg_s2m

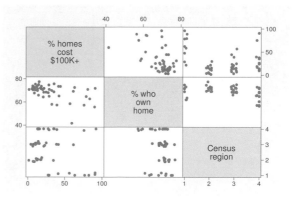

```
graph matrix propval100 ownhome region, scale(1.5)
```

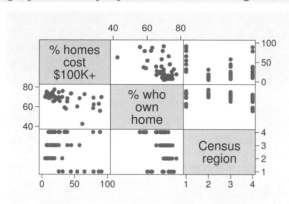

We can use the `scale()` option to magnify the contents of the graph, including the markers, labels, and lines, but not the overall size of the graph. Here we increase the size of these items, making them 1.5 times their normal size. Unlike other similar options, this option does not take an asterisk preceding the value; i.e., we specify 1.5, not *1.5.

Uses allstates.dta & scheme vg_s2m

4.4 Graphing by groups

This section looks at the use of the `by()` option for showing separate graphs based on the levels of a `by()` variable. For more information, see Options: By (346) and [G-3] *by_option*. This section uses the `vg_brite` scheme.

```
graph matrix propval100 ownhome borninstate, by(north)
```

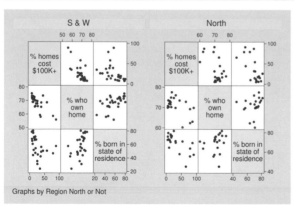

We can use the `by()` option with `graph matrix` to show separate scatterplot matrices by a particular variable. Here we show separate scatterplot matrices for households in northern states and nonnorthern states.

Uses allstates.dta & scheme vg_brite

```
graph matrix propval100 ownhome borninstate, by(north, compact)
```

To display the graphs closer together,
we use the `compact` option.
Uses allstates.dta & scheme vg_brite

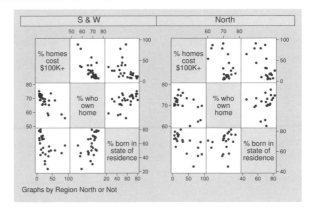

```
twoway scatter propval100 ownhome, by(north, compact)
```

If we compare the previous scatterplot
matrix with this twoway scatterplot, we
see that the `compact` option does not
make the scatterplot matrix as compact
as it does with a `twoway scatter`
command, which joins the two graphs
on their edges by omitting the *y* labels
between the two graphs.
Uses allstates.dta & scheme vg_brite

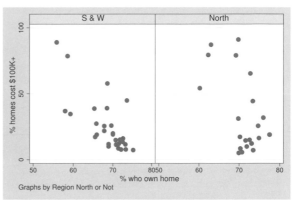

```
graph matrix propval100 ownhome borninstate, by(north, compact)
    maxes(ylabel(, nolabels))
```

We can make the `graph matrix` display
more compactly with the `by()` option
by using the `maxes(ylabel(,`
`nolabels))` option to suppress the
labels on all the *y* axes. Then, when we
use the `compact` option, the edges of
the plots are pushed closer together.
Uses allstates.dta & scheme vg_brite

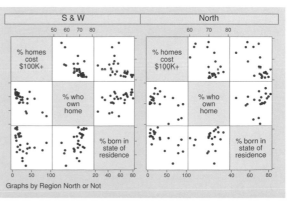

Introduction   Editor   Twoway   Matrix   Bar   Box   Dot   Pie   Options   Standard options   Styles   Appendix

Marker options   Axes   Matrix options   By

```
graph matrix propval100 ownhome borninstate, by(north, compact scale(*1.3))
    maxes(ylabel(, nolabels))
```

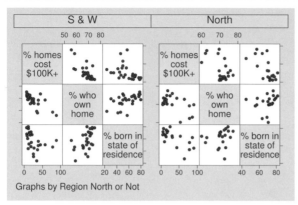

Graphs by Region North or Not

We use `scale()` within the `by()` option to increase the size of the markers, labels, and text to make them more readable. `scale()` is especially useful for the display of small graphs.

Uses allstates.dta & scheme vg_brite

5 Bar graphs

This chapter explores how to create bar charts by using the **graph bar** command. It shows how to use **graph bar** to graph one or more continuous y variables and shows how you can break them down by one or more categorical variables. This chapter also illustrates how to control the display of each of the axes, the legend, and the bars, and how to use the **by()** option. This chapter begins by looking at features related to graphing one or more y variables. For this entire chapter, we will use the **nlsw** data file.

5.1 Y variables

A bar chart graphs one or more continuous variables broken down by one or more categorical variables. The continuous variables are graphed on the y axis and are referred to as y variables. This section shows you how to specify the y variables using the **graph bar** command, how to include one or more y variables, and how to obtain different summary statistics for the y variables. For more information, see [G-2] **graph bar**. This section begins by using the **vg_past** scheme.

```
graph bar ttl_exp
```

This is probably the most basic bar chart that we can make (and perhaps the most boring, as well). It shows the average total work experience for all observations in the file. It graphs one y variable by using the default summary statistic, the mean.

Uses nlsw.dta & scheme vg_past

`graph bar prev_exp tenure ttl_exp`

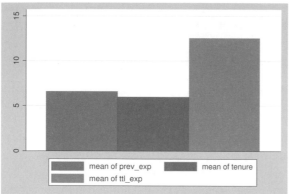

We can specify multiple y variables to be plotted simultaneously. Here we graph the mean of previous, current, and total work experience in the same plot. The bars are plotted touching each other, and a legend indicates which bar corresponds to which variable.
Uses nlsw.dta & scheme vg_past

`graph bar (median) prev_exp tenure ttl_exp`

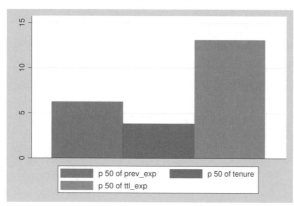

This graph is much like the last one, but it shows the median of these y variables. We only specified (`median`) before `prev_exp` but it applied to all the y variables that follow. You can summarize the y variables by using any of the summary statistics permitted by the `collapse` command (e.g., `mean`, `sd`, `sum`, `median`, and `p10`); see [D] **collapse**.
Uses nlsw.dta & scheme vg_past

`graph bar (median) prev_exp tenure (mean) ttl_exp`

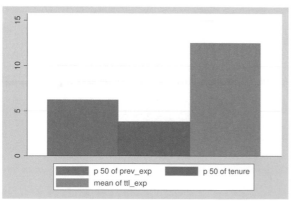

Here we get the median of the first two y variables and then the mean of the last y variable.
Uses nlsw.dta & scheme vg_past

graph bar (mean) wage (median) wage

We can plot different summary
statistics for the same *y* variable. In
this example, we plot the mean and
median wage.

Uses nlsw.dta & scheme vg_past

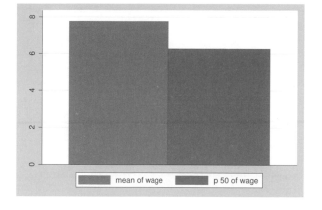

Let's now consider a handful of options that are useful when we have multiple *y* variables. These options allow us to display the *y* variables as though they were categories of the
same variable, to create stacked bar charts, and to display the *y* variables as percentages of
the total *y* variables. These options are illustrated in the following graphs using the vg_s1m
scheme.

graph bar prev_exp tenure ttl_exp hours

First, consider this bar chart showing
four *y* variables. Each *y* variable is
shown with a different colored bar and
with a legend indicating which *y*
variable corresponds to which bar. See
the next example for another way to
differentiate these four bars.

Uses nlsw.dta & scheme vg_s1m

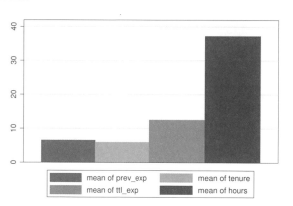

`graph bar prev_exp tenure ttl_exp hours, ascategory`

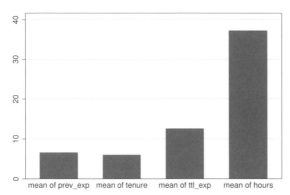

The `ascategory` option indicates that we want Stata to graph multiple y variables by using the style that would be used for the levels of an `over()` variable. Comparing this graph with the previous graph, note how the bars for the different variables are the same color and labeled on the x axis rather than using a legend.
Uses nlsw.dta & scheme vg_s1m

`graph bar prev_exp tenure, over(occ5)`

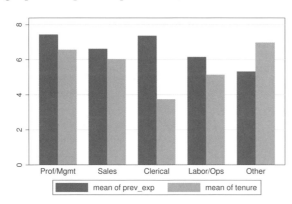

Consider this graph, where we show work experience prior to one's current job (`prev_exp`) and work experience at one's current job (`tenure`) broken down by `occ5`. The total of previous and current work experience represents total work experience, and we might want to show each bar as a percentage of total work experience. The next example shows how we can do that.
Uses nlsw.dta & scheme vg_s1m

`graph bar prev_exp tenure, over(occ5) percentages`

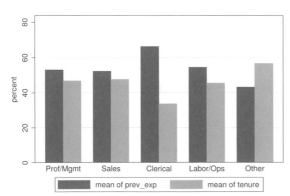

Here we use the `percentages` option to show the time worked before one's current job, `prev_exp`, and the time at the current job, `tenure`, as a percentage of total work experience.

In the Object Browser, double-click on **bar region** and then check **Graph percentages**.
Uses nlsw.dta & scheme vg_s1m

```
graph bar prev_exp tenure, over(occ5) stack
```

The `stack` option shows the *y* variables as a stacked bar chart. This option allows us to see the mean of each *y* variable, as well as the mean of the total *y* variables.

In the Object Browser, double-click on **bar region** and then check **Stack bars**.

Uses nlsw.dta & scheme vg_s1m

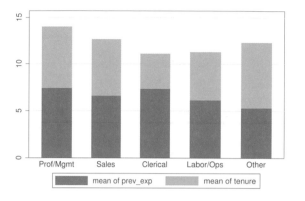

```
graph bar prev_exp tenure, over(occ5) percentages stack
```

We can also combine the `stack` and `percentages` options to create a stacked bar chart in terms of percentages.

In the Object Browser, double-click on **bar region** and then check **Stack bars** and **Graph percentages**.

Uses nlsw.dta & scheme vg_s1m

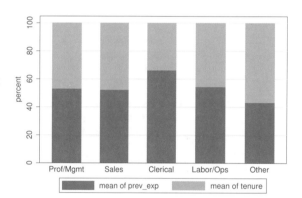

5.2 Graphing bars over groups

This section focuses on using the `over()` option for showing bar charts by one or more categorical variables. It illustrates using the `over()` option with one or multiple *y* variables. We will also look at some basic options, including those that do the following: display the `over()` variable as though its levels were multiple *y* variables, include missing values on the `over()` variable, and suppress empty combinations of multiple `over()` variables. See the *group_options* and *over_subopts* tables of [G-2] **graph bar** for more details.

Introduction Editor Twoway Matrix Bar Box Dot Pie Options Standard options Styles Appendix

Y variables Over Bar gaps Bar sorting Cat axis Legends/Labels Y axis Lookofbar options By

`graph hbar wage, over(occ5)`

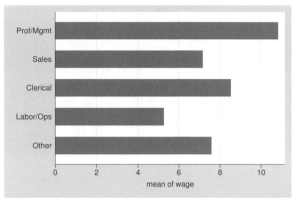

In this example, we use the `over()` option to show the average wages broken down by occupation. We are using `graph hbar` to produce horizontal, rather than vertical, bar charts.

Uses nlsw.dta & scheme vg_brite

`graph hbar wage, over(occ5) over(collgrad)`

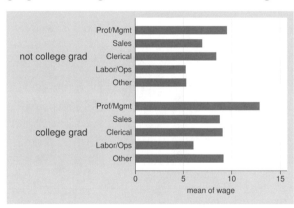

Here we use the `over()` option twice to show the wages broken down by occupation and whether one graduated college. Note the appropriate way to produce this graph is to use two `over()` options, rather than using one `over()` option with two variables. As we will see later, each `over()` can have its own options, allowing us to customize the display of each `over()` variable.

Uses nlsw.dta & scheme vg_brite

`graph hbar wage, over(urban2) over(occ5) over(collgrad)`

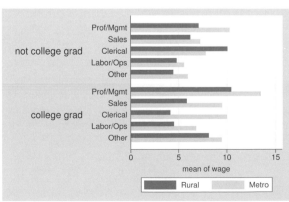

We can even add a third `over()` option, here, using `over(urban2)` to compare those living in rural versus urban areas. The look of the graph changes when we add the third `over()` variable. This is because Stata is now treating the first `over()` variable as though it were multiple *y* variables. Because of this, we can only specify one *y* variable when we have three `over()` options.

Uses nlsw.dta & scheme vg_brite

Now let's look at examples of using multiple y variables with the **over()** option. Let's start with a simple bar graph with multiple y variables. These examples will use the **vg_lgndc** scheme, which places the legend to the left of the graph and displays it in one, stacked column.

graph hbar prev_exp tenure ttl_exp

This graph shows the overall mean of previous, current, and total work experience.
Uses nlsw.dta & scheme vg_lgndc

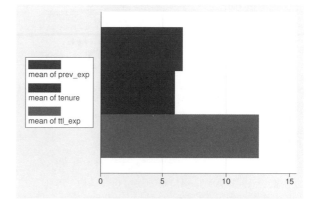

graph hbar prev_exp tenure ttl_exp, over(occ5)

We can take the graph from above and break down the means by occupation by adding the **over(occ5)** option.
Uses nlsw.dta & scheme vg_lgndc

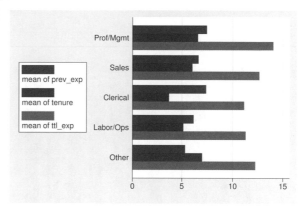

Introduction   Editor   Twoway   Matrix   Bar   Box   Dot   Pie   Options   Standard options   Styles   Apperdix

Y variables   Over   Bar gaps   Bar sorting   Cat axis   Legends/Labels   Y axis   Lookofbar options   By

```
graph hbar prev_exp tenure ttl_exp, over(occ5) over(union)
```

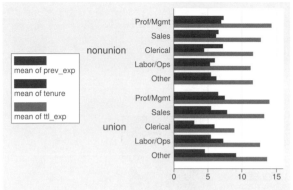

We can take the previous graph and further break down the results by whether one belongs to a union. However, we cannot add a third `over()` option when we have multiple *y* variables.

Uses nlsw.dta & scheme vg_lgndc

Now let's consider options we can use in combination with the `over()` option to customize the graphs. These examples will illustrate how we can treat the levels of the variable in the first `over()` option as though they were multiple *y* variables and can even graph those levels as percentages or stacked bar charts. We can also request that missing values for the levels of the `over()` variables be displayed, and we can suppress empty levels when multiple `over()` options are used. These examples are shown below using the **vg_rose** scheme.

```
graph bar wage, over(occ5) over(union)
```

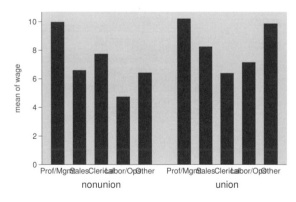

Consider this graph, where we show wages broken down by occupation and whether one belongs to a union. The labels for the levels of **occ5** overlap, but this is corrected in the next example.

Uses nlsw.dta & scheme vg_rose

`graph bar wage, over(occ5) over(union) asyvars`

If we add the `asyvars` option, the first
`over()` variable (`occ5`) is graphed as if
there were five y variables
corresponding to the five levels of `occ5`.
The levels of `occ5` are shown as
differently colored bars pushed next to
each other and labeled using the legend.

In the Object Browser, click on **bar
region**. Then, in the Contextual
Toolbar, click on **Rotate** until you see
the graph you desire.

Uses nlsw.dta & scheme vg_rose

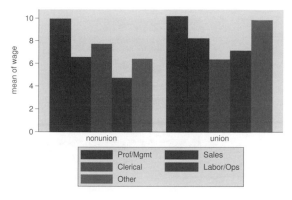

`graph bar wage, over(occ5) over(union) asyvars percentages`

With the levels of `occ5` considered as y
variables, we can use some of the
options that apply when we have
multiple y variables. Here we request
that the values be plotted as
`percentages`.

In the Object Browser, double-click
on **bar region** and then check **Graph
percentages**.

Uses nlsw.dta & scheme vg_rose

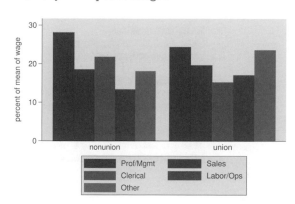

`graph bar wage, over(occ5) over(union) asyvars percentages stack`

Again, because we are treating the
levels of `occ5` as though they were
multiple y variables, we can add the
`stack` option to view the graph as a
stacked bar chart.

In the Object Browser, double-click
on **bar region** and then check **Stack
bars**.

Uses nlsw.dta & scheme vg_rose

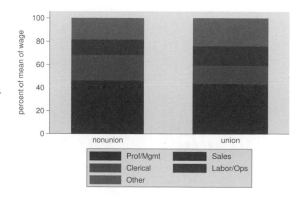

Introduction Editor Twoway Matrix Bar Box Dot Pie Options Standard options Styles Appendix

Y variables Over Bar gaps Bar sorting Cat axis Legends/Labels Y axis Lookofbar options By

`graph hbar wage, over(urban3) over(union)`

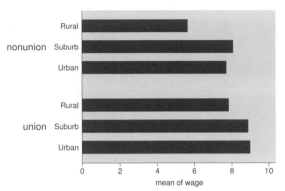

Consider this graph, where we use the `over(union)` option to compare the mean wages of union workers with nonunion workers. One aspect this graph hides is that there are several missing values on the variable `union`. *Uses nlsw.dta & scheme vg_rose*

`graph hbar wage, over(urban3) over(union) missing`

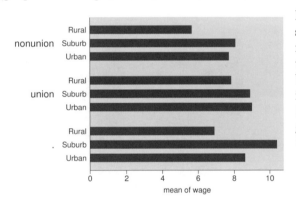

By adding the `missing` option, we then see a category for those who are missing on the `union` variable, shown as the third set of bars. The label for this bar is one dot, which is the Stata indicator of missing values. The section `Bar:Cat axis` (193) shows how we can give this bar a more meaningful label. *Uses nlsw.dta & scheme vg_rose*

`graph bar wage, over(grade) over(collgrad)`

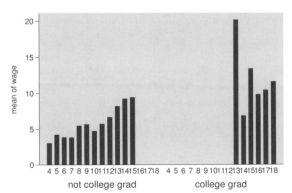

Consider this bar chart, which breaks down wages by two variables: the last grade that one completed and whether one is a college graduate. By default, Stata shows all possible combinations for these two variables. Usually, all combinations are possible, but not here. *Uses nlsw.dta & scheme vg_rose*

```
graph bar wage, over(grade) over(collgrad) nofill
```

If we want to display only the
combinations of the `over()` variables
that exist in the data, we use the
`nofill` option.

📊 In the Object Browser, double-click
on **bar region** and then uncheck
Include missing categories.

Uses nlsw.dta & scheme vg_rose

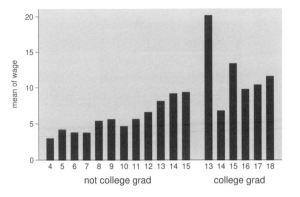

5.3   Options for controlling gaps between bars

This section illustrates how to control gaps between bars and the gaps outside the bars.
There are no options to control the actual width of the bars. These are controlled by specifying the size of the gaps, and whatever is leftover determines the width of the bars (i.e.,
specifying big gaps yields skinny bars, and specifying small gaps yields wide bars).

```
graph hbar wage, over(grade4)
```

We start with a simple graph showing
the average wages broken down by
`grade4`. This graph is annotated using
the Graph Editor to show the gaps that
we can control. The default gap
between levels of `grade4` bars
(controlled by `over(grade4, gap(#))`
is 67% of the width of a bar. The
default value for the `outergap(#)` is
20% the width of a bar. The following
examples will illustrate altering these
defaults.

Uses nlsw.dta & scheme vg_s2c

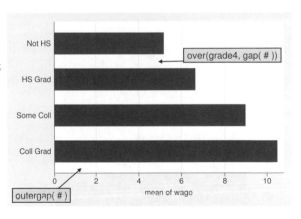

`graph hbar wage, over(grade4, gap(10))`

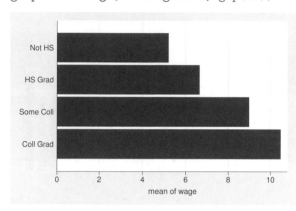

Here we add the `gap(10)` option to make the gap between the bars be 10% of the width of a bar.
 In the Object Browser, double-click on **bar region** to bring up the *Bar region properties* dialog box. Change the bar spacing for **Categories** to 10pct. The following Graph Editor examples using this graph will assume we are still within the *Bar region properties* dialog box.
Uses nlsw.dta & scheme vg_s2c

`graph hbar wage, over(grade4, gap(10)) outergap(200)`

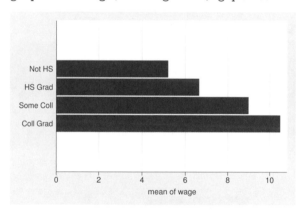

We now add the `outergap(200)` option to increase the outer gap between the peripheral bar and the edge of the graph from 20% of the width of a bar to 200% times the width of a bar.
 Continuing from the previous graph, change the bar spacing for **Outer** to 200pct.
Uses nlsw.dta & scheme vg_s2c

`graph hbar wage, over(grade4, gap(-50))`

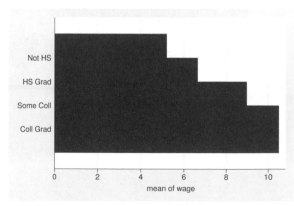

Specifying a negative gap might seem illogical, but this is how to indicate that you want the bars to overlap. Specifying −50 indicates that the bars should overlap by 50% of the width of a bar.
 In the Object Browser, double-click on **bar region** and change the bar spacing for **Categories** to neg50pct.
Uses nlsw.dta & scheme vg_s2c

```
graph hbar wage, over(grade4) over(married)
```

We now look at a graph with two
`over()` options. This graph is
annotated using the Graph Editor to
show the gaps that we can control. The
default gap between levels of `grade4`
bars (controlled by `over(grade4,`
`gap(#))` is 67% of the width of a bar,
whereas the default gap between the
levels of `married` is 200% times the
width of a bar. The default value for
the `outergap(#)` is 20% the width of a
bar. The following examples will
illustrate altering these defaults.
Uses nlsw.dta & scheme vg_s2c

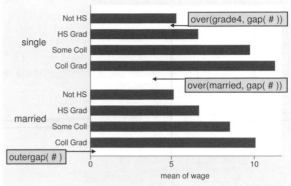

```
graph hbar wage, over(grade4, gap(20)) over(married)
```

By adding the `gap(20)` option, we
shrink the gap among the education
categories from 67% to 20%.

  🔖 In the Object Browser, double-click
on **bar region** to bring up the *Bar
region properties* dialog box. You can
change the bar spacing for **Categories**
to 20pct. The following examples using
this graph will assume we are still
within the *Bar region properties* dialog
box.
Uses nlsw.dta & scheme vg_s2c

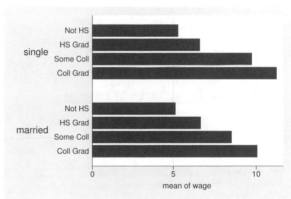

```
graph hbar wage, over(grade4, gap(20)) over(married, gap(50))
```

The gap between single and married
would normally be 200% times the
width of a bar, but here we reduce it to
50%.

  🔖 Continuing from the previous graph,
change the bar spacing for **Super
categories** to 50pct.
Uses nlsw.dta & scheme vg_s2c

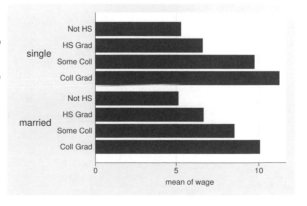

Introduction Editor Twoway Matrix Bar Box Dot Pie Options Standard options Styles Appendix

Y variables Over Bar gaps Bar sorting Cat axis Legends/Labels Y axis Lookofbar options By

```
graph hbar wage, over(grade4, gap(20)) over(married, gap(50)) outergap(200)
```

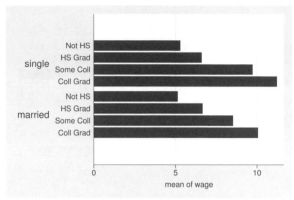

The outer gap is normally 20% of the width of a bar, but here we expand it to 200% times the width of a bar by specifying the `outergap(200)` option.
📊 Continuing from the previous graph, change the bar spacing for **Outer** to 200pct.
Uses nlsw.dta & scheme vg_s2c

```
graph hbar wage, over(collgrad) over(married) over(union)
```

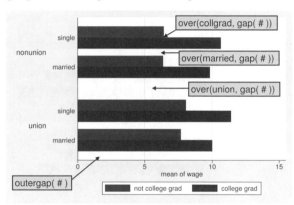

We now look at a graph with three `over()` options. This graph is annotated using the Graph Editor to show the gaps that we can control. The default gap between levels of `collgrad` bars (controlled by `over(collgrad, gap(#))`) is zero. The default gap between levels of married is 67% of the width of a bar, whereas the default gap between the levels of `union` is 200% times the width of a bar. The default value for the `outergap(#)` is 20% the width of a bar. The following examples will illustrate altering these defaults.
Uses nlsw.dta & scheme vg_s2c

```
graph hbar wage, over(collgrad, gap(25))
    over(married) over(union)
```

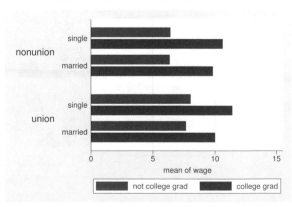

Normally, the gap among the levels of `collgrad` would be 0%, but here we make the gap 25% of the width of a bar.
📊 In the Object Browser, double-click on **bar region** to bring up the *Bar region properties* dialog box. Change the bar spacing for **Variables** to 25pct. The following examples using this graph will assume we are still within the *Bar region properties* dialog box.
Uses nlsw.dta & scheme vg_s2c

```
graph hbar wage, over(collgrad, gap(25))
   over(married, gap(200)) over(union)
```

Normally, the gap between single and married would be two-thirds the width of a bar (67%), but here we increase that gap to 200%.

Continuing from the previous graph, change the bar spacing for **Categories** to 200pct.

Uses nlsw.dta & scheme vg_s2c

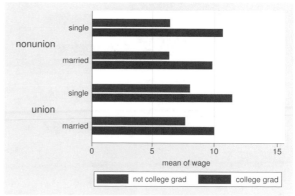

```
graph hbar wage, over(collgrad, gap(25))
   over(married, gap(200)) over(union, gap(400))
```

The gap between union and nonunion would normally be twice the width of a bar (200%), but here we increase it to 400%.

Continuing from the previous graph, change the bar spacing for **Super categories** to 400pct.

Uses nlsw.dta & scheme vg_s2c

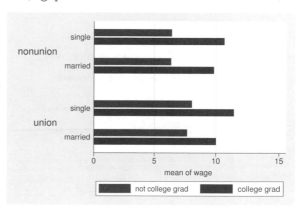

```
graph hbar wage, over(collgrad, gap(25))
   over(married, gap(200)) over(union, gap(400)) outergap(400)
```

The outer gap is normally one-fifth (20%) of the width of a bar, but here we increase it to four times the width of a bar (400%).

Continuing from the previous graph, change the bar spacing for **Outer** to 400pct.

Uses nlsw.dta & scheme vg_s2c

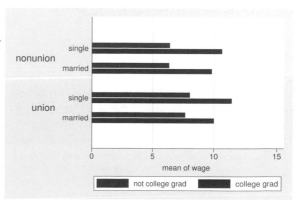

Introduction   Editor   Twoway   Matrix   Bar   Box   Dot   Pie   Options   Standard options   Styles   Appendix

Y variables   Over   Bar gaps   Bar sorting   Cat axis   Legends/Labels   Y axis   Lookofbar options   By

`graph hbar prev_exp ttl_exp, over(married) over(union)`

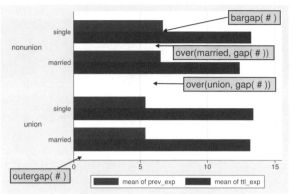

The final examples of this section look at a graph that has multiple y variables (here, two) along with two `over()` options. This graph is annotated using the Graph Editor to show the gaps that we can control. The default gaps are all the same as the previous example, but because this graph involves multiple y variables, the gap between `prev_exp` and `ttl_exp` is controlled by the `bargap()` option, as illustrated in the next example.

Uses nlsw.dta & scheme vg_s2c

`graph hbar prev_exp ttl_exp, bargap(25)`
`    over(married, gap(200)) over(union, gap(400)) outergap(400)`

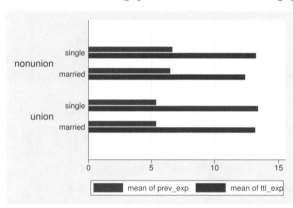

Here we use the `bargap()` option to make the gap between `prev_exp` and `ttl_exp` 25% of the width of a bar. We also see that the `gap()` and `outergap()` options work as they did in the previous examples.

In the Object Browser, double-click on **bar region** to bring up the *Bar region properties* dialog box. Then change the bar spacing for **Variables** to 25pct.

Uses nlsw.dta & scheme vg_s2c

5.4   Options for sorting bars

This section considers some of the options we can use to control the order in which the bars are displayed. This is an area where the control afforded to commands is far greater than the control you can obtain with the Graph Editor. This section will focus on graph commands without concern for the Graph Editor. By default, the bars formed by `over()` variables are ordered in ascending sequence according to the values of the `over()` variable. However, Stata gives us considerable flexibility in the ordering of the bars, as illustrated in the following examples using the `vg_outc` scheme.

```
graph hbar wage, over(occ7, gap(-25))
```

Let's start with this graph showing the
average wages broken down by
occupation. The bars are ordered by
the levels of occ7, going from 1 to 7,
where 1 is *Prof* and 7 is *Other*. The
first bar (*Prof*) overlaps the second bar
(*Mgmt*), and the second bar overlaps
the third (*Sales*), and so forth.
Uses nlsw.dta & scheme vg_outc

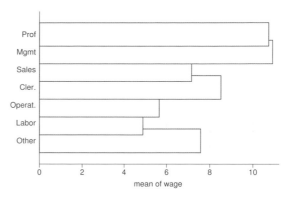

```
graph hbar wage, over(occ7, gap(-25) descending)
```

By using the descending option, the
order of the bars is switched. The bars
are still ordered according to the seven
levels of occupation, but they are
ordered going from 7 to 1.
Uses nlsw.dta & scheme vg_outc

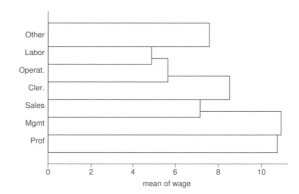

```
graph hbar wage, over(occ7, gap(-25) reverse)
```

Using the reverse option not only
switches the order of the bars, but it
also changes the order of overlapping.
Here *Prof* overlaps *Mgmt*, whereas the
overlapping was the opposite in the
previous graph.
Uses nlsw.dta & scheme vg_outc

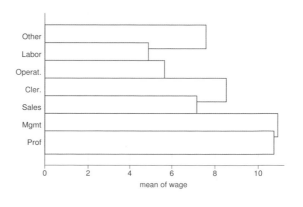

Introduction   Editor   Twoway   Matrix   Bar   Box   Dot   Pie   Options   Standard options   Styles   Appendix

Y variables   Over   Bar gaps   Bar sorting   Cat axis   Legends/Labels   Y axis   Lookofbar options   By

```
graph hbar wage, over(occ7, gap(-25) reverse descending)
```

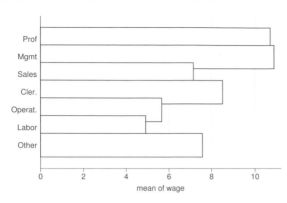

Although it might seem counterintuitive, we can combine the **reverse** and **descending** options. They counteract each other with respect to the ordering of the categories (the categories are ordered from 1 to 7), but the net effect is that the pattern of overlapping is reversed. In this graph *Mgmt* overlaps *Prof*, compared with the first graph in this section, where *Prof* overlaps *Mgmt*.

Uses nlsw.dta & scheme vg_outc

```
graph hbar wage, over(occ7, sort(occ7alpha))
```

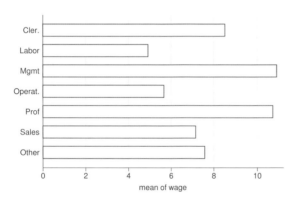

We might want to put these bars in alphabetical order (but with *Other* still appearing last). We can do this by recoding **occ7** into a new variable (say, **occ7alpha**) such that as **occ7alpha** goes from 1 to 7, the occupations are alphabetical. We recoded **occ7** with these assignments: $4 = 1$, $6 = 2$, $2 = 3$, $5 = 4$, $1 = 5$, $3 = 6$, and $7 = 7$; see [D] **recode**. Then the **sort(occ7alpha)** option has the effect of alphabetizing the bars.

Uses nlsw.dta & scheme vg_outc

```
graph hbar wage, over(occ7, sort(1))
```

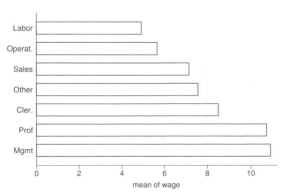

Here we sort the variables on the height of the bars (in ascending order). The **sort(1)** option sorts the bars according to the height of the first y variable, the mean of **wage**.

Uses nlsw.dta & scheme vg_outc

```
graph hbar wage, over(occ7, sort(1) descending)
```

Adding the `descending` option yields
bars in descending order.
Uses nlsw.dta & scheme vg_outc

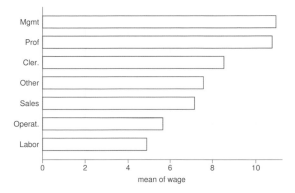

```
graph hbar wage hours, over(occ7, sort(1))
```

Here we plot two *y* variables. In
addition to wages, we show the average
hours worked per week. The `sort(1)`
option sorts the bars according to the
mean of `wage` because that is the first *y*
variable.
Uses nlsw.dta & scheme vg_outc

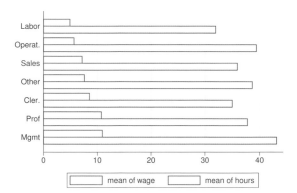

```
graph hbar wage hours, over(occ7, sort(2))
```

Changing `sort(1)` to `sort(2)` sorts the
bars according to the second *y* variable,
the mean of `hours`.
Uses nlsw.dta & scheme vg_outc

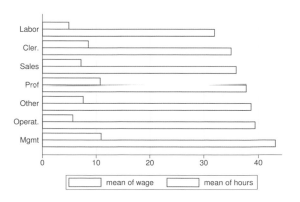

Introduction Editor Twoway Matrix Bar Box Dot Pie Options Standard options Styles Appendix

Y variables Over Bar gaps Bar sorting Cat axis Legends/Labels Y axis Lookofbar options By

`graph hbar wage hours, over(occ7, sort(2)) over(married)`

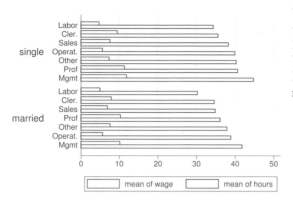

We can use the `sort()` option when there are multiple `over()` variables. Here the `sort(2)` option orders the bars according to the mean number of hours worked within each level of `married`.

Uses nlsw.dta & scheme vg_outc

`graph hbar wage hours, over(occ7, sort(2)) over(married, descending)`

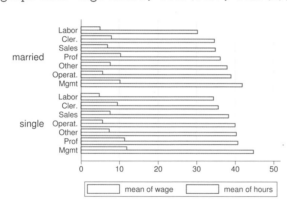

Each `over()` option can have its own separate sorting options. In this example, we add the `descending` option to the second `over()` option, and the levels of `married` are now shown with those who are married appearing first.

Uses nlsw.dta & scheme vg_outc

`graph hbar (sum) wage, over(collgrad) over(occ7) asyvars stack`

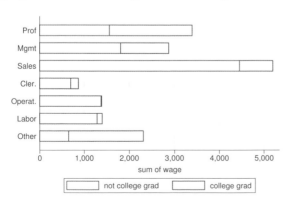

Say that we were to graph the sum of `wage` broken down by `collgrad` and `occ7`. We further treat the levels of `collgrad` as y variables and form a stacked bar chart. We might want to sort these bars on the sum of wages for each occupation. See the next example for how we can do that.

Uses nlsw.dta & scheme vg_outc

```
graph hbar (sum) wage,
   over(collgrad) over(occ7, sort((sum) wage)) asyvars stack
```

Here we add `sort((sum) wage)` to the
`over()` option for `occ7`. The bars are
sorted on the sum of wages at each
level of `occ7`, sorting the bars on their
total height.
Uses nlsw.dta & scheme vg_outc

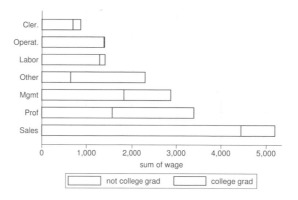

```
graph hbar (sum) wage,
   over(collgrad) over(occ7, sort((sum) wage) descending) asyvars stack
```

In this example, we add the
`descending` option to change the sort
order from highest to lowest. Note the
placement of the `descending` option
outside the `sort()` option.
Uses nlsw.dta & scheme vg_outc

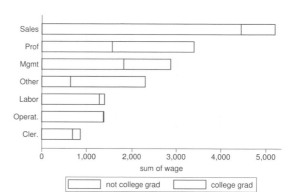

5.5   Controlling the categorical axis

This section describes ways that we can label categorical axes. Bar charts are special
because their x axis is formed by categorical variables. This section describes options we can
use to customize these categorical axes. For more details, see [G-3] **cat_axis_label_options**
and [G-3] **cat_axis_line_options**.

Let's start by exploring how we can change the labels for the bars on the x axis.

```
graph hbar wage, over(union) over(smsa) over(south) missing
```

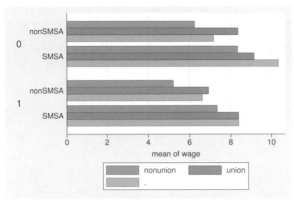

This bar chart breaks down wages by union status, whether one lives in the south, and whether one lives in a metropolitan area (smsa). Each variable could use a better label to make the graph clearer. The following examples will illustrate how to change the labels for these variables.

Uses nlsw.dta & scheme vg_palec

```
graph hbar wage, over(union) over(smsa)
    over(south, relabel(1 "N & W" 2 "South")) missing
```

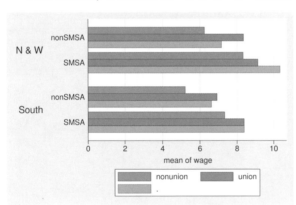

The relabel() option changes the labels displayed for the levels of south. We wrote relabel(1 "N & W") and not relabel(0 "N & W") because these numbers do not represent the actual levels of south but the ordinal position of the levels, i.e., first and second.

See the next graph.

Uses nlsw.dta & scheme vg_palec

```
graph hbar wage, over(union) over(smsa) over(south) missing
```

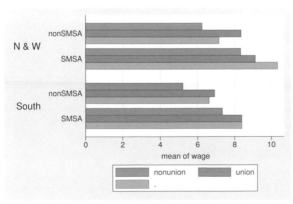

Here we show how to use the Graph Editor to change the labels for south as illustrated in the previous example. Double-click on the label for south (e.g., 0 or 1) and click on the **Edit or add individual ticks and labels** button. Select the tick labeled 1 and click on **Edit** and change the **Label** to South and click on **OK**. Next select the tick labeled 0 and click on **Edit** and change the **Label** to N & W and click on **OK**.

Uses nlsw.dta & scheme vg_palec

```
graph hbar wage, over(union)
   over(smsa, relabel(1 "Nonmetro" 2 "Metro"))
   over(south, relabel(1 "N & W" 2 "South")) missing
```

We now add a `relabel()` option to change the labels for the levels of `smsa`. Each `over()` option can have its own `relabel()` option.

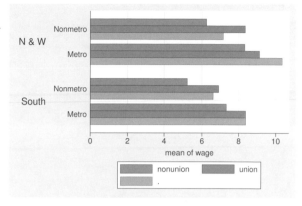 See the next graph.

Uses nlsw.dta & scheme vg_palec

```
graph hbar wage, over(union)
   over(smsa) over(south, relabel(1 "N & W" 2 "South")) missing
```

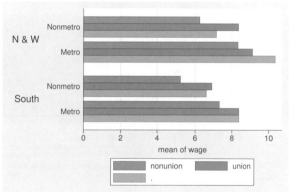 In the Graph Editor, double-click on the label for `smsa` (e.g., *nonSMSA* or *SMSA*) and click on **Edit or add individual ticks and labels**. Select one of the ticks labeled *SMSA* and click on **Edit** and change the **Label** to `Metro`. Repeat this step for the other *SMSA* tick. Likewise edit the ticks labeled *nonSMSA*, changing their labels to `Nonmetro`.

Uses nlsw.dta & scheme vg_palec

```
graph hbar wage, over(union, relabel(3 "missing"))
   over(smsa, relabel(1 "Nonmetro" 2 "Metro"))
   over(south, relabel(1 "N & W" 2 "South")) missing
```

We can even use the `relabel()` option on the first `over()` option. Because this variable is labeled with a legend, we could also use the `legend()` option. See Bar : Legend (201) for more details.

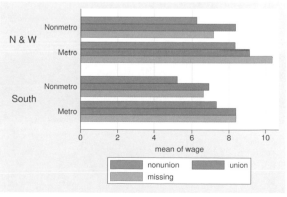 Point on the dot in the legend and double-click. In the *Textbox properties* dialog box, change the **Text** to `missing`.

Uses nlsw.dta & scheme vg_palec

Introduction   Editor   Twoway   Matrix   Bar   Box   Dot   Pie   Options   Standard options   Styles   Appendix

Y variables   Over   Bar gaps   Bar sorting   Cat axis   Legends/Labels   Y axis   Lookofbar options   By

```
graph hbar prev_exp tenure ttl_exp, ascategory
```

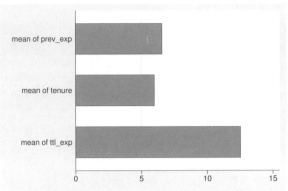

This bar chart shows three y variables, but we use the `ascategory` option to plot the different y variables as categorical variables on the x axis. The default labels on the x axis are not bad, but we might want to change them.
Uses nlsw.dta & scheme vg_palec

```
graph hbar prev_exp tenure ttl_exp, ascategory
    yvaroptions(relabel(1 "Previous Exp" 2 "Current Exp" 3 "Total Exp"))
```

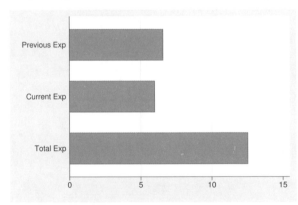

If the three level-of-experience variables were indicated by an `over()` option, we would use the `over(, relabel())` option to change the labels. Instead, because we have treated the multiple y variables as categories, we then use `yvaroptions(relabel())` to modify the labels on the x axis.
📖 See the next graph.
Uses nlsw.dta & scheme vg_palec

```
graph hbar prev_exp tenure ttl_exp, ascategory
```

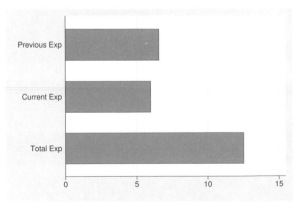

📖 We can customize the graph as shown in the previous example using the Graph Editor. Double-click on the vertical axis (e.g., mean of `prev_exp`) and click on **Edit or add individual ticks and labels**. Select *mean of ttl_exp* and click on **Edit** and change the **Label** to Total Exp. We can likewise change the labels for the other two variables.
Uses nlsw.dta & scheme vg_palec

```
graph hbar prev_exp tenure ttl_exp, ascategory yalternate
    yvaroptions(relabel(1 "Previous Exp" 2 "Current Exp" 3 "Total Exp"))
```

We move the y axis to the opposite side of the graph by using the `yalternate` option. You can also use the `xalternate` option to move the x axis to its opposite side.

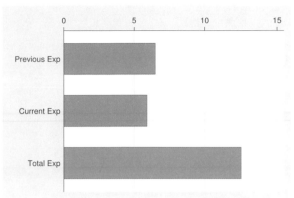

Click the Grid Edit tool. Click and hold the vertical axis and drag it to the rightmost part of the graph. Then select the Pointer tool. Double-click on the vertical axis, select *Advanced*, and change the **Position** to `Right`.

Uses nlsw.dta & scheme vg_palec

Previously, we saw that we can use the `relabel()` option within the `over()` option to control the labeling of `over()` variables and within `yvaroptions()` to control the labeling of multiple y variables (provided that we use the `ascategory` option to convert the multiple y variables into categories). Let's now explore other `over()` options, which we can use with either `over()` or `yvaroptions()`.

```
graph bar wage, over(occ7, label(nolabels))
```

The `label(nolabels)` option suppresses the display of the labels associated with the levels of `occ7`. The `label(nolabels)` option is generally not useful alone but is very useful in combination with other means to label the bars. Consider the next example.

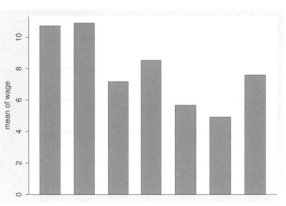

Right-click on the bottom axis and click on **Hide**. To reshow it, in the Object Browser, right-click on **grpaxis** and click on **Show**.

Uses nlsw.dta & scheme vg_palec

Introduction   Editor   Twoway   Matrix   Bar   Box   Dot   Pie   Options   Standard options   Styles   Appendix

Y variables   Over   Bar gaps   Bar sorting   Cat axis   Legends/Labels   Y axis   Lookofbar options   By

`graph bar wage, over(occ7, label(nolabels)) blabel(group)`

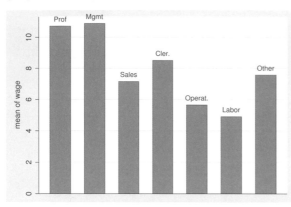

By adding the `blabel(group)` (bar label) option, we label the bars with the name of the group to which the bar belongs. See Bar : Legend (201) for more about `blabel()`.

In the Object Browser, double-click on **bar region** and in the *Labels* tab change the **Label** to `Group`.

Uses nlsw.dta & scheme vg_palec

`graph bar wage, over(occ7, label(angle(45))) over(collgrad)`

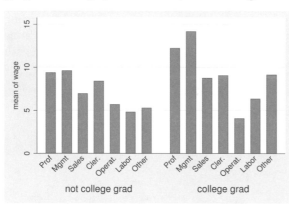

This graph shows wages broken down by occupation and by whether one graduated college. We add the `label(angle(45))` option to rotate the labels for occupation by 45 degrees. If this had been omitted, the labels would have overlapped each other.

Click on any occupation and, in the Contextual Toolbar, change the **Label angle** to `45 degrees`.

Uses nlsw.dta & scheme vg_palec

`graph bar wage, over(occ7, label(alternate)) over(collgrad)`

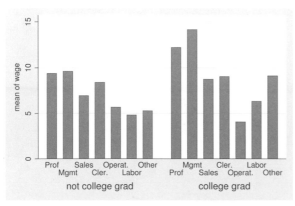

Compare this graph with the previous example. This example uses the `label(alternate)` option to avoid overlapping by alternating the labels for occupation.

Double-click on any of the occupation labels, then click on the **Label properties** button and check **Alternate spacing of adjacent labels**.

Uses nlsw.dta & scheme vg_palec

```
graph bar wage hours ttl_exp, ascategory over(collgrad)
    yvaroptions(label(alternate))
```

This example illustrates using the
`label(alternate)` option to alternate
labels created by multiple *y* variables
converted to categories using the
`ascategory` option. Then the option is
specified as
`yvaroptions(label(alternate))`.

Double-click on any of the labels for
the *y* variables, click on the **Label
properties** button, and check
**Alternate spacing of adjacent
labels**.

Uses nlsw.dta & scheme vg_palec

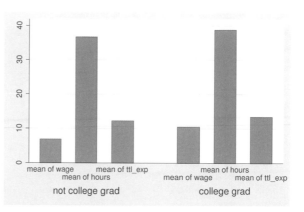

```
graph bar wage hours ttl_exp, ascategory over(union) nolabel
```

If we add the `nolabel` option, the
names of the variables are shown
instead of the value labels.

Uses nlsw.dta & scheme vg_palec

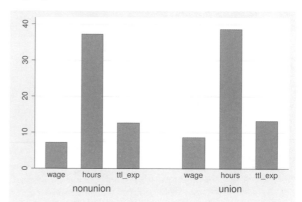

```
graph hbar wage, over(occ5, label(labsize(small)))
    over(collgrad, label(labcolor(maroon)))
```

Here we make the labels for `occ5` small
and the color of the labels for `collgrad`
maroon. See **Styles**: **Colors** (412) and
Styles: **Textsize** (428) for more details
about other values.

Click on the label for `occ5` and, in
the Contextual Toolbar, change the
Label size to `Small`. Double-click on
the label for `collgrad`, click on the
Label properties button, and change
the **Color** to `Maroon`.

Uses nlsw.dta & scheme vg_palec

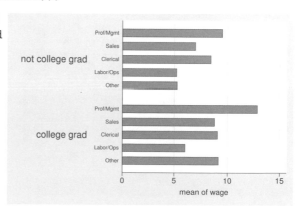

Introduction   Editor   Twoway   Matrix   Bar   Box   Dot   Pie   Options   Standard options   Styles   Appendix

Y variables   Over   Bar gaps   Bar sorting   Cat axis   Legends/Labels   Y axis   Lookofbar options   By

```
graph bar wage, over(age3, label(ticks tlwidth(thick)
    tlength(medium) tposition(crossing))) over(collgrad)
```

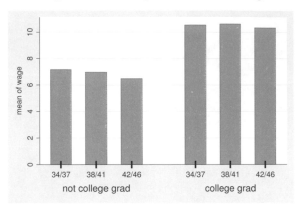

The `ticks` option adds tick marks. Here we make them thick in width and medium in length, and they cross the x axis. See [G-3] *cat_axis_label_options* for more details.

🖳 Double-click on the labels for `age3` and click on the **Tick properties** button. Check **Show ticks** and change the **Length** to Medium, the **Width** to Thick, and the **Placement** to Crossing.

Uses nlsw.dta & scheme vg_palec

```
graph bar wage, over(age3, label(labgap(zero))) over(collgrad)
```

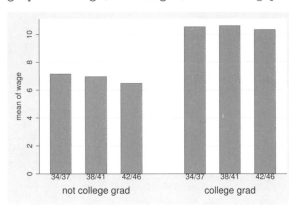

We use the `labgap(zero)` option to make no gap between the label and the axis.

🖳 Double-click on the label for `age3`, then click on the **Label properties** button and change the **Text gap** to Zero.

Uses nlsw.dta & scheme vg_palec

```
graph bar wage, over(age3) over(collgrad, label(labgap(huge)))
```

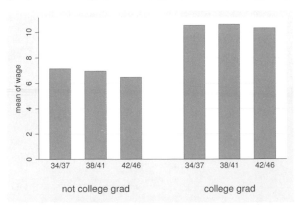

We use the `label(labgap(huge))` option to control the gap between the labels for `age3` and `collgrad`, making that gap huge.

🖳 Double-click on the label for `collgrad`, click on the **Label properties** button, and change the **Text gap** to Huge.

Uses nlsw.dta & scheme vg_palec

```
graph bar wage, over(age3) over(collgrad, axis(outergap(*20)))
```

The `axis(outergap(*20))` option
controls the gap between the labels of
the x axis and the outside of the graph.
As you can see, this increases the space
below the labels for `collgrad` and the
bottom of the graph.
Uses nlsw.dta & scheme vg_palec

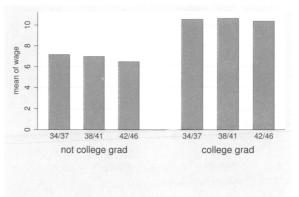

```
graph bar wage, over(union) over(grade4) asyvars
    b1title("Education Level in Four Categories")
```

The `b1title()` option adds a title to
the bottom of the graph, in effect
labeling the x axis. We can add a
second title below that by using the
`b2title()` option. If you used `graph`
`hbar`, you could label the left axis using
the `l1title()` and `l2title()` options.

In the Object Browser, double-click
on **bottom 1** (which is below
positional titles) and enter a title
under **text**.
Uses nlsw.dta & scheme vg_palec

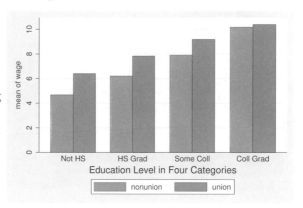

5.6   Legends and labeling bars

This section discusses the use of legends for bar charts, emphasizing the features that are
unique to bar charts. The section Options : Legend (361) goes into great detail about legends,
as does [G-3] *legend_options*. You can use legends for multiple y variables or when the
first `over()` variable is treated as a y variable with the `asyvars` option. See Bar : Y-variables
(173) for more information about the use of multiple y variables and Bar : Over (177) for
more examples of treating the first `over()` variable as a y variable. The following examples
show the different kinds of labels that we can create by using the `blabel()` option. We
can create labels that display the name of y variable, the name of the first `over()` group,
the height of the bar, or the overall height of the bar (when used with the `stack` option).
These examples begin using the `vg_s1c` scheme.

Introduction   Editor   Twoway   Matrix   Bar   Box   Dot   Pie   Options   Standard options   Styles   Appendix

Y variables   Over   Bar gaps   Bar sorting   Cat axis   Legends/Labels   Y axis   Lookofbar options   By

`graph bar wage hours tenure ttl_exp age`

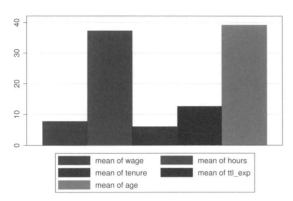

Consider this bar graph of five different y variables. The bars for the different y variables are shown with different colors, and a legend is used to identify the y variables.

Uses nlsw.dta & scheme vg_s1c

`graph bar wage, over(occ7) asyvars`

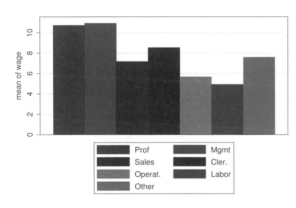

This is another example of where a legend can arise in a Stata bar graph by specifying the **asyvars** option, which treats an **over()** variable as though the levels were different y variables.

Uses nlsw.dta & scheme vg_s1c

Unless otherwise mentioned, the legend options described below work the same regardless of whether we derived the legend from multiple y variables or from an **over()** variable that was combined with the **asyvars** option. These next examples use the **vg_s2m** scheme.

`graph hbar wage hours tenure ttl_exp age, nolabel`

The `nolabel` option only works when we have multiple *y* variables. When it is used, the variable names (not the variable labels) are used in the legend. For example, instead of showing the variable label `hourly wage`, it shows the variable name `wage`.

Uses nlsw.dta & scheme vg_s2m

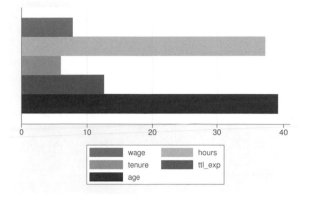

`graph hbar wage hours tenure ttl_exp age, showyvars`

The `showyvars` option puts the labels on the axis, beside or "under" the bars.

Uses nlsw.dta & scheme vg_s2m

`graph bar wage, over(occ7) asyvars showyvars`

Even though the `showyvars` option sounds like it would work only with multiple *y* variables, it also works when we combine the `over()` and `asyvars` options. As we can see, the legend is now redundant and could be suppressed.

Uses nlsw.dta & scheme vg_s2m

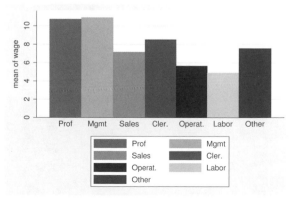

Introduction   Editor   Twoway   Matrix   Bar   Box   Dot   Pie   Options   Standard options   Styles   Appendix

Y variables   Over   Bar gaps   Bar sorting   Cat axis   Legends/Labels   Y axis   Lookofbar options   By

```
graph bar wage, over(occ7) asyvars showyvars legend(off)
```

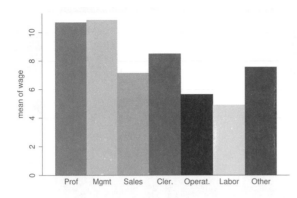

This example is similar to the previous one, but here we use the `legend(off)` option to suppress the display of the legend.

In the Object Browser, right-click on **legend** and select **Hide**. (You can redisplay the legend by right-clicking on **legend** and selecting **Show**.)
Uses nlsw.dta & scheme vg_s2m

```
graph bar wage, over(occ7) asyvars legend(label(1 "Professional")
    label(2 "Management"))
```

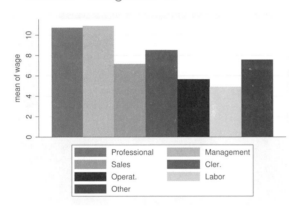

We use the `legend(label())` option to change the labels for the first two bars in the graph. We use a separate `label()` option for each bar.

Click on *Prof* and, in the Contextual Toolbar, change the **Text** to `Professional`. Likewise, click on *Mgmt* and, in the Contextual Toolbar, change the **Text** to `Management`.
Uses nlsw.dta & scheme vg_s2m

```
graph bar wage, over(occ7) asyvars legend(rows(2) colfirst)
```

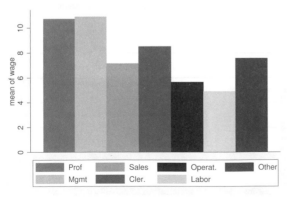

The `rows(2)` option combined with `colfirst` displays the legend in two rows with the keys ordered by column (instead of by row, the default). This yields keys that are more adjacent to the bars that they label.

In the Object Browser, double-click on **legend**. Change **Rows/Columns** to Rows and change **Rows** to 2. Then change the **Key Sequence** to Down first.
Uses nlsw.dta & scheme vg_s2m

As we can see, the default placement for the legend is below the x axis. However, Stata gives us flexibility in the placement of the legend. Let's now consider options that control the placement of the legend, along with options useful for controlling the placement of the items within the legend. The following examples use the **vg_blue** scheme.

`graph bar wage, over(occ7) asyvars exclude0 legend(position(1))`

We can use the `legend(position(1))` option to place the legend in the top right corner of the graph. The values supplied for `position()` are like the numbers on a clock face, where twelve o'clock is the top, six o'clock is the bottom, and 0 represents the center of the clock face. Specifying one o'clock places the legend in the top right; see Styles : Clockpos (414) for more details.

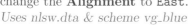 See the next graph.
Uses nlsw.dta & scheme vg_blue

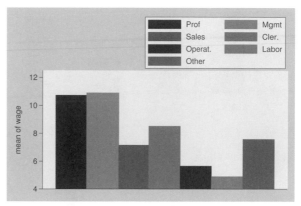

`graph bar wage, over(occ7) asyvars exclude0`

Using the Graph Editor, select the Grid Edit tool and then, in the Object Browser, click on **legend** to select it. Click and hold the legend and drag it to the top of the graph. Then, in the Object Browser, double-click **legend** and, in the *Advanced* tab, change the **Alignment** to East.
Uses nlsw.dta & scheme vg_blue

`graph bar wage, over(occ7) asyvars exclude0 legend(position(1) ring(0))`

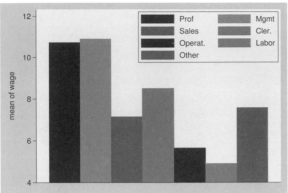

By adding the `ring(0)` option, we can tuck the legend inside the top right corner of the plot area. Think of the `ring()` option as specifying concentric rings around the graph, where 0 is a position inside the plot region, 1 is just outside the plot region, and increasing values are farther from the center of the plot region.

See the next graph.

Uses nlsw.dta & scheme vg_blue

`graph bar wage, over(occ7) asyvars exclude0`

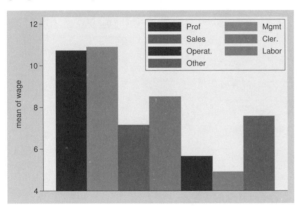

Using the Graph Editor, select the Grid Edit tool and then, in the Object Browser, click on **legend** to select it. Click and hold the legend and drag it to the center of the graph. Then to tuck it in the top right corner, go to the Object Browser, and double-click **legend** and, in the *Advanced* tab, change the **Alignment** to Northeast.

Uses nlsw.dta & scheme vg_blue

`graph hbar wage, over(occ7) asyvars legend(cols(1) position(9))`

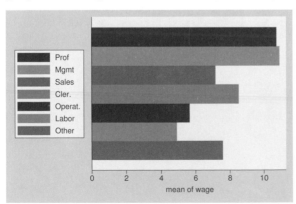

By using `graph hbar`, we switch this to a horizontal bar chart. The legend is placed in the nine o'clock position with the `position(9)` option and displays in one column with the `cols(1)` option.

See the next graph.

Uses nlsw.dta & scheme vg_blue

```
graph hbar wage, over(occ7) asyvars
```

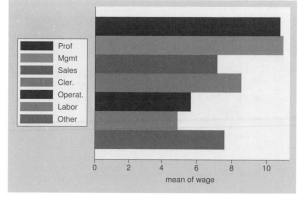

Using the Graph Editor, select the Grid Edit tool and then, in the Object Browser, click on **legend** to select it. Click and hold the legend and drag it to the left edge of the graph. Then, in the Object Browser, double-click on **legend** and change **Rows/Columns** to Columns and 1. Finally, in the *Advanced* tab, change the **Alignment** to Center.
Uses nlsw.dta & scheme vg_blue

```
graph hbar wage, over(occ7) asyvars legend(cols(1) position(9) textfirst)
```

Adding the `textfirst` option places the description of the key before the symbol in the legend.

In the Object Browser, double-click on **legend** and change the **Symbol order** to Labels first.
Uses nlsw.dta & scheme vg_blue

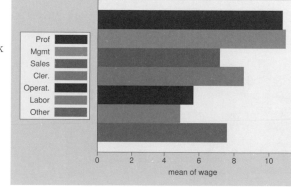

```
graph hbar wage, over(occ7) asyvars legend(cols(1) position(9) stack)
```

The `stack` option stacks the keys and their labels. Other options for controlling the legend include `rowgap()`, `keygap()`, `symxsize()`, `symysize()`, `textwidth()`, and `symplacement()`. See Options:Legend (361) and [G-3] *legend_options* for more details.

In the Object Browser, double-click on **legend** and change **Stack symbols and text** to Yes.
Uses nlsw.dta & scheme vg_blue

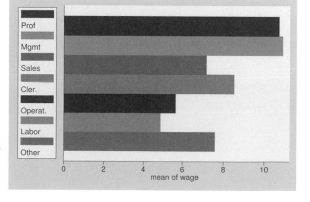

Introduction Editor Twoway Matrix Bar Box Dot Pie Options Standard options Styles Appendix

Y variables Over Bar gaps Bar sorting Cat axis Legends/Labels Y axis Lookofbar options By

```
graph hbar wage, over(occ7) asyvars
```

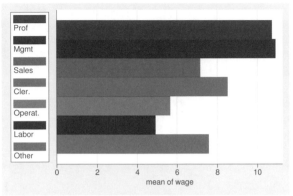

This example uses the vg_lgndc scheme. Notice how it positions and customizes the legend, as in the previous example. With this scheme, the legend defaults to the nine o'clock position, in a single column, with the keys and symbols stacked.

Uses nlsw.dta & scheme vg_lgndc

Now let's look at how we can use the blabel() (bar label) option to add labels to the bars. These labels can show the name of the over() option, the name of *y* variables, or the height of the bar. These options are illustrated below along with other related options we might use in conjunction with blabel() for identifying the bars. These examples begin using the vg_past scheme.

```
graph bar wage hours tenure, over(collgrad)
```

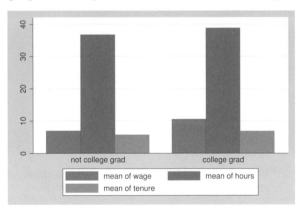

Consider this graph, where we look at wage, hours, and tenure broken down by the levels of collgrad. The legend identifies the bars for us. There are ways, in addition to the legend, to label these bars, as we will see in the upcoming examples.

Uses nlsw.dta & scheme vg_past

`graph bar wage hours tenure, over(collgrad) blabel(name)`

The `blabel(name)` (bar label) option
places labels on each bar with the name
of the *y* variable. Here each label is
preceded with *mean of* because each
bar represents the mean of *y* variable.

In the Object Browser, double-click
on **bar region** and, in the *Labels* tab,
change the **Label** to `Name`.

Uses nlsw.dta & scheme vg_past

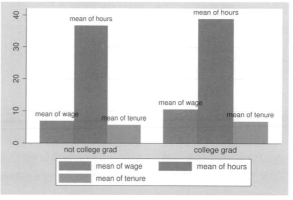

`graph bar wage hours tenure, over(collgrad) blabel(name) nolabel`

The `nolabel` option specifies to show
only the name of the *y* variable. For
example, instead of showing the label
mean of wage, it shows the variable
name `wage`.

Uses nlsw.dta & scheme vg_past

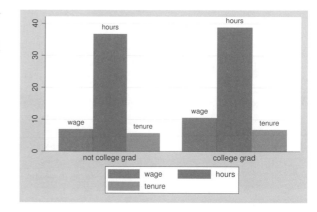

`graph bar wage hours tenure, over(collgrad) blabel(name) nolabel`
    `legend(off)`

Now the legend is no longer needed, so
we can suppress the display of the
legend with the `legend(off)` option.
See **Options**: Legend (361) for more
information about legend options.

In the Object Browser, right-click
on **legend** and select **Hide**. (You can
redisplay the legend by right-clicking on
legend and selecting **Show**.)

Uses nlsw.dta & scheme vg_past

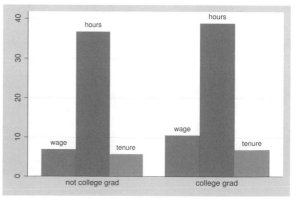

Introduction   Editor   Twoway   Matrix   Bar   Box   Dot   Pie   Options   Standard options   Styles   Appendix

Y variables   Over   Bar gaps   Bar sorting   Cat axis   Legends/Labels   Y axis   Lookofbar options   By

```
graph bar tenure, over(occ7) exclude0 blabel(group)
```

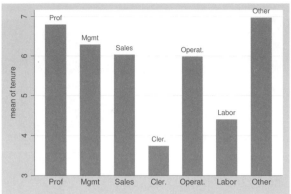

The `blabel(group)` option shows the label for the first `over()` group at the top of each bar. Here the label at the bottom of the bar becomes unnecessary.

In the Object Browser, double-click on **bar region** and change the **Label** to Group in the *Label* tab.

Uses nlsw.dta & scheme vg_past

```
graph bar tenure, over(occ7, label(nolabels)) exclude0
    blabel(group)
```

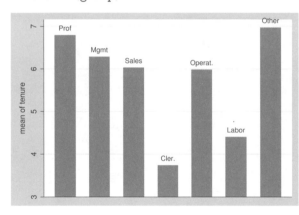

The `label(nolabels)` option suppresses the display of the labels below each bar.

Right-click on the labels for the x axis and select **Hide**. (Reshow the labels by right-clicking on **grpaxis** in the Object Browser and selecting **Show**.)

Uses nlsw.dta & scheme vg_past

```
graph bar tenure, over(occ5, label(nolabels)) exclude0 blabel(group)
    yscale(range(7.2)) over(union)
```

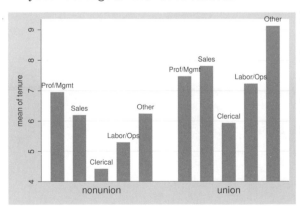

Even if we add a second `over()` option, the levels of the first `over()` variable are labeled at the top of each bar because of the `blabel()` option, and the levels of the second `over()` variable are labeled, as usual, at the bottom of the bars. The `blabel()` option does not work this way when you have three `over()` options or multiple y variables.

Uses nlsw.dta & scheme vg_past

```
graph hbar prev_exp tenure ttl_exp, over(grade4) blabel(bar)
```

Consider this graph showing previous, current, and total work experience broken down by education. Here we use the `blabel(bar)` option to display the bar height (i.e., the mean of y variables).

📊 In the Object Browser, double-click **bar region** and, in the *Labels* tab, change the **Label** to Bar.

Uses nlsw.dta & scheme vg_past

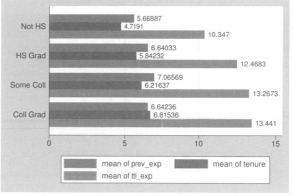

```
graph bar (sum) prev_exp tenure, stack over(grade4) blabel(bar)
```

Using the (`sum`) statistic, this graph shows the total experience for all individuals in a grade level before their current job (`prev_exp`) and the sum of experience for all individuals in a grade level in their current job (`tenure`) and then uses `stack` to stack these two totals. With the `blabel(bar)` option, the bar labels are the totals for each y variable broken down by `grade4`.

Uses nlsw.dta & scheme vg_past

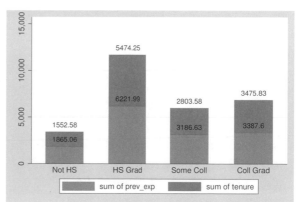

```
graph bar (sum) prev_exp tenure, stack over(grade4) blabel(total)
```

As compared with the prior example, this example uses the `blabel(total)` option to display the results as totals. Now the labels represent the cumulative total height of the bar.

📊 In the Object Browser, double-click on **bar region** and, in the *Labels* tab, change the **Label** to Total.

Uses nlsw.dta & scheme vg_past

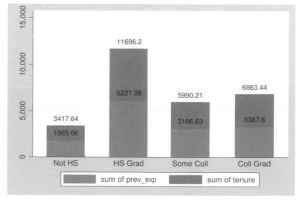

Introduction Editor Twoway Matrix Bar Box Dot Pie Options Standard options Styles Appendix

Y variables Over Bar gaps Bar sorting Cat axis Legends/Labels Y axis Lookofbar options By

We have seen several ways that we can use the `blabel()` option to label the bars. Stata also offers other options to control the display of these labels. Next let's consider some of these options to customize the way these labels are displayed. These examples begin using the `vg_palec` scheme.

`graph hbar hours, over(occ7, label(nolabels)) blabel(group)`

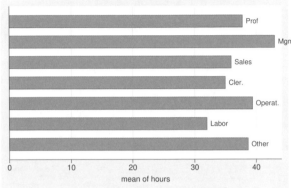

Consider this graph of the average hours worked by occupation. We add labels of the occupation at the top of each bar but suppress the label at the bottom of each bar. Note how the label for the second bar runs off the right of the graph.

In the Object Browser, double-click on **bar region** and, in the *Labels* tab, change the **Label** to `Group`.

Uses nlsw.dta & scheme vg_palec

`graph hbar hours, over(occ7, label(nolabels))`
`    blabel(group, position(inside))`

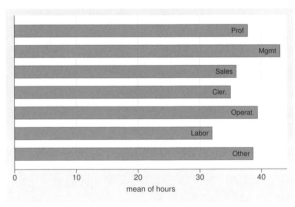

The `position(inside)` option places the group label inside the bar. By default, `inside` refers to the "top" inside the bar. Because we chose the `vg_palec` scheme, the bar colors are pale, so the labels within the bars are readable.

In the Object Browser, double-click on **bar region** and, in the *Labels* tab, change the **Position** to `Inside`.

Uses nlsw.dta & scheme vg_palec

```
graph hbar hours, over(occ7, label(nolabels))
   blabel(group, position(base))
```

The `position(base)` option places the label at the "base" inside of the bar. You can also specify `position(center)` to place the label in the center of the bar.

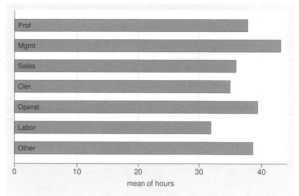

In the Object Browser, double-click on **bar region** and, in the *Labels* tab, change the **Position** to `base`.

Uses nlsw.dta & scheme vg_palec

```
graph hbar hours, over(occ7, label(nolabels))
   blabel(group, position(base) gap(huge))
```

We can use the `gap()` option to fine-tune the placement of the label. Here we position the label at the base but make the gap between the label and the base huge. We can also use the `gap()` option with `position(inside)` to position the label with respect to the top of the bar.

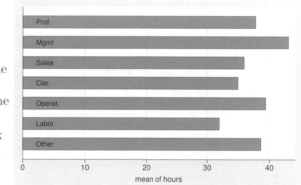

In the Object Browser, double-click on **bar region** and, in the *Labels* tab, change the **Gap** to `Huge`.

Uses nlsw.dta & scheme vg_palec

5.7   Controlling the y axis

This section describes options to control the y axis in bar charts. To be precise, when Stata refers to the y axis on a bar chart, it refers to the axis with the continuous variable, whether that be the left axis when using `graph bar` or the bottom axis when using `graph hbar`. This section emphasizes the features that are particularly relevant to bar charts. For more details, see Options: Axis titles (327), Options: Axis labels (330), and Options: Axis scales (339). Also see [G-3] *axis_title_options*, [G-3] *axis_label_options*, and [G-3] *axis_scale_options*. This section uses the `vg_s2c` scheme.

```
graph hbar wage, over(occ5) over(married) asyvar
```

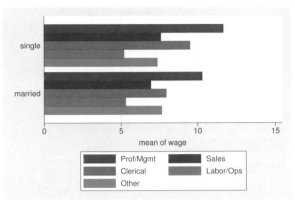

Consider this graph showing the mean hourly wage broken down by occupation and marital status.
Uses nlsw.dta & scheme vg_s2c

```
graph hbar wage, over(occ5) over(married) asyvar
    ytitle("Average Wages")
```

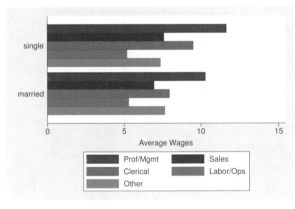

We can use the `ytitle()` option to add a title to the y axis. See Options: Axis titles (327) and [G-3] *axis_title_options* for more details, but disregard any references to `xtitle()` because that option is not valid when using `graph bar`.

Double-click on the y title (i.e., *Mean of Wage*) and change the **Text** to `Average Wages`.
Uses nlsw.dta & scheme vg_s2c

```
graph hbar wage, over(occ5) over(married) asyvar
    ytitle("Average Wages", size(vlarge) box bexpand)
```

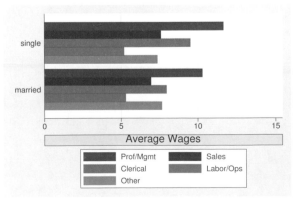

Because this title is considered a textbox, we can use a variety of textbox options to control the look of the title. Here the title is made very large with a box around it; the box is expanded (bexpand) to fill the width of the plot area. See Options: Textboxes (379) for more examples of how to use textbox options to control the display of text.

See the next graph.
Uses nlsw.dta & scheme vg_s2c

```
graph hbar wage, over(occ5) over(married) asyvar
   ytitle("Average Wages")
```

📊 Using the Graph Editor,
double-click on the title and change the
Size to v Large. Next select the *Box*
tab and check **Place box around
text**. Finally, select the *Format* tab
and check **Expand area to fill cell**.
Uses nlsw.dta & scheme vg_s2c

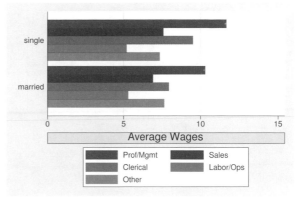

```
graph hbar wage, over(occ5) over(married) asyvar
   yline(10, lwidth(thick) lcolor(red) lpattern(dash))
```

Here the yline() option is used to
place a thick, red, dashed line on the
graph where *y* equals 10. This option is
still called yline() because the *y* axis
is the axis with the continuous variable.
📊 See the next graph.
Uses nlsw.dta & scheme vg_s2c

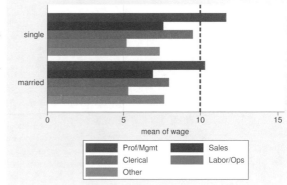

```
graph hbar wage, over(occ5) over(married) asyvar
```

📊 In the Graph Editor's Object
Browser, double-click on **bar region**.
In the *Reference lines* tab, click on
Add vertical reference line and for
X axis value type in 10. Second,
double-click on the reference line and
change the **color** to Red, the **Width** to
Thick, and the **Pattern** to Dash.
Uses nlsw.dta & scheme vg_s2c

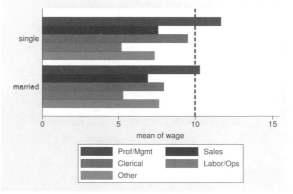

Introduction Editor Twoway Matrix Bar Box Dot Pie Options Standard options Styles Appendix

Y variables Over Bar gaps Bar sorting Cat axis Legends/Labels Y axis Lookofbar options By

`graph bar hours, over(occ7) asyvar ylabel(30(5)45)`

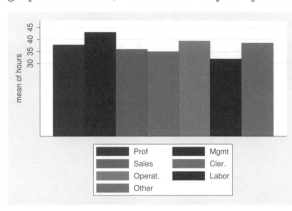

We use the `ylabel()` option to label the y axis from 30 to 45, incrementing by 5. See Options: Axis labels (330) and [G-3] **axis_label_options** for more details. The y axis still begins at 0. See the following example to see how this is controlled.

Double-click on the y axis and click on **Range/Delta** and enter a **Minimum value** of 30, a **Maximum value** of 45, and a **Delta** of 5.

Uses nlsw.dta & scheme vg_s2c

`graph bar hours, over(occ7) asyvar ylabel(30(5)45) exclude0`

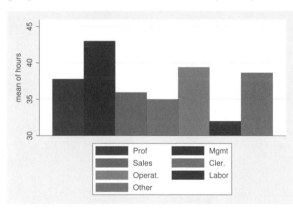

By default, bar charts include 0 on the y axis. The `exclude0` option removes the 0, as shown here.

In the Object Browser, double-click on **bar region** and uncheck **Include zero in scale**.

Uses nlsw.dta & scheme vg_s2c

`graph bar hours, over(occ7) asyvar ylabel(30(5)45, angle(0)) exclude0`

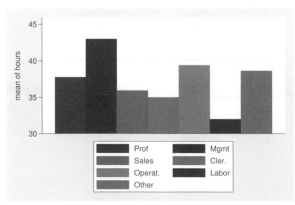

We can add the `angle()` option to modify the angle of the y label, making the labels for the y axis horizontal (0 degrees).

Double-click on the y axis, click on **Label properties**, and change the **Angle** to Horizontal.

Uses nlsw.dta & scheme vg_s2c

```
graph bar hours, over(occ7) asyvar ylabel(30(5)45, nogrid) exclude0
```

The `nogrid` option placed within the `ylabel()` option suppresses the grid for the y axis. (With bar charts, there is never a grid with respect to the x axis.) If the grid were absent and you wanted to include it, you could add the `grid` option. For more details, see Options : Axis labels (330).

📊 Double-click on the y axis, click on **Grid Lines**, and uncheck **Show grid**.

Uses nlsw.dta & scheme vg_s2c

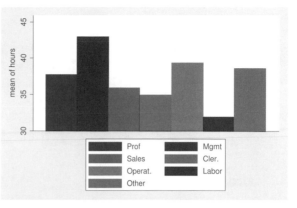

```
graph bar prev_exp tenure, over(occ7) yscale(off)
```

If we want to suppress the display of the y axis entirely, we can use the `yscale(off)` option. See Options : Axis scales (339) and [G-3] *axis_scale_options* for more details. Disregard any references to `xscale()` because that option is not valid when using `graph bar` or `graph hbar`.

📊 In the Object Browser, find the **scaleaxis**, right-click on it, and select **Hide**. We can reverse this by right-clicking on **scaleaxis** and selecting **Show**.

Uses nlsw.dta & scheme vg_s2c

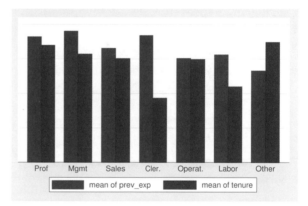

```
graph bar prev_exp tenure, over(occ7) yalternate
```

The `yalternate` option places the y axis on the opposite side, in this example, on the right side of the graph.

📊 Select the Grid Edit 📊 tool, click and hold the y axis, and drag it to the right side of the graph. Then select the Pointer ↖ tool, double-click on the y axis, click on **Advanced**, and change **Position** to Right.

Uses nlsw.dta & scheme vg_s2c

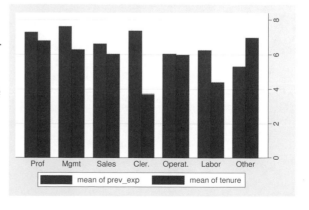

```
graph hbar prev_exp tenure, over(occ7) xalternate yreverse
```

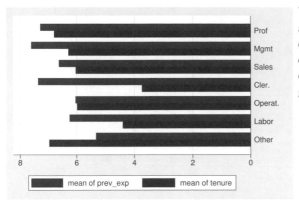

We can reverse the direction of the y axis with the `yreverse` option. We combine this with the `xalternate` option to place the labels for the bars on the alternate (right) side of the graph.

📊 See the next graph.

Uses nlsw.dta & scheme vg_s2c

```
graph hbar prev_exp tenure, over(occ7)
```

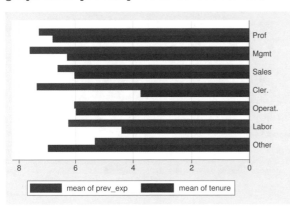

📊 First, in the Graph Editor's Object Browser, double-click on **bar region** and check **Reverse scale**. Second, select the Grid Edit 📋 tool, click and hold the vertical axis at the left, and drag it to the right side of the graph. Then select the Pointer ↖ tool, double-click on the right axis, click on **Advanced**, and change **Position** to Right.

Uses nlsw.dta & scheme vg_s2c

5.8   Changing the look of bars

This section shows how to control the color of the bars and the characteristics of the line outlining the bars. For more information, see the *lookofbar_options* table in [G-2] **graph bar** and [G-3] ***barlook_options***. This section begins using the `vg_rose` scheme.

```
graph bar wage hours ttl_exp tenure, over(collgrad)
```

Consider this bar chart. It shows the
mean wages, hours worked per week,
total experience, and job tenure broken
down by whether one graduated college.
Uses nlsw.dta & scheme vg_rose

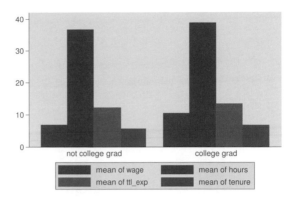

```
graph bar wage hours ttl_exp tenure, over(collgrad)
   intensity(*.5)
```

We use the intensity() option to
control the intensity of the color within
the bars. In this example, we request
that the color be 50% as intense as it
normally would be.

🔲 Double-click on the bar for wage and
change the **Fill intensity** to 50%. We
can repeat the same steps for the other
three bars.
Uses nlsw.dta & scheme vg_rose

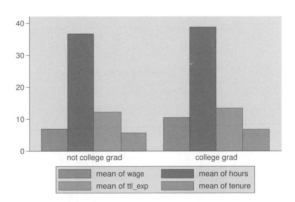

```
graph bar wage hours ttl_exp tenure, over(collgrad)
   intensity(*2)
```

Here we use the intensity() option to
make the colors within the bars 2 times
more intense than they would normally
be. Stata also has an option called
lintensity() that works the same way
but controls the intensity of the line
surrounding the bar. (This option is
not illustrated.)

🔲 Double-click on the bar for wage and
change the **Fill intensity** to 200%. We
can repeat the same steps for the other
three bars.
Uses nlsw.dta & scheme vg_rose

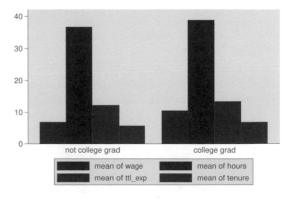

Introduction   Editor   Twoway   Matrix   Bar   Box   Dot   Pie   Options   Standard options   Styles   Appendix

Y variables   Over   Bar gaps   Bar sorting   Cat axis   Legends/Labels   Y axis   Lookofbar options   By

So far, all the options that we have examined determine the overall behavior and look of all the bars as a group. The `bar()` option controls the look of the bars for each y variable, as illustrated below. These graphs use the `vg_s2c` scheme.

```
graph bar wage hours ttl_exp tenure, over(collgrad)
   bar(1, bcolor(dkgreen))
```

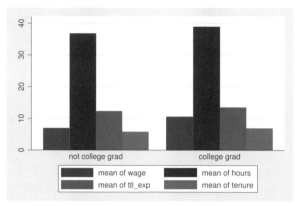

Here we use the `bar()` option to make the color of the first bar dark green. See Styles : Colors (412) for more information about available colors.

📷 Double-click on the first bar (e.g., the bar `wage`) and change the **Color** to **Dark green.**

Uses nlsw.dta & scheme vg_s2c

```
graph bar wage hours ttl_exp tenure, over(collgrad)
   bar(1, fcolor(ltblue) lcolor(blue) lwidth(vthick))
```

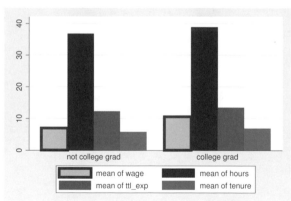

In this example, we make the fill color of the first bar light blue and the outline very thick and blue. See Styles : Linewidth (422) for more details on controlling the thickness of lines. We could also use the `lpattern()` option to control the pattern of the line surrounding the bar; see Styles : Linepatterns (420) for more details.

📷 See the next graph.

Uses nlsw.dta & scheme vg_s2c

```
graph bar wage hours ttl_exp tenure, over(collgrad)
```

🖬 Using the Graph Editor,
double-click on the first bar and change
the **Color** to Light blue.
Next change the **Outline width** to
v Thick. Finally, check **Different
outline color** and change the **Outline
color** to Blue.
Uses nlsw.dta & scheme vg_s2c

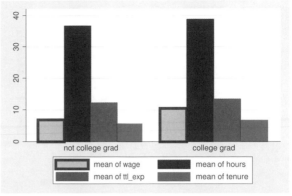

```
graph bar wage hours ttl_exp tenure, over(collgrad)
```

Although we can use the **bar()** option
to control the look of each bar,
selecting a different scheme allows us to
control the look of all the bars. For
example, this graph uses the **vg_palec**
scheme. See Intro : Schemes (15) for
other schemes and
Appendix : Customizing schemes (479) for
tips on customizing our own schemes.
Uses nlsw.dta & scheme vg_palec

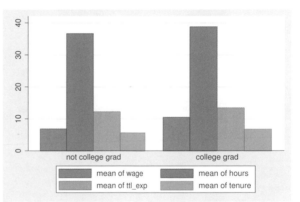

5.9   Graphing by groups

This section discusses the use of the **by()** option in combination with **graph bar**. Normally, you would use the **over()** option instead of the **by()** option, but there are cases
where the **by()** option is either necessary or more advantageous. For example, a **by()**
option is useful if you exceed the maximum number of **over()** options (three if you have
a single *y* variable or two if you have multiple *y* variables). Thus the **by()** option allows
you to break down your data by more categorical variables. Also, **by()** gives you more
flexibility in the placement of the separate panels. For more information about the **by()**
option, see Options : By (346); for more information about the **over()** option, see Bar : Over
(177). These examples use the **vg_s1c** scheme.

Introduction   Editor   Twoway   Matrix   Bar   Box   Dot   Pie   Options   Standard options   Styles   Appendix

Y variables   Over   Bar gaps   Bar sorting   Cat axis   Legends/Labels   Y axis   Lookofbar options   By

`graph bar wage, over(urban2) over(married) over(union)`

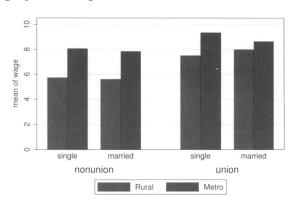

Consider this bar graph that breaks down wages by three categorical variables. If we wanted to break down this by another categorical variable, we could not use another `over()` option because we can have a maximum of three `over()` options with one y variable.

Uses nlsw.dta & scheme vg_s1c

`graph bar wage, over(urban2) over(married) over(union) by(collgrad)`

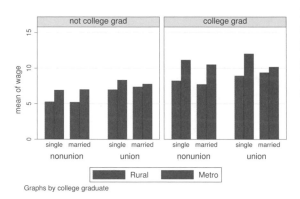

If we want to show the previous graph separately by `collgrad`, we can use the `by()` option. This option gives us two graphs side by side: one for those who are not college graduates and one for college graduates.

Uses nlsw.dta & scheme vg_s1c

`graph bar ttl_exp tenure, over(married) over(urban2)`

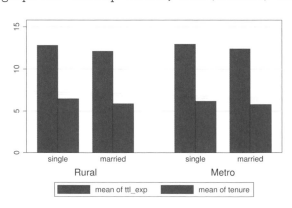

Consider this bar graph with multiple y variables broken down by two categorical variables by using two `over()` options. When we have multiple y variables, we can have a maximum of two `over()` options.

Uses nlsw.dta & scheme vg_s1c

```
graph bar ttl_exp tenure, over(married) over(urban2)
   by(union)
```

If we want to show the previous graph
by another categorical variable, say,
union, we can use the by() option.
Uses nlsw.dta & scheme vg_s1c

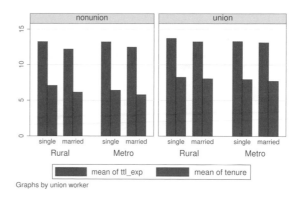

```
graph bar ttl_exp tenure, over(married) over(urban2)
   by(union, missing)
```

We can add the missing option to
include a panel for the missing values of
union.
Uses nlsw.dta & scheme vg_s1c

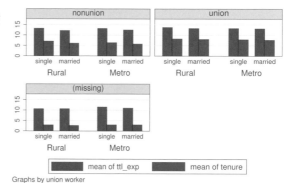

```
graph bar ttl_exp tenure, over(married) over(urban2)
   by(union, missing total)
```

We can add the total option to include
a panel for all observations.
Uses nlsw.dta & scheme vg_s1c

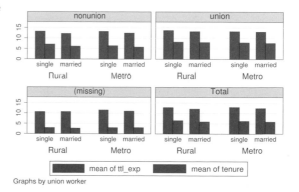

```
graph hbar ttl_exp tenure, over(married) over(urban2)
   by(union, cols(1))
```

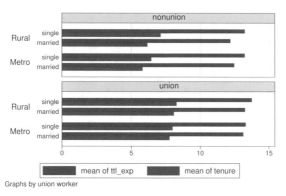

We remove the total and missing options and flip the graph to make a horizontal bar chart. We then use the cols(1) option to show these graphs in one column. This makes the graph pretty cramped. Let's explore several options we can add to this graph to make it less cramped, adding the options a few at a time.

In the Object Browser, double-click on **Graph**, select the *Organization* tab, and change the **Columns** to 1.

Uses nlsw.dta & scheme vg_s1c

```
graph hbar ttl_exp tenure, over(married) over(urban2)
   by(union, cols(1) note(""))
```

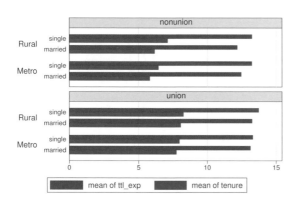

We add the note("") option within the by() option to suppress the note in the left corner, leaving more room for the graph.

Right-click on the note and then click on **Hide** to hide the note. We can reshow it by going to the Object Browser and right-clicking on **note** and selecting **Show**.

Uses nlsw.dta & scheme vg_s1c

```
graph hbar ttl_exp tenure, over(married) over(urban2)
   by(union, cols(1) note("") legend(position(3)))
```

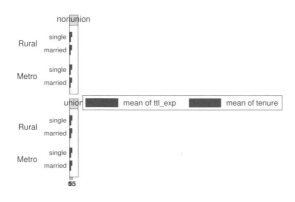

We add the legend(position(3)) option to put the legend at the right. legend(position(3)) is contained within the by() option because it changes the position of the legend. If we could make the legend narrow (instead of wide), it would work well in this position.

See the next graph.

Uses nlsw.dta & scheme vg_s1c

```
graph hbar ttl_exp tenure, over(married) over(urban2)
   by(union, cols(1) note(""))
```

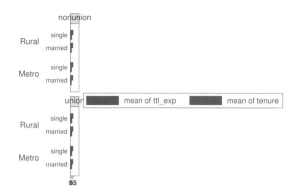 Using the Graph Editor, select the Grid Edit 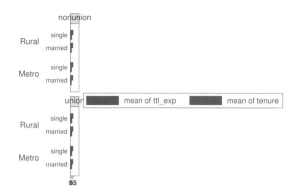 tool and then, in the Object Browser, click on **legend** to select the entire legend. Click and hold the legend, drag it to the rightmost part of the graph, and release the mouse. Then, in the Object Browser, double-click on **legend** and, in the *Advanced* tab, change the **Alignment** to Center.

Uses nlsw.dta & scheme vg_s1c

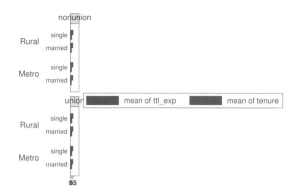

```
graph hbar ttl_exp tenure, over(married) over(urban2)
   by(union, cols(1) note("") legend(position(3)))
   legend(cols(1) stack label(1 "Tot Exp") label(2 "Curr Exp"))
```

We add the legend(cols(1) stack) option to make the legend narrow and the label() option to change the labels in the legend. Here the legend() option appears outside the by() option. See Options : By (346) and Options : Legend (361) for more information about the interactions of by() and legend().

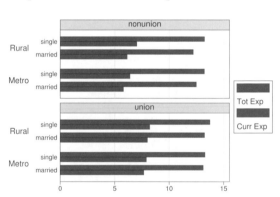 See the next graph.

Uses nlsw.dta & scheme vg_s1c

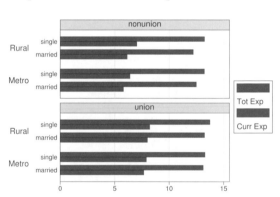

```
graph hbar ttl_exp tenure, over(married) over(urban2)
   by(union, cols(1) note("") legend(position(3)))
```

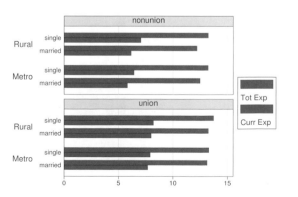 From the Graph Editor, go to the Object Browser and double-click on **legend**. In the *Organization* tab, change the **Columns** to 1 and **Stack symbols and text** to Yes and click on **OK**. Then, using the Pointer ▶ tool, double-click on *mean of ttl_exp* and change the **Text** to Tot Exp. Likewise, double-click on *mean of tenure* and change the **Text** to Curr Exp.

Uses nlsw.dta & scheme vg_s1c

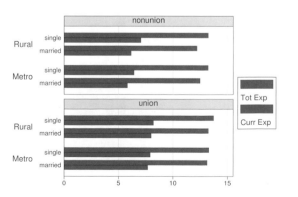

Introduction   Editor   Twoway   Matrix   Bar   Box   Dot   Pie   Options   Standard options   Styles   Appendix

Y variables   Over   Bar gaps   Bar sorting   Cat axis   Legends/Labels   Y axis   Lookofbar options   By

```
graph hbar ttl_exp tenure, over(married) over(urban2)
  by(union, cols(1) note("") legend(position(3)))
  legend(cols(1) stack label(1 "Tot Exp") label(2 "Curr Exp"))
  subtitle(, position(5) ring(0) nobexpand)
```

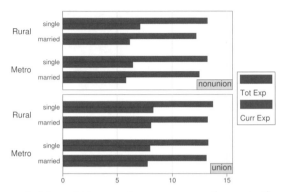

We can add the subtitle() option to position the title for each graph in the lower right corner. The position(5) option puts the title in the five o'clock position, and the ring(0) option puts the title inside the plot area. The nobexpand (no box expand) option keeps the title from expanding to fill the entire plot area.

📈 See the next graph.

Uses nlsw.dta & scheme vg_s1c

```
graph hbar ttl_exp tenure, over(married) over(urban2)
  by(union, cols(1) note("") legend(position(3)))
  legend(cols(1) stack label(1 "Tot Exp") label(2 "Curr Exp"))
```

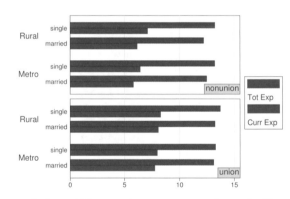

📈 Using the Graph Editor, select the Grid Edit 🔲 tool and select the subtitle *nonunion* and drag it to the center of the plotregion. Then select the Pointer 🔺 tool, double-click on *nonunion*, and, in the *Format* tab, uncheck the box **Expand area to fill cell**. Click and hold the subtitle and drag it to the bottom right corner. Do likewise for the *union* title.

Uses nlsw.dta & scheme vg_s1c

```
graph bar ttl_exp tenure, over(married) over(urban2)
  by(union collgrad)
```

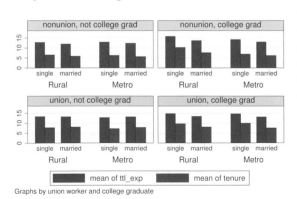

We can include multiple variables within the by() option. Here, in addition to breaking down these variables by two over() variables, we break them down by two more variables using the by(union collgrad) option.

Uses nlsw.dta & scheme vg_s1c

6   Box plots

A box plot displays a box(es) bordered at the 25th and 75th percentiles of the y variable with a *median line* at the 50th percentile. Whiskers extend from the box to the upper and lower *adjacent values* and are capped with an *adjacent line*. Values exceeding the upper and lower adjacent values are called *outside values* and are displayed as markers. This chapter starts by showing the use of the `over()` option to break down box plots by categorical variables and continues by showing how we can specify multiple y variables to display plots for multiple variables. The next two sections show more options we can use to customize the display of the `over()` option and show options that control the display of categorical axes. The chapter then discusses options for legends and options that control the display of the y axis. The final two sections take a look at options that control the look of boxes before ending with the `by()` option.

6.1   Specifying variables and groups

This section introduces the use of box plots, illustrating the use of the `over()` option for showing box plots by one or more grouping variables. We will look at examples showing how we can graph multiple variables at once by specifying more y variables, followed by some examples of general options for controlling the display of multiple y variables and the behavior of the `over()` options. See the *group_options* table in [G-2] **graph box** for more details. This section begins with the `vg_s2c` scheme.

`graph box wage, over(grade4)`

This box plot uses the `over(grade4)` option to break down wages by education level (in four categories). By default, the separate levels of `grade4` are graphed using the same color, and the levels are labeled on the x axis. The graph shows many outside values that are displayed as markers beyond the whiskers. The following example shows how we can suppress the display of the outside values.

Uses nlsw.dta & scheme vg_s2c

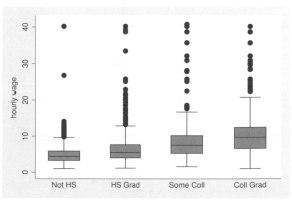

`graph box wage, over(grade4) nooutsides`

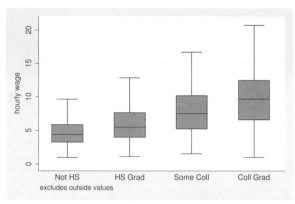

By adding the `nooutsides` option, we suppress the display of the outside values. Graphs using this option have a note in the bottom left corner indicating that the outside values have been excluded from display in the graph. For most of the graphs in this chapter, there would be many outside values, which would make the graphs cluttered, so most of the graphs will use the `nooutsides` option.

Uses nlsw.dta & scheme vg_s2c

`graph box wage, nooutsides over(grade4) over(union)`

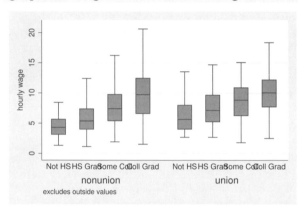

In this example, we add the `over(union)` option to show wages broken down by education and whether one is a member of a union. However, the labels for `grade4` overlap each other. See the next example for one solution.

Uses nlsw.dta & scheme vg_s2c

`graph hbox wage, nooutsides over(grade4) over(union)`

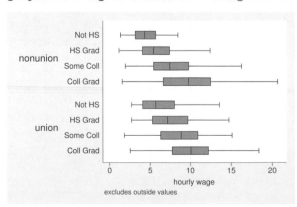

Here we use `graph hbox` to make a horizontal box plot. This eliminates the overlapping of the labels for `grade4`. The next example will show another possible solution.

In the Object Browser, double-click on **box region** and choose whether the orientation is **Horizontal** or **Vertical**.

Uses nlsw.dta & scheme vg_s2c

```
graph box wage, nooutsides over(grade4) over(union) asyvars
```

With the `asyvars` option, the first
`over()` variable, `grade4`, is treated as
though it were multiple y variables. As
a result, the levels of `grade4` are shown
in multiple colors and labeled with a
legend. We can only use `asyvars` when
we have one y variable.

◨ In the Object Browser, double-click
on **box region** and click on the
Rotate Categories button until the
graph appears as we wish.
Uses nlsw.dta & scheme vg_s2c

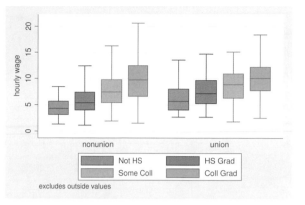

```
graph box wage, nooutsides over(grade4) over(union) over(urban2)
```

Here we add a third `over()` option,
comparing people who live in rural and
metropolitan areas. The first `over()`
variable, `grade4`, is now treated as
though it were multiple y variables.
Because of this, we can specify only one
y variable when we have three `over()`
options.
Uses nlsw.dta & scheme vg_s2c

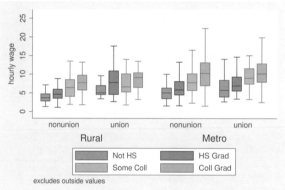

Now let's look at examples of using multiple y variables with the `over()` option, starting
with a graph with multiple y variables. These examples use the `vg_outc` scheme.

Introduction   Editor   Twoway   Matrix   Bar   Box   Dot   Pie   Options   Standard options   Styles   Appendix

Yvars and over   Box gaps   Box sorting   Cat axis   Legend   Y axis   Boxlook options   By

`graph hbox prev_exp tenure, nooutsides`

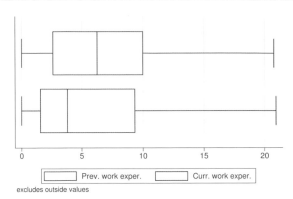

This graph shows work experience before one's current job and work experience at one's current job.
Uses nlsw.dta & scheme vg_outc

`graph hbox prev_exp tenure, nooutsides over(married)`

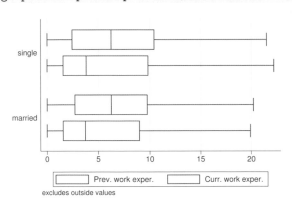

We can further break down these variables by marital status.
Uses nlsw.dta & scheme vg_outc

`graph hbox prev_exp tenure, nooutsides over(married) over(union)`

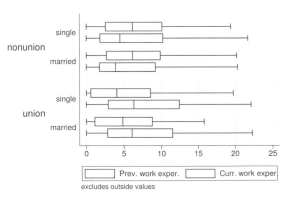

We can take the last graph and add another `over()` option to even further break down these variables by whether one belongs to a union. However, we cannot add a third `over()` option when we have multiple *y* variables, but we could add the `by()` option; see Box : By (258).
Uses nlsw.dta & scheme vg_outc

Now let's consider options that may be used in combination with the `over()` option to customize the behavior of the graphs. We will first explore how you can treat the levels of the first `over()` option as though they were multiple y variables. We will then see how you can request that missing values for the levels of the `over()` variables be displayed and how you can suppress empty categories when multiple `over()` options are used. These examples are shown below using the `vg_s2m` scheme.

`graph hbox wage, nooutsides over(grade4) over(union)`

Consider this graph where we show wages broken down by education level and whether one belongs to a union.
Uses nlsw.dta & scheme vg_s2m

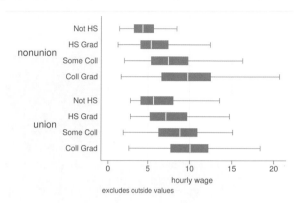

`graph hbox wage, nooutsides over(grade4) over(union) asyvars`

If we add the `asyvars` option, the first `over()` variable (`grade4`) is graphed as if there were four y variables corresponding to each level of `grade4`. Each level of `grade4` is shown as a differently colored/shaded box and labeled with the legend.

In the Object Browser, double-click on **box region** and click on the **Rotate Categories** button until the graph appears as we wish.
Uses nlsw.dta & scheme vg_s2m

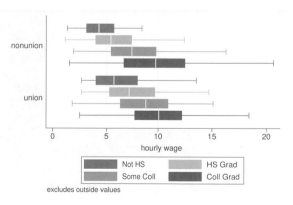

`graph hbox wage, nooutsides over(grade4) over(union) asyvars missing`

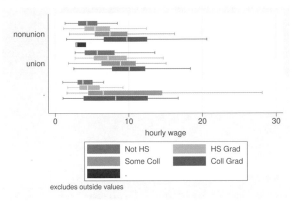

By adding the `missing` option to the previous graph, we see a category for those who are missing on the `union` variable, shown as the third group, which is labeled with a dot to indicate that those values are missing; see Box:Cat axis (238) to see how we could label this differently (e.g., labeling it with the word "Missing").

Uses nlsw.dta & scheme vg_s2m

`graph box wage, nooutsides over(grade) over(collgrad)`

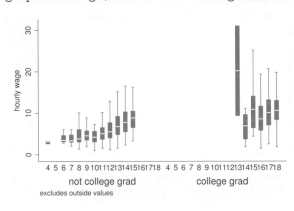

Consider this box chart that breaks down wages by two variables: the last grade that one completed and whether one is a college graduate. By default, Stata shows all possible combinations for these two variables. Usually, all combinations are possible, but not here.

Uses nlsw.dta & scheme vg_s2m

`graph box wage, nooutsides over(grade) over(collgrad) nofill`

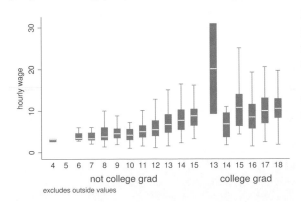

If we want to display only the combinations of the `over()` variables that exist in the data, we can use the `nofill` option.

📖 In the Object Browser, double-click on **box region** and uncheck **Include missing categories**.

Uses nlsw.dta & scheme vg_s2m

6.2   Options for controlling gaps between boxes

This section considers some of the options that we can use to control the gaps (spacing) between boxes. The options for controlling gaps between boxes work exactly the same as the options to control gaps between bars with the `graph bar` command. These options were illustrated in more detail in Bar : Gaps (183), so I refer you to that section for further exploration. These graphs are shown using the `vg_past` scheme.

```
graph hbox tenure, nooutsides over(occ5) over(collgrad)
```

Consider this graph that shows box plots of `tenure` broken down by `occ5` and `collgrad`.
Uses nlsw.dta & scheme vg_past

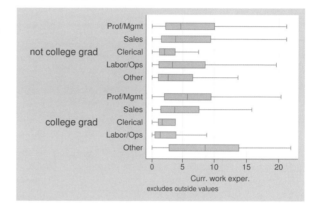

```
graph hbox tenure, nooutsides over(occ5, gap(300)) over(collgrad)
```

By default, the gap among levels `occ5` would be two-thirds (67%) the width of a box. With the `gap(300)` option, that gap is increased to three times the width of a box.

In the Object Browser, double-click on **box region** and under **Spacing** change the spacing for **Categories** to 300pct.
Uses nlsw.dta & scheme vg_past

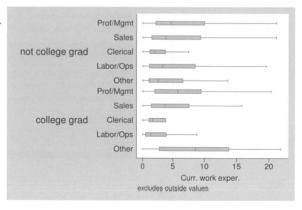

```
graph hbox tenure, nooutsides over(occ5) over(collgrad, gap(25))
```

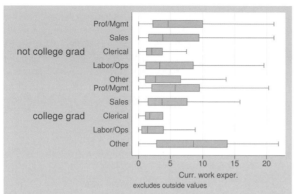

Here we shrink the gap between the levels of `collgrad`, making the gaps 25% of the width of a box, yielding boxes that are wider than they normally would be.

In the Object Browser, double-click on **box region** and change the spacing for **Categories** to 25pct.

Uses nlsw.dta & scheme vg_past

```
graph hbox tenure, nooutsides over(occ5, gap(25)) over(collgrad, gap(400))
```

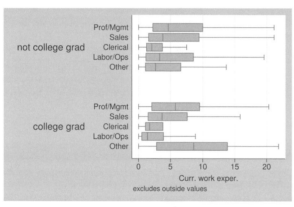

We can control the gap with respect to each `over()` variable. In this example, we make the gap among the `occ5` categories small (25% of the width of a box) and the gap between the levels of `collgrad` large (four times the width of a box).

In the Object Browser, double-click on **box region** and change the spacing for **Categories** to 25pct and the spacing for **Super categories** to 400pct.

Uses nlsw.dta & scheme vg_past

```
graph box prev_exp tenure ttl_exp, nooutsides over(collgrad)
    outergap(150)
```

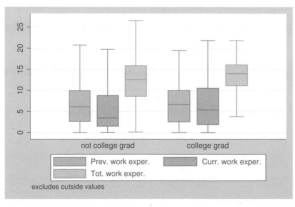

We can change the outer gap between the boxes and the edge of the plot area with the `outergap()` option. Here the gap is 1.5 times the size of a box.

In the Object Browser, double-click on **box region** and change the spacing for **Outer** to 150pct.

Uses nlsw.dta & scheme vg_past

```
graph box prev_exp tenure ttl_exp, nooutsides over(collgrad)
   boxgap(10)
```

The `boxgap()` option controls the size of the gap among the boxes formed by the multiple *y* variables. The default value is 33, meaning that the distance between the boxes is 33% of the width of the boxes. Here we make the gap only 10% of the width of a box.

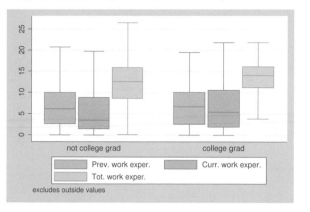

🔖 In the Object Browser, double-click on **box region** and change the spacing for **Variables** to `10pct`.

Uses nlsw.dta & scheme vg_past

6.3   Options for sorting boxes

This section considers some of the options that we can use to control the order in which the boxes are displayed. This is an area where the control afforded with commands is far greater than the control you can obtain with the Graph Editor. This section will focus on graph commands without concern for the Graph Editor. By default, the boxes formed by `over()` variables are ordered in ascending sequence according to the values of the `over()` variable. Stata allows us to control the order of the boxes by allowing us to put them in descending order, order them according to the values of another variable, or sort the boxes according to their medians. This section uses the `vg_rose` scheme.

```
graph hbox tenure, nooutsides over(occ7, descending)
```

Consider this graph showing `tenure` broken down by the seven levels of occupation. The boxes would normally be ordered by levels of `occ7`, going from 1 to 7. The `descending` option switches the order of the boxes. They still are ordered according to the seven levels of occupation, but the boxes are ordered going from 7 to 1.

Uses nlsw.dta & scheme vg_rose

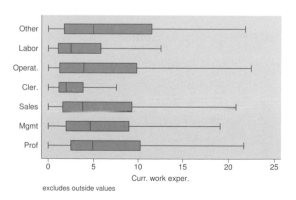

```
graph hbox tenure, nooutsides over(occ7, sort(occ7alpha))
```

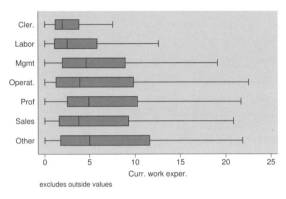

excludes outside values

We might want to put these boxes in alphabetical order, but with *Other* still appearing last. We can do this by recoding occ7 into a new variable (say, occ7alpha) such that, as occ7alpha goes from 1 to 7, the occupations are alphabetically ordered. We recoded occ7 with these assignments: $4 = 1$, $6 = 2$, $2 = 3$, $5 = 4$, $1 = 5$, $3 = 6$, and $7 = 7$. Then the sort(occ7alpha) option alphabetizes the boxes but with *Other* still appearing last.
Uses nlsw.dta & scheme vg_rose

```
graph hbox tenure, nooutsides over(occ7, sort(1))
```

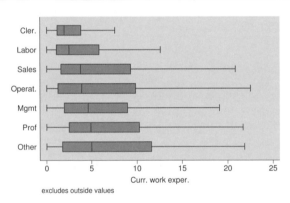

excludes outside values

Here we sort the variables on the median of tenure, yielding boxes with medians in ascending order. The sort(1) option sorts the boxes according to the median of the first y variable, meaning to sort on the median of tenure.
Uses nlsw.dta & scheme vg_rose

```
graph hbox tenure, nooutsides over(occ7, sort(1) descending)
```

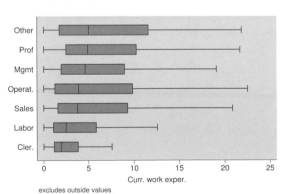

excludes outside values

Adding the descending option yields boxes in descending order, going from highest median tenure to lowest median tenure.
Uses nlsw.dta & scheme vg_rose

```
graph hbox prev_exp tenure, nooutsides over(occ7)
```

Here we plot two *y* variables: the
number of years of work experience
before one's current job and the years
in one's current job. Because we have
removed all `sort()` options, the boxes
are sorted according to the values of
`occ7`.
Uses nlsw.dta & scheme vg_rose

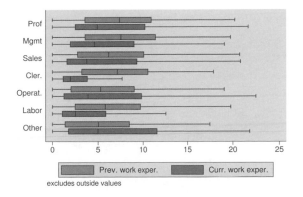

```
graph hbox prev_exp tenure, nooutsides over(occ7, sort(1))
```

Adding the `sort(1)` option now sorts
the boxes according to the median of
`prev_exp` because that is the first *y*
variable.
Uses nlsw.dta & scheme vg_rose

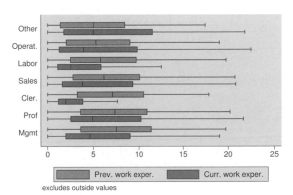

```
graph hbox prev_exp tenure, nooutsides over(occ7, sort(2))
```

Changing `sort(1)` to `sort(2)` now
sorts the boxes according to the median
of the second *y* variable, `tenure`.
Uses nlsw.dta & scheme vg_rose

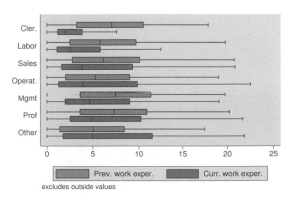

```
graph hbox tenure, nooutsides over(occ7, sort(1)) over(collgrad)
```

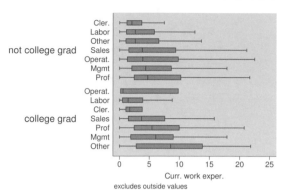

We can use the `sort()` option when there are more `over()` variables. Here the boxes are ordered according to the median of `tenure` across `occ7` but within each level of `collgrad`.
Uses nlsw.dta & scheme vg_rose

```
graph hbox tenure, nooutsides over(occ7, sort(1)) over(collgrad, descending)
```

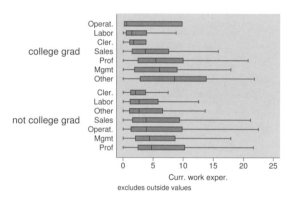

We add the `descending` option to the second `over()` option, and the levels of `collgrad` are now shown with college graduates appearing first.
Uses nlsw.dta & scheme vg_rose

6.4   Controlling the categorical axis

This section describes ways that we can label categorical axes. Box plots are similar to bar charts, but they are different from other graphs because their x axes are represented by categorical variables. This section describes options we can use to customize these categorical axes. For more details, see [G-3] *cat_axis_label_options* and [G-3] *cat_axis_line_options*.

This section begins by showing examples of how we can change the labels for the x axis for these categorical variables. The next set of examples will use the `vg_teal` scheme.

```
graph box wage, nooutsides over(south)
```

This is an example of a box plot with one `over()` variable graphing wages broken down by whether one lives in the South. The variable `south` is a dummy variable that does not have any value labels, so the x axis is not labeled well.
Uses nlsw.dta & scheme vg_teal

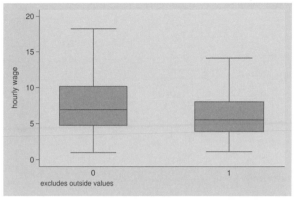

```
graph box wage, nooutsides over(south, relabel(1 "N & W" 2 "South"))
```

The `relabel()` option changes the labels displayed for the levels of `south`, giving the x axis more meaningful labels. Note that we wrote `relabel(1 "N & W")`, not `relabel(0 "N & W")`, because these numbers do not represent the actual levels of `south` but the ordinal position of the levels, i.e., first and second.

☞ See the next graph.
Uses nlsw.dta & scheme vg_teal

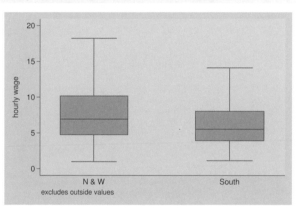

```
graph box wage, nooutsides over(south)
```

☞ Using the Graph Editor, double-click on the label for `south` (e.g., 0 or 1) and click on the **Edit or add individual ticks and labels** button. Select the tick labeled 1 and click on **Edit**, and change the **Label** to South and click on **OK**. Next select the tick labeled 0 and click on **Edit** and change the **Label** to N & W.
Uses nlsw.dta & scheme vg_teal

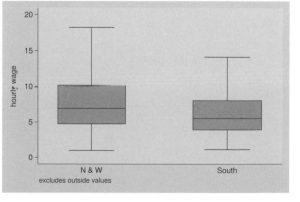

```
graph box wage, nooutsides over(smsa, relabel(1 "Nonmetro" 2 "Metro"))
    over(south, relabel(1 "N & W" 2 "South"))
```

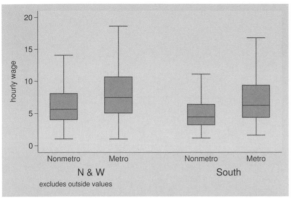

This is an example of a box plot with two `over()` variables. Here we use the `relabel()` option to change the labels displayed for the levels of `south` and `smsa`.

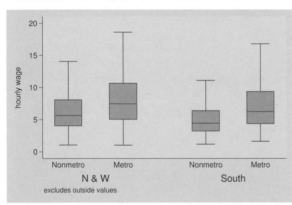 See the next graph.

Uses nlsw.dta & scheme vg_teal

```
graph box wage, nooutsides over(smsa)
    over(south, relabel(1 "N & W" 2 "South"))
```

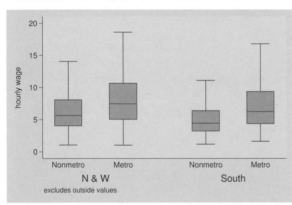

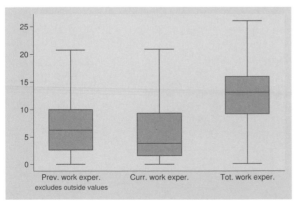 You can use the Graph Editor to change the labels of the categorical variable. Double-click on a label for `smsa` (i.e., *nonSMSA* or *SMSA*) and click on **Edit or add individual ticks and labels**. Select one of the ticks labeled *SMSA* and click on **Edit** and change the **Label** to `Metro`. Likewise edit the ticks labeled *nonSMSA*, changing their labels to `Nonmetro`.

Uses nlsw.dta & scheme vg_teal

```
graph box prev_exp tenure ttl_exp, nooutsides ascategory
```

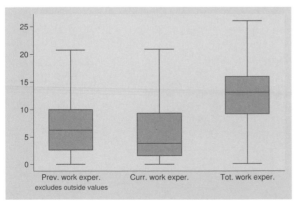

This box plot shows multiple y variables, but we use the `ascategory` option to plot the different y variables as if they were categorical variables. The boxes for the different variables are the same color, and the categories are labeled on the x axis rather than with a legend. The default labels on the x axis are not bad, but we might want to change them.

Uses nlsw.dta & scheme vg_teal

```
graph box prev_exp tenure ttl_exp, nooutsides ascategory
    yvaroptions(relabel(1 "Prev Exp" 2 "Curr Exp" 3 "Tot Exp"))
```

If we had an `over()` option, we would
use the `relabel()` option to change the
labels on the *x* axis. But because we
had multiple *y* variables that we have
treated as categories, we use the
`yvaroptions(relabel())` option to
modify the labels on the *x* axis.
📊 See the next graph.
Uses nlsw.dta & scheme vg_teal

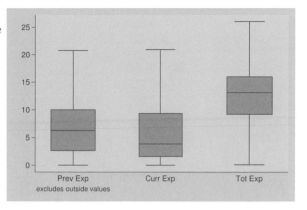

```
graph box prev_exp tenure ttl_exp, nooutsides ascategory
```

📊 Using the Graph Editor,
double-click on the vertical axis (e.g.,
mean of `prev_exp`) and click on **Edit
or add individual ticks and labels**.
Select *mean of ttl_exp*, click on **Edit**,
and change the **Label** to `Tot Exp`.
Likewise, you can change the labels for
the other two variables.
Uses nlsw.dta & scheme vg_teal

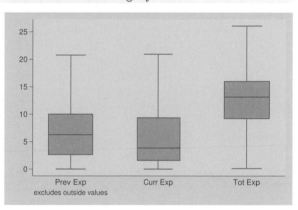

```
graph box prev_exp tenure ttl_exp, nooutsides ascategory
    yvaroptions(relabel(1 "Prev Exp" 2 "Curr Exp" 3 "Tot Exp")) xalternate
```

We move the *x* axis to the opposite side
of the graph using the `xalternate`
option. We can also use the
`yalternate` option to move the *y* axis
to its opposite side.
📊 Click the Grid Edit 📊 tool and,
pointing to the horizontal axis, click
and hold and then drag the axis to the
top of the graph. Then select the
Pointer 🢑 tool, double-click on the
horizontal axis, select *Advanced*, and
change the **Position** to `Above`.
Uses nlsw.dta & scheme vg_teal

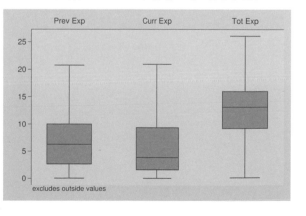

Introduction   Editor   Twoway   Matrix   Bar   Box   Dot   Pie   Options   Standard options   Styles   Appendix

Yvars and over   Box gaps   Box sorting   Cat axis   Legend   Y axis   Boxlook options   By

We have seen that although the `relabel()` option is called an `over()` option, we can use it within `yvaroptions()` to control the labeling of multiple y variables (provided that the `ascategory` option is used to convert the multiple y variables into categories). We next explore other `over()` options that we can use with `yvaroptions()`. These examples will use the `vg_rose` scheme.

`graph box wage, nooutsides over(occ7, label(angle(45))) over(collgrad)`

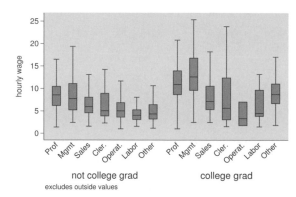

This graph shows wages broken down by occupation and by whether one graduated college. We add the `label(angle(45))` option to rotate the labels for occupation by 45 degrees. If this had been omitted, the labels would have overlapped each other.

Click on any occupation and, in the Contextual Toolbar, change the **Label angle** to **45 degrees**.

Uses nlsw.dta & scheme vg_rose

`graph box wage, nooutsides over(occ7, label(alternate)) over(collgrad)`

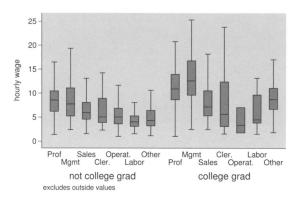

Compare this graph with the previous example. This example uses the `label(alternate)` strategy to avoid overlapping by alternating the labels for occupation.

Double-click on any of the occupation labels and click on the **Label properties** button and check **Alternate spacing of adjacent labels**.

Uses nlsw.dta & scheme vg_rose

```
graph hbox wage, nooutsides over(occ5, label(labsize(small)))
   over(collgrad, label(labcolor(maroon)))
```

Here we make the labels for `occ5` small and the color of the labels for `collgrad` maroon. See Styles : Colors (412) and Styles : Textsize (428) for more details about other values we could choose.

🖺 Click on the label for `occ5` and, in the Contextual Toolbar, change the **Label size** to Small. Double-click on the label for `collgrad`, click on the **Label properties** button, and change the **Color** to Maroon.

Uses nlsw.dta & scheme vg_rose

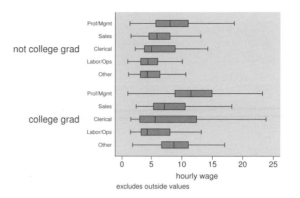

```
graph hbox wage, nooutsides
   over(occ5, label(ticks tlwidth(thick) tlength(medium)))
   over(collgrad)
```

The `ticks` option adds tick marks. Here we make them thick in width and medium in length. See [G-3] ***cat_axis_label_options*** for more details.

🖺 Double-click on the labels for `occ5` and click on the **Tick properties** button. Check **Show ticks** and change the **Length** to Medium and the **Width** to Thick.

Uses nlsw.dta & scheme vg_rose

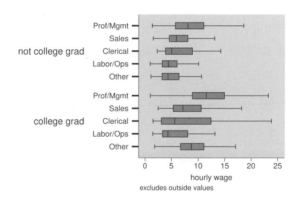

```
graph hbox wage, nooutsides over(occ5, label(labgap(huge))) over(collgrad)
```

The `label(labgap(huge))` option controls the gap between the label and the axis line. Here we increase the gap between the label for the levels of `occ5` and the axis line to huge.

🖺 Double-click on the label for `occ5`, click on the **Label properties** button, and change the **Text gap** to Huge.

Uses nlsw.dta & scheme vg_rose

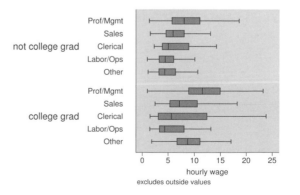

Introduction Editor Twoway Matrix Bar Box Dot Pie Options Standard options Styles Appendix

Yvars and over Box gaps Box sorting Cat axis Legend Y axis Boxlook options By

```
graph hbox wage, nooutsides over(occ5) over(collgrad, label(labgap(huge)))
```

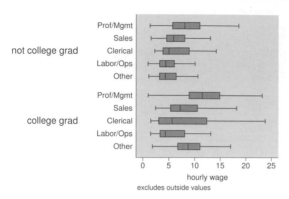

By using the `label(labgap(huge))` option, we increase the gap associated with `collgrad`. This example makes the gap between `collgrad` and `occ5` huge.

📊 Double-click on the label for `collgrad`, click on the **Label properties** button, and change the **Text gap** to Huge.

Uses nlsw.dta & scheme vg_rose

```
graph box wage, nooutsides over(union) over(grade4) asyvars
    b1title("Education Level in Four Categories")
```

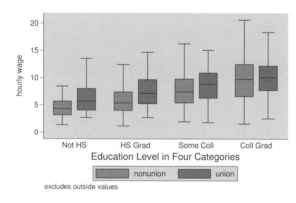

The `b1title()` option adds a title to the bottom of the graph, in effect labeling the *x* axis. We can add a second title below that by using the `b2title()` option. If we used `graph hbox`, we could label the left axis using the `l1title()` and `l2title()` options.

📊 In the Object Browser, double-click on **bottom 1** (which is below **positional titles**) and enter a title under **text**.

Uses nlsw.dta & scheme vg_rose

6.5   Controlling legends

This section discusses the use of legends for box charts, emphasizing the features that are unique to box charts. The section Options : Legend (361) goes into great detail about legends, as does [G-3] *legend_options*. We can use legends for multiple *y* variables or when the first `over()` variable is treated as a *y* variable with the `asyvars` option. See Box : Yvars and over (227) for more information about using multiple *y* variables and more examples of treating the first `over()` variable as a *y* variable. These first examples use the `vg_brite` scheme.

```
graph box prev_exp tenure ttl_exp, nooutsides
```

Consider this box plot of three different variables. These variables are shown with different colors, and a legend is used to identify the variables.

Uses nlsw.dta & scheme vg_brite

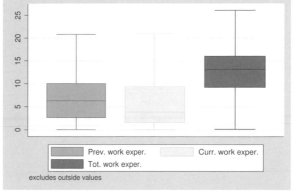

```
graph box wage, nooutsides over(occ7) asyvars
```

This is another example of where a legend can arise in a Stata box plot by using the `asyvars` option, which treats an `over()` variable as though the levels were different *y* variables.

Uses nlsw.dta & scheme vg_brite

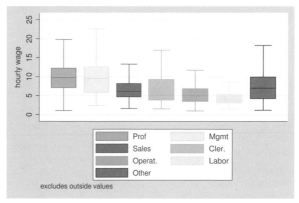

Unless otherwise mentioned, the `legend()` option described below works the same whether we derived the legend from multiple *y* variables or from an `over()` option that we combined with the `asyvars` option. These examples use the `vg_teal` scheme.

Introduction   Editor   Twoway   Matrix   Bar   Box   Dot   Pie   Options   Standard options   Styles   Appendix

Yvars and over   Box gaps   Box sorting   Cat axis   Legend   Y axis   Boxlook options   By

```
graph box prev_exp tenure ttl_exp, nooutsides nolabel
```

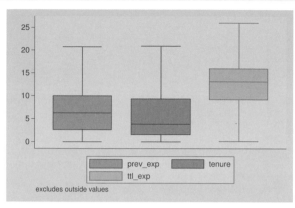

The `nolabel` option works only when we have multiple y variables. When we use this option, the variable names (not the variable labels) are used in the legend. For example, instead of showing the variable label `Prev. work exper.`, it shows the variable name `prev_exp`.
Uses nlsw.dta & scheme vg_teal

```
graph box prev_exp tenure ttl_exp, nooutsides showyvars
```

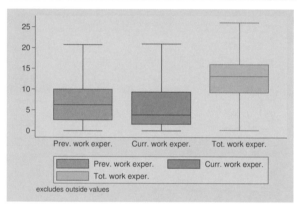

The `showyvars` option puts the labels in the legend.
Uses nlsw.dta & scheme vg_teal

```
graph box prev_exp tenure ttl_exp, nooutsides showyvars legend(off)
```

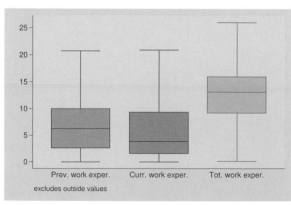

One instance when the `showyvars` option would be useful is when we want separately colored boxes labeled at the bottom. Here we use `showyvars` to show the labels at the bottom of the boxes and the `legend(off)` option to suppress the display of the legend.

Hide the legend by going to the Object Browser, right-clicking on **legend**, and selecting **Hide**. (Redisplay it by right-clicking on **legend** and selecting **Show**.)
Uses nlsw.dta & scheme vg_teal

```
graph box wage, nooutsides over(occ7) asyvars showyvars legend(off)
```

Even though the `showyvars` option sounds like it would work only with multiple *y* variables, it also works when we combine the `over()` and `asyvars` options. As before, we suppress the legend by using the `legend(off)` option.

Uses nlsw.dta & scheme vg_teal

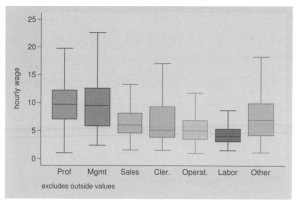

```
graph box wage, nooutsides over(occ7) asyvars
    legend(label(1 "Professional") label(2 "Management"))
```

We use the `legend(label())` option to change the labels for the first and second variables in the legend. We use a separate `label()` option for each box. 📊 Click on *Prof* and, in the Contextual Toolbar, change the **Text** to `Professional`. Likewise, click on *Mgmt* and, in the Contextual Toolbar, change the **Text** to `Management`.

Uses nlsw.dta & scheme vg_teal

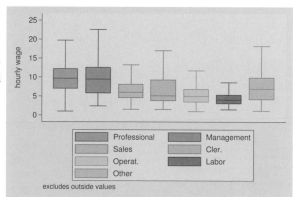

```
graph box wage, nooutsides over(occ7) asyvars legend(rows(2) colfirst)
```

The `rows(2)` option combined with `colfirst` displays the legend in two rows with the keys ordered by column (instead of by row, the default). This yields keys that are more adjacent to the bars that they label.

📊 In the Object Browser, double-click on **legend**. Change **Rows/Columns** to `Rows` and **Rows** to 2. Then change the **Key Sequence** to `Down first`.

Uses nlsw.dta & scheme vg_teal

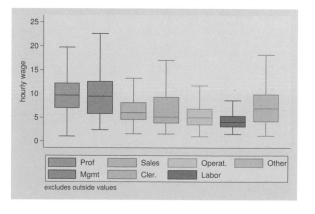

Introduction   Editor   Twoway   Matrix   Bar   Box   Dot   Pie   Options   Standard options   Styles   Appendix

Yvars and over   Box gaps   Box sorting   Cat axis   Legend   Y axis   Boxlook options   By

As we can see, the default placement for the legend is below the x axis. However, Stata gives us tremendous flexibility in the placement of the legend. Let's now consider options that control the placement of the legend, along with options useful for controlling the placement of the items within the legend. The following examples use the **vg_rose** scheme.

```
graph box wage, nooutsides over(occ7) asyvars
    legend(position(1))
```

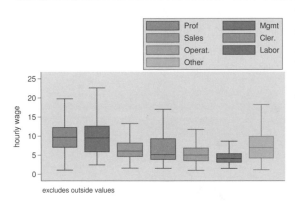

We can use the legend(position(1)) option to place the legend in the top right corner of the graph. The values we supply for position() are like the numbers on a clock face, where twelve o'clock is the top, six o'clock is the bottom, and 0 represents the center of the clock face. Specifying one o'clock places the legend in the top right; see Styles : Clockpos (414) for more details.

📊 See the next graph.

Uses nlsw.dta & scheme vg_rose

```
graph box wage, nooutsides over(occ7) asyvars
```

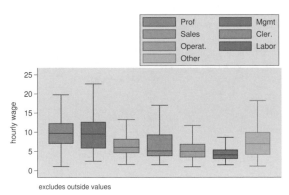

📊 We can replicate the previous customizations with the Graph Editor. Select the Grid Edit 📊 tool and then, in the Object Browser, click on **legend** to select it. Click and hold the legend and drag it to the top of the graph. Then, in the Object Browser, double-click **legend** and, in the *Advanced* tab, change the **Alignment** to East.

Uses nlsw.dta & scheme vg_rose

```
graph hbox wage, nooutsides over(occ7) asyvars
    legend(cols(1) position(9))
```

We switch to making this a horizontal
box plot and move the legend by using
the position(9) option to place the
legend in the nine o'clock position. We
also use the cols(1) option to display
the legend as one column.

📖 See the next graph.

Uses nlsw.dta & scheme vg_rose

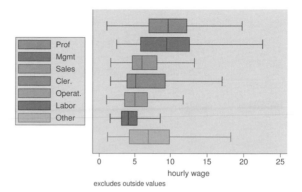

```
graph hbox wage, nooutsides over(occ7) asyvars
```

📖 To reposition the legend by using
the Graph Editor, select the Grid Edit
🔧 tool and, in the Object Browser,
click on **legend** to select it. Click and
hold the legend and drag it to the
complete left edge of the graph. Then,
in the Object Browser, double-click on
legend and change **Rows/Columns**
to Columns and 1. Finally, in the
Advanced tab, change the **Alignment**
to Center.

Uses nlsw.dta & scheme vg_rose

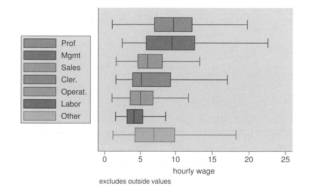

```
graph hbox wage, nooutsides over(occ7) asyvars
    legend(cols(1) position(9) stack)
```

The stack option stacks the keys and
their labels. Other options for
controlling the legend include
rowgap(), keygap(), symxsize(),
symysize(), textwidth(), and
symplacement(). See Options:Legend
(361) and [G-3] *legend_options* for
more details.

📖 In the Object Browser, double-click
on **legend** and change **Stack symbols
and text** to Yes.

Uses nlsw.dta & scheme vg_rose

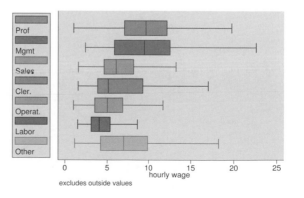

```
graph hbox wage, nooutsides over(occ7) asyvars
```

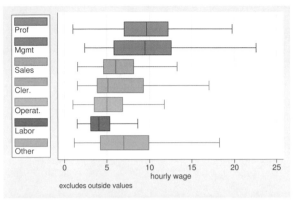

We now switch to the vg_lgndc scheme. This scheme positions the legend at the left in one column.

Uses nlsw.dta & scheme vg_lgndc

6.6   Controlling the y axis

This section describes options we can use with respect to the y axis with box charts. To be precise, when Stata refers to the y axis on a box chart, it refers to the axis with the continuous variable, whether that be the left axis when using **graph box** or the bottom axis when using **graph hbox**. This section emphasizes the features that are particularly relevant to box charts. For more details, see Options : Axis titles (327), Options : Axis labels (330), and Options : Axis scales (339). See also [G-3] *axis_title_options*, [G-3] *axis_label_options*, and [G-3] *axis_scale_options*. These examples are shown using the vg_lgndc scheme, which places the legend to the left in one column.

```
graph box prev_exp tenure, nooutsides over(occ5)
```

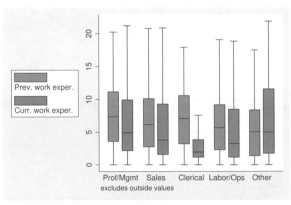

Consider this graph showing previous and current work experience broken down by occupation. We use the **nooutsides** option to suppress the display of outside values. For the rest of the graphs in this section, there would be many outside values, which would make the graphs cluttered, so we will continue to use the **nooutsides** option for each example.

Uses nlsw.dta & scheme vg_lgndc

```
graph box prev_exp tenure, nooutsides over(occ5)
    ytitle("Years of experience")
```

Looking at previous and current work experience over occupations, we can use the ytitle() option to add a title to the *y* axis. See Options : Axis titles (327) and [G-3] ***axis_title_options*** for more details.

🔲 Double-click on **Graph** in the Object Browser, select the *Titles* tab, and enter Years of Experience for the **Title**.

Uses nlsw.dta & scheme vg_lgndc

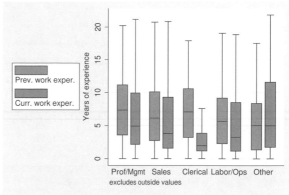

```
graph box prev_exp tenure, nooutsides over(occ5)
    ytitle("Years of experience", size(vlarge) box bexpand)
```

Because this title is considered a textbox, we can use a variety of textbox options to control the look of the title. This example makes the title large, surrounds it with a box, and uses the bexpand (box expand) option to stretch the box to fill the width of the plot area. See Options : Textboxes (379) for more examples of how to use textbox options to control the display of text.

🔲 See the next graph.

Uses nlsw.dta & scheme vg_lgndc

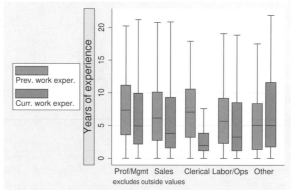

```
graph box prev_exp tenure, nooutsides over(occ5)
    ytitle("Years of experience")
```

🔲 We can use the Graph Editor to customize the title as illustrated in the previous example. Double-click on the title and change the **Size** to v Large. Next select the *Box* tab and check **Place box around text**. Finally, select the *Format* tab and check **Expand area to fill cell**.

Uses nlsw.dta & scheme vg_lgndc

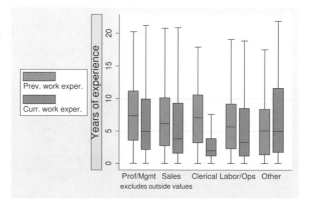

Introduction   Editor   Twoway   Matrix   Bar   Box   Dot   Pie   Options   Standard options   Styles   Appendix

Yvars and over   Box gaps   Box sorting   Cat axis   Legend   Y axis   Boxlook options   By

```
graph box wage, nooutsides over(occ5) over(collgrad) asyvar
   yline(12, lwidth(thick) lcolor(maroon) lpattern(dash))
```

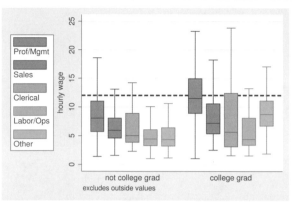

Here we use the `yline()` option to add a thick, maroon, dashed line to the points in the graph where wages equal 12. We would still use `yline()`, even if we used `graph hbox`, placing the *y* axis at the bottom.

📊 See the next graph.

Uses nlsw.dta & scheme vg_lgndc

```
graph box wage, nooutsides over(occ5) over(collgrad) asyvar
```

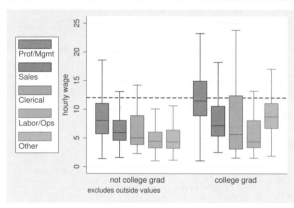

📊 First, in the Graph Editor's Object Browser, double-click on **box region**. In the *Reference lines* tab, click on **Add horizontal reference line** and for **Y axis value** type in 12. Second, double-click on the reference line and change the **color** to Maroon, the **Width** to Thick, and the **Pattern** to Dash.

Uses nlsw.dta & scheme vg_lgndc

```
graph box wage, nooutsides over(occ5) over(collgrad) asyvar
   ylabel(5(10)25)
```

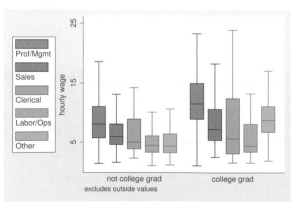

We use the `ylabel()` option to label the *y* axis going from 5 to 25, incrementing by 10. The *y* axis still starts at 0, but we could override this by using the `exclude0` option. See Options:Axis labels (330) for more details.

📊 Double-click on the *y* axis and click on **Range/Delta** and enter a **Minimum value** of 5, a **Maximum value** of 25, and a **Delta** of 10.

Uses nlsw.dta & scheme vg_lgndc

```
graph box wage, nooutsides over(occ5) over(collgrad) asyvar
    ylabel(5(10)25, angle(0))
```

The `angle()` option modifies the angle of the y labels. Here we display them horizontally.

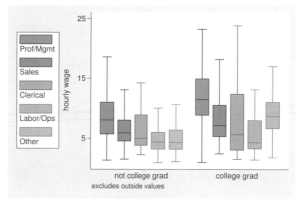

Double-click on the y axis and click on **Label properties** and change the **Angle** to Horizontal.

Uses nlsw.dta & scheme vg_lgndc

```
graph box wage, nooutsides over(occ5) over(collgrad) asyvar
    ylabel(5(10)25, nogrid)
```

We use the `nogrid` option to suppress the display of the grid for the y axis. (With box plots, there is never a grid with respect to the x axis.) If the grid were absent and you wanted to include it, you could add the `grid` option. For more details, see Options : Axis labels (330).

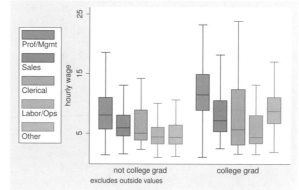

Double-click on the y axis, click on **Grid Lines**, and uncheck **Show grid**.

Uses nlsw.dta & scheme vg_lgndc

```
graph box wage, nooutsides over(occ5) over(collgrad) asyvar yscale(off)
```

We can use `yscale(off)` to turn off the y axis. See Options : Axis scales (339) and [G-3] *axis_scale_options* for more details. Disregard any references to `xscale()` because this option is not valid when using `graph box`

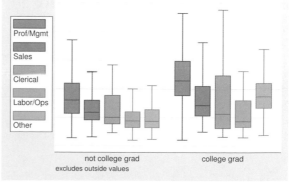

In the Object Browser, right-click on **scaleaxis** and select **Hide**. You can redisplay the y axis by right-clicking on **scaleaxis** and selecting **Show**.

Uses nlsw.dta & scheme vg_lgndc

Introduction Editor Twoway Matrix Bar Box Dot Pie Options Standard options Styles Appendix

Yvars and over Box gaps Box sorting Cat axis Legend Y axis Boxlook options By

`graph box wage, nooutsides over(occ5) over(collgrad) asyvar yalternate`

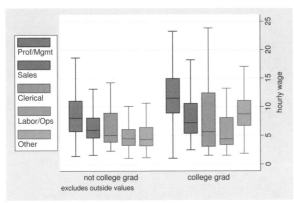

We can put the y axis on the opposite side, which here is on the right side of the graph, by using the `yalternate` option.

▨ Select the Grid Edit ▨ tool, click and hold the y axis, and drag it to the right side of the graph. Then select the Pointer ▨ tool, double-click on the y axis, click on **Advanced**, and change **Position** to Right.

Uses nlsw.dta & scheme vg_lgndc

`graph box wage, nooutsides over(occ5) over(collgrad) asyvar yreverse`

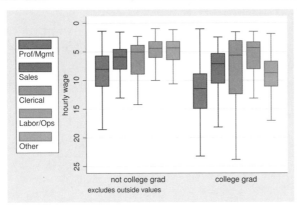

You can reverse the direction of the y axis, in effect turning your boxes upside down, with the `yreverse` option.

▨ In the Object Browser, double-click on **box region** and check **Reverse scale**.

Uses nlsw.dta & scheme vg_lgndc

6.7   Changing the look of boxes

This section shows how we can control the color of the boxes and the characteristics of the line outlining the boxes. We will first look at options that control the overall intensity of the color for all the boxes and then show how you can control the color of each box. The following examples use the `vg_s2c` scheme.

```
graph box wage, over(occ5) over(collgrad) asyvars nooutsides intensity(10)
```

The `intensity()` option controls the intensity of the color within the boxes. Here we request that the color be displayed with 10% intensity (contrasted with the default, which is 50%).

🖳 Double-click on one of the boxes for *Prof/Mgmt* and change the **Fill intensity** to 10%. We can repeat the same steps for the other four boxes.
Uses nlsw.dta & scheme vg_s2c

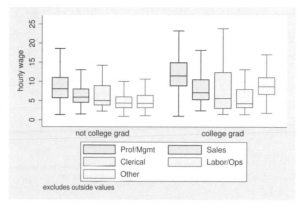

```
graph box wage, over(occ5) over(collgrad) asyvars nooutsides intensity(80)
```

In this example, we use the `intensity()` option to make the intensity of the colors 80%.

🖳 Double-click on one of the boxes for *Prof/Mgmt* and change the **Fill intensity** to 80%. We can repeat the same steps for the other four boxes.
Uses nlsw.dta & scheme vg_s2c

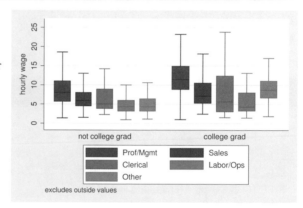

```
graph box wage, over(occ5) over(collgrad) asyvars nooutsides
    box(1, bcolor(sand))
```

Here we add `box(1, bcolor(sand))` to specify that the box color for the first bar be sand. See **Styles : Colors** (412) for more information about the available colors.

🖳 Double-click on one of the boxes for *Prof/Mgmt* and change the **Color** to Sand.
Uses nlsw.dta & scheme vg_s2c

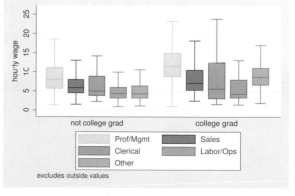

Introduction   Editor   Twoway   Matrix   Bar   Box   Dot   Pie   Options   Standard options   Styles   Appendix

Yvars and over   Box gaps   Box sorting   Cat axis   Legend   Y axis   Boxlook options   By

```
graph box wage, over(occ5) over(collgrad) asyvars nooutsides
   box(1, bcolor(sand) blcolor(brown) blwidth(thick))
```

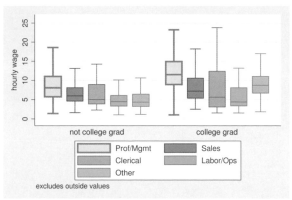

We add the `blcolor()` (box line color) and `blwidth()` (box line width) options to make the outline for the first box brown and thick.

Double-click on one of the boxes for *Prof/Mgmt*, check **Different outline color**, and change the **Outline color** to Brown. Then change the **Outline Width** to Thick.

Uses nlsw.dta & scheme vg_s2c

Now let's consider options that allow us to control the display of the median, whiskers, caps, and outside markers. These examples use the `vg_s1m` scheme.

```
graph box prev_exp tenure ttl_exp, nooutsides
   medtype(cline) medline(lwidth(thick) lcolor(black))
```

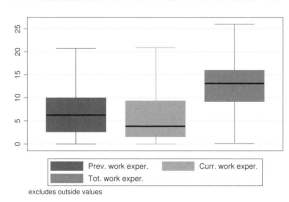

The `medtype(cline)` option sets the median type to be a custom line. We then customize the median line by using the `medline()` option to specify that the line width be thick and the line color be black.

Double-click on any box and, within the *Median* tab, change the **Median type** to Custom line. Then change the **Color** to Black and the **Width** to Thick.

Uses nlsw.dta & scheme vg_s1m

```
graph box prev_exp tenure ttl_exp, nooutsides
    medtype(marker) medmarker(msymbol(+) msize(large))
```

The `medtype(marker)` option specifies
that we want to use a marker symbol to
label the median, and the `medmarker()`
option controls the display of the
median marker. Here we make the
marker symbol a plus sign and the
marker size large.

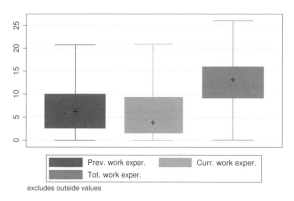

📖 Double-click on any box and, within
the *Median* tab, change the **Median
type** to `Marker`, the **Symbol** to `Plus`,
and the **Size** to `Large`.

Uses nlsw.dta & scheme vg_s1m

```
graph box prev_exp tenure ttl_exp, nooutsides
    cwhiskers lines(lwidth(thick) lcolor(black))
```

To customize the whiskers, we need to
specify the `cwhiskers` (customize
whiskers) option, and then we can add
the `lines()` option to specify how we
want the whiskers customized. In this
example, we make the whiskers thick
and black.

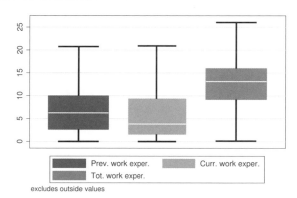

📖 Double-click on any box and, within
the *Box/Whisker* tab, check **Custom
whiskers**. Then change the **Width** to
`Thick` and the **Color** to `Black`.

Uses nlsw.dta & scheme vg_s1m

```
graph box prev_exp tenure ttl_exp, nooutsides alsize(20)
```

The `alsize()` (adjacent line size)
option allows us to control the size
(width) of the adjacent line. By
default, the adjacent line is 67% of the
width of the box. Here we make the
adjacent line much smaller, 20% of the
width of the box.

Uses nlsw.dta & scheme vg_s1m

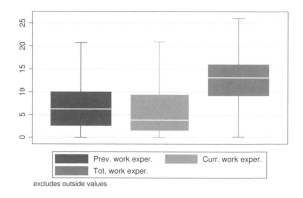

Introduction

Editor

Twoway

Matrix

Bar

Box

Dot

Pie

Options

Standard options

Styles

Appendix

Yvars and over

Box gaps

Box sorting

Cat axis

Legend

Y axis

Boxlook options

By

```
graph box prev_exp tenure ttl_exp, nooutsides capsize(5)
```

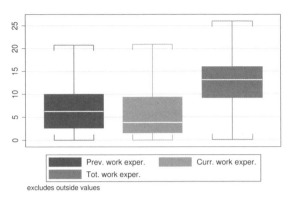

excludes outside values

The `capsize()` option allows us to specify the size of the caps (if any) on the adjacent line. The default value is 0, meaning that no cap is displayed. In this example, we add a small cap to the adjacent line.

Uses nlsw.dta & scheme vg_s1m

```
graph box prev_exp tenure ttl_exp, marker(2, msymbol(Oh) msize(vlarge))
```

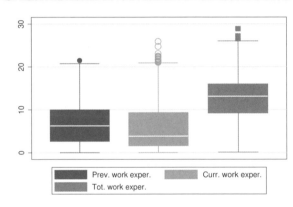

The `marker()` option allows us to control the markers used to display the outside values. We can control this separately for each *y* variable. Here we make the outside value for `tenure` display as large, hollow circles.

Double-click on any of the markers for the second box. In the *Markers* tab, change the **Symbol** to `Hollow circle` and the **Size** to `v Large`.

Uses nlsw.dta & scheme vg_s1m

6.8   Graphing by groups

This section discusses the use of the `by()` option in combination with `graph box`. Normally, you would use the `over()` option instead of the `by()` option, but sometimes the `by()` option is either necessary or more advantageous. For example, a `by()` option is useful if you exceed the maximum number of `over()` options (three if you have one *y* variable or two if you have multiple *y* variables). Thus the `by()` option allows you to break down your data by more categorical variables. Also, `by()` gives you more flexibility in the placement of the separate panels. For more information about the `by()` option, see **Options : By** (346); for more information about the `over()` option, see **Box : Yvars and over** (227).

```
graph hbox wage, nooutsides
    over(collgrad) over(urban2) over(married)
```

Consider this box graph, which breaks
down wages by three categorical
variables. If we wanted to further break
down this by another categorical
variable, we could not use another
`over()` option because we can have a
maximum of three `over()` options with
one y variable.
Uses nlsw.dta & scheme vg_s1m

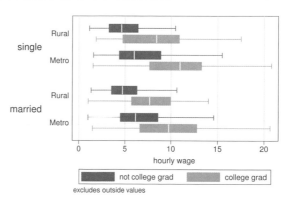

```
graph hbox wage, nooutsides note("")
    over(collgrad) over(urban2) over(married) by(union)
```

If we want to further break down
`prev_exp` by `union`, we can use the
`by(union)` option. We also add the
`note("")` option to suppress the note
"excludes outside values".
Uses nlsw.dta & scheme vg_s1m

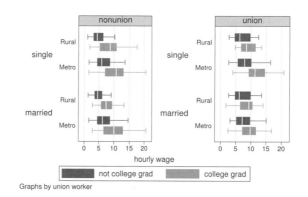

```
graph hbox prev_exp tenure, nooutsides note("")
    over(urban2) over(married)
```

Consider this box graph with multiple y
variables breaking down them by two
categorical variables using two `over()`
options. When we have multiple y
variables, we can have a maximum of
two `over()` options.
Uses nlsw.dta & scheme vg_s1m

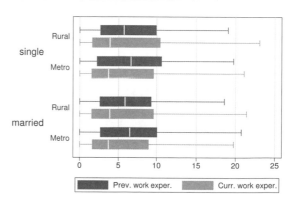

```
graph box prev_exp tenure, nooutsides note("")
   over(urban2) over(married) by(union)
```

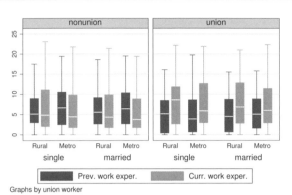

Graphs by union worker

We can add the by(union) option to further break down these by plots by union status. We can include multiple variables within by(), although this can make some small graphs.
Uses nlsw.dta & scheme vg_s1m

```
graph hbox ttl_exp tenure, nooutsides note("")
   over(urban2) over(married) by(union, missing)
```

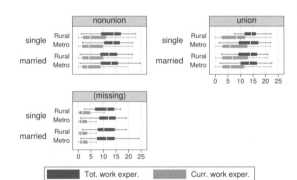

Graphs by union worker

We can use the missing option to include a panel for the missing values of union.
Uses nlsw.dta & scheme vg_s1m

```
graph hbox ttl_exp tenure, nooutsides note("")
   over(urban2) over(married) by(union, total)
```

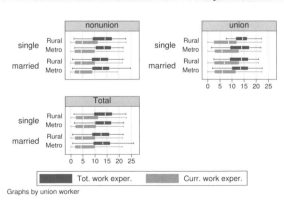

Graphs by union worker

We can add the total option to include a panel for all observations.
Uses nlsw.dta & scheme vg_s1m

```
graph hbox ttl_exp tenure, nooutsides note("")
   over(urban2) over(married) by(union, total rows(1))
```

We can use the `rows(1)` option to show
the multiple graphs in one row.

🖳 In the Object Browser, double-click
on **Graph** and, in the *Organization*
tab, change **Rows/Columns** to Rows
and then change **Rows** to 1.
Uses nlsw.dta & scheme vg_s1m

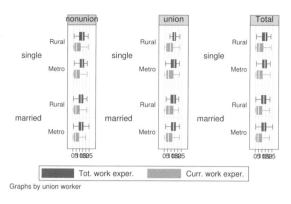

Graphs by union worker

```
graph hbox ttl_exp tenure, nooutsides note("")
   over(urban2) over(married) by(union, cols(1))
```

Here we omit the `total` option and use
the `cols(1)` option to show both
graphs in one column.

🖳 In the Object Browser, double-click
on **Graph** and, in the *Organization*
tab, change the **Columns** to 1.
Uses nlsw.dta & scheme vg_s1m

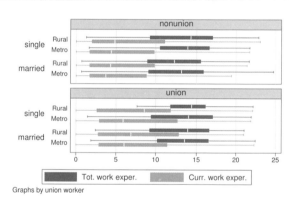

Graphs by union worker

```
graph hbox ttl_exp tenure, nooutsides note("")
   over(urban2) over(married) by(union, cols(1) legend(position(9)))
```

We use the `legend(position(9))`
option within the `by()` option to put
the legend at nine o'clock.

🖳 Select the Grid Edit 🖳 tool and
then, in the Object Browser, click on
legend to select the entire legend.
Then click and hold the legend, drag it
to the leftmost part of the graph, and
release the mouse. Then, in the Object
Browser, double-click on **legend** and,
in the *Advanced* tab, change the
Alignment to Center.
Uses nlsw.dta & scheme vg_s1m

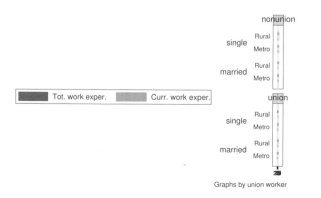

Graphs by union worker

Introduction Editor Twoway Matrix Bar Box Dot Pie Options Standard options Styles Appendix

Yvars and over Box gaps Box sorting Cat axis Legend Y axis Boxlook options By

```
graph hbox ttl_exp tenure, nooutsides note("")
    over(urban2) over(married) by(union, cols(1) legend(position(9)))
    legend(cols(1) stack)
```

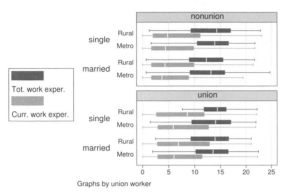

We now add the `legend(cols(1) stack)` option to make the legend one stacked column.

In the Object Browser, double-click on **legend**. In the *Organization* tab, change the **Columns** to 1 and **Stack symbols and text** to Yes and click on **OK**.

Uses nlsw.dta & scheme vg_s1m

7   Dot plots

This chapter discusses the use of dot plots in Stata. The chapter begins by showing how we can specify multiple y variables to display plots for multiple variables and by showing how we can use the `over()` option to break down dot plots by categorical variables. The next section discusses the `over()` option, which we can use to customize the display of these categorical variables, followed by a section on options concerning the display of categorical axes. Next the chapter covers options that control legends, followed by options that control the y axis. The chapter draws to a close with options that control the look of the lines and dots that form the dot plot and, lastly, with the `by()` option.

7.1   Specifying variables and groups

This section introduces the use of dot plots. It shows how we can use the `over()` option for displaying dot plots by one or more grouping variables. It then shows how we can specify one or more y variables in a plot and control the summary statistic used for collapsing the y variable(s). See the *group_options* table in [G-2] **graph dot** for more details. This section uses the `vg_s1c` scheme.

`graph dot tenure, over(occ7)`

Here we use the `over()` option to show the average current work experience broken down by occupation. By default, the y variable (`tenure`) is placed on the bottom axis and is considered the y axis. Likewise, the levels of `occ7` are placed on the left axis and are considered to form the x axis, or categorical axis.

Uses nlsw.dta & scheme vg_s1c

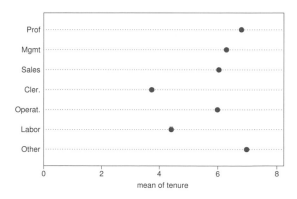

263

`graph dot tenure, over(occ7) over(collgrad)`

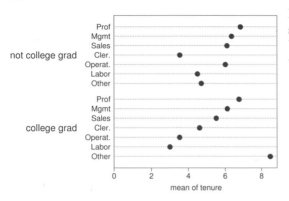

Here we use a second `over()` option to show the mean of work experience broken down by occupation and whether one graduated college.
Uses nlsw.dta & scheme vg_s1c

`graph dot tenure, over(occ7) over(collgrad) over(married)`

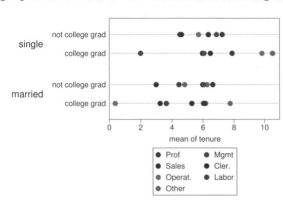

We can add a third `over()` option to further break down `tenure` by whether one is married. The first `over()` variable (`occ7`) is now treated as multiple *y* variables. When we use three `over()` options, the first variable is then treated as multiple *y* variables, as though we had specified the `asyvars` option. This graph can be difficult to read with `occ7` forming the multiple *y* variables.
Uses nlsw.dta & scheme vg_s1c

`graph dot tenure, over(married) over(occ7) over(collgrad)`

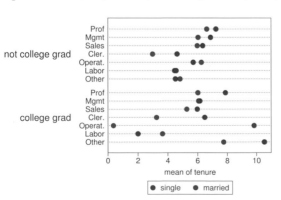

This graph shows the same data as the last one, except that we have switched the order of the `over()` options, making `over(married)` come first and thus forming the multiple *y* variables. This might be easier to read than the previous graph.
Uses nlsw.dta & scheme vg_s1c

Let's now consider examples with multiple y variables. These examples are shown using the `vg_outc` scheme.

graph dot prev_exp tenure, over(occ7)

This graph shows the average previous experience and average current tenure broken down by occupation. Although we do not need to use the `over()` option, omitting it may make a fairly boring graph.

Uses nlsw.dta & scheme vg_outc

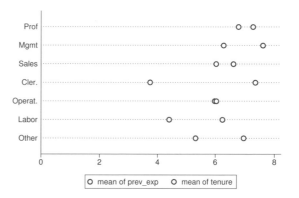

graph dot prev_exp tenure, over(occ7) over(collgrad)

This graph also shows whether one is a college graduate as another grouping level. The command has multiple y variables, so we cannot include another `over()` option because dot plots support three levels of nesting and the multiple y variables account for a level.

Uses nlsw.dta & scheme vg_outc

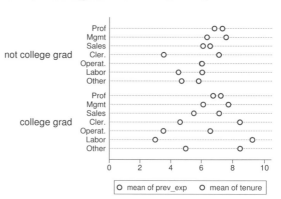

```
graph dot (median) prev_exp tenure, over(occ7) over(collgrad)
```

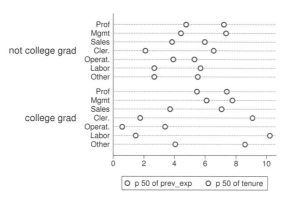

So far, all the examples we have seen have graphed the mean of y variable(s). Here we preface the y variables with (median), plotting the median for each y variable.

Uses nlsw.dta & scheme vg_outc

```
graph dot (p10) wage (p25) wage (p50) wage (p75) wage (p90) wage,
    over(occ7)
```

We can request different statistics for the same variable, such as here, which shows the 10th, 25th, 50th, 75th, and 90th percentiles of wages broken down by occupation.

Uses nlsw.dta & scheme vg_outc

Now let's consider options that we can use with the over() option to customize the behavior of the graphs. We will see how to treat the levels of the first over() option as though they were multiple y variables. We can also request that missing values for the levels of the over() variables be displayed, and we can suppress empty categories when multiple over() options are used. These examples are shown using the vg_s2m scheme.

`graph dot tenure, over(collgrad) over(occ7)`

Consider this graph showing the
average current work experience broken
down by whether one is a college
graduate and by occupation.
Uses nlsw.dta & scheme vg_s2m

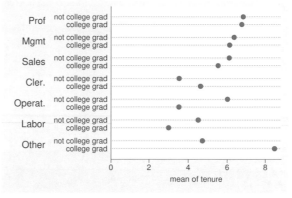

`graph dot tenure, over(collgrad) over(occ7) asyvars`

With the `asyvars` option, the first
`over()` variable (`collgrad`) is graphed
as if there were two *y* variables. The
two levels of `collgrad` are shown as
different markers on the same line, and
they are labeled using the legend.

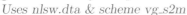 In the Object Browser, double-click
on **dot region** and click on the **Rotate
Categories** button until the graph
appears as you wish.
Uses nlsw.dta & scheme vg_s2m

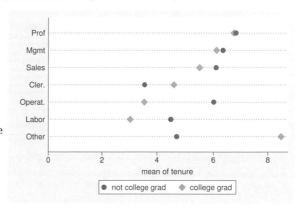

`graph dot tenure, over(occ5) over(union) missing`

Consider this graph in which we use the
`over()` option to show `tenure` broken
down by `occ5` and `union`. By including
the `missing` option, we see the category
for those who are missing on the `union`
variable, shown as the third group
labeled with a dot. See **Dot : Cat axis**
(274) for examples showing how we
could change the label (.) to something
more meaningful, e.g., "Missing".
Uses nlsw.dta & scheme vg_s2m

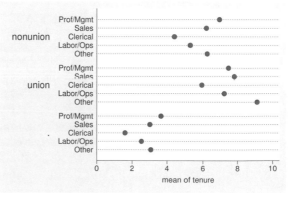

Introduction Editor Twoway Matrix Bar Box Dot Pie Options Standard options Styles Appendix

Yvars and over   Dot gaps   Dot sorting   Cat axis   Legend   Y axis   Dotlook options   By

`graph dot tenure, over(grade) over(collgrad)`

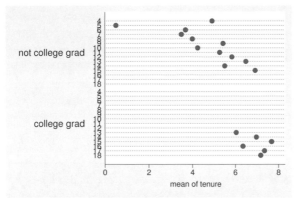

Consider this dot plot, which breaks down `tenure` by two variables: the last grade that one completed and whether one is a college graduate. By default, Stata shows all possible combinations for these two variables. Usually, all combinations are possible, but not here, and including them has caused the labels for grade to overlap.

Uses nlsw.dta & scheme vg_s2m

`graph dot tenure, over(grade) over(collgrad) nofill`

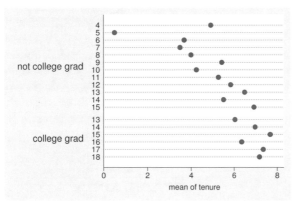

If we want to display only the combinations of the `over()` variables that exist in the data, we can use the `nofill` option.

📊 Double-click on **dot region** in the Object Browser and then uncheck **Include missing categories**.

Uses nlsw.dta & scheme vg_s2m

7.2   Options for controlling gaps between dots

This section considers some of the options that we can use for controlling the spacing between the markers. For more information on the options covered in this section, see the *over_subopts* table in [G-2] **graph dot**.

When dealing with bar and box plots, the spacing was measured for the width of a bar or box. The gap could be increased or decreased a fraction of the width of a bar/box. With a dot plot, the same concept applies except the width is measured not in terms of the dot itself, but in terms of the width of the entire line the dot occupies.

These examples use the **vg_blue** scheme.

`graph dot tenure, over(occ5) over(collgrad)`

Consider this graph in which we show a dot plot of `tenure` broken down by `occ5` and `collgrad`. Note the width of each line for each occupation.
Uses nlsw.dta & scheme vg_blue

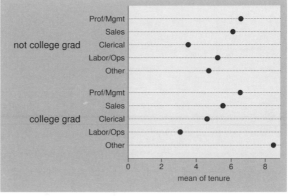

`graph dot tenure, over(occ7) over(collgrad, gap(300))`

Suppose that we wanted to make the gap between the levels of `collgrad` larger. Here we use the `gap(300)` option to make this gap 300% (three times) the width of an individual line.

In the Graph Editor's Object Browser, double-click on **dot region** to bring up the *Dot region properties* dialog box. We can change the spacing for **Super Categories** to 300pct.
Uses nlsw.dta & scheme vg_blue

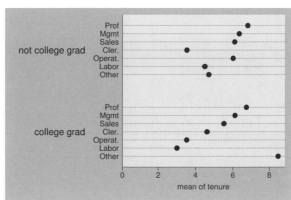

`graph dot prev_exp tenure, over(occ7) linegap(30)`

We use the `linegap()` option to display the *y* variables on different lines and specify the gap between these lines. The gap is 30% of the width of the entire line. The default value is 0, meaning that all *y* variables are displayed on the same line.

See the next graph.
Uses nlsw.dta & scheme vg_blue

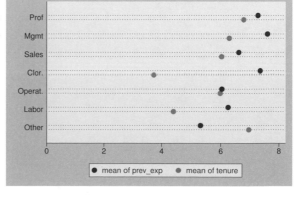

Introduction
Editor
Twoway
Matrix
Bar
Box
Dot
Pie
Options
Standard options
Styles
Appendix

Yvars and over
Dot gaps
Dot sorting
Cat axis
Legend
Y axis
Dotlook options
By

`graph dot prev_exp tenure, over(occ7) linegap(30)`

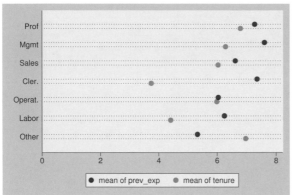

In the Object Browser, double-click on **dot region**. In the spacing section, the **Variables** option controls the gap among the variables. The scaling is `linegap()` minus 100, so a value of `Neg70pct` yields the same spacing as the previous graph. A value of `Neg100pct` means that the dots are parallel, and a value of `0pct` means that the dots are on either edge of the dot line.

Uses nlsw.dta & scheme vg_blue

`graph dot tenure, over(occ5) over(collgrad) over(married)`

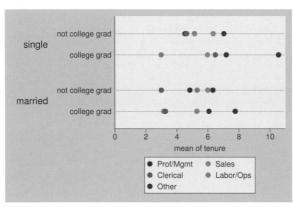

Consider this graph. Because we have used three `over()` options, the levels of the first `over()` variable are displayed as though they were different *y* variables. We may want to use the `linegap()` option to display the different *y* variables on different lines to make the graph more readable; see the next example.

Uses nlsw.dta & scheme vg_blue

`graph dot tenure, over(occ5) over(collgrad) over(married) linegap(70)`
`    legend(rows(1) span)`

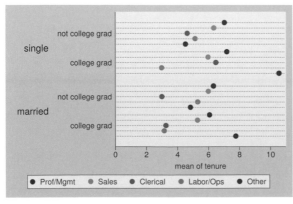

This example is similar to the previous one, but we have added the `linegap(70)` option to make the levels of `occ5` display on separate lines, making the results more readable. We have also added a `legend()` option to make the legend display in one line and span the width of the graph.

In the Object Browser, double-click on **dot region** and change the spacing for **Variables** to `Neg30pct` (i.e., 70 minus 100).

Uses nlsw.dta & scheme vg_blue

7.3   Options for sorting dots

This section considers some of the options that we can use to change the order in which the markers are sorted. This is an area where the control afforded with commands is far greater than the control you can obtain with the Graph Editor. This section will focus on graph commands without concern for the Graph Editor. These examples use the `vg_brite` scheme.

`graph dot tenure, over(occ7)`

Consider this graph showing `tenure` broken down by the seven levels of occupation. The markers are ordered by levels of `occ7`, going from 1 to 7.
Uses nlsw.dta & scheme vg_brite

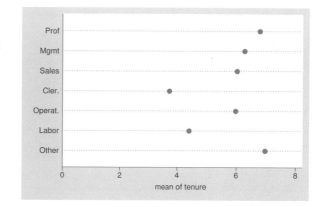

`graph dot tenure, over(occ7, descending)`

The `descending` option switches the order of the markers. They are still ordered according to the seven levels of occupation, but the markers are ordered from 7 to 1.
Uses nlsw.dta & scheme vg_brite

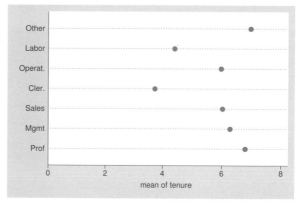

`graph dot tenure, over(occ7, sort(occ7alpha))`

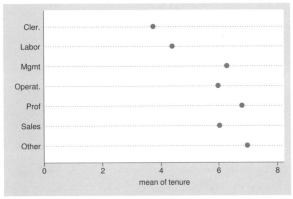

We might want to put these markers in alphabetical order but with *Other* appearing last. We can do this by recoding `occ7` into a new variable (say, `occ7alpha`) such that as `occ7alpha` goes from 1 to 7, the occupations are alphabetical. We recoded `occ7` with these assignments: $4 = 1$, $6 = 2$, $2 = 3$, $5 = 4$, $1 = 5$, $3 = 6$, and $7 = 7$; see [D] **recode**. Then the `sort(occ7alpha)` option alphabetizes the markers but with *Other* still appearing last.

Uses nlsw.dta & scheme vg_brite

`graph dot tenure, over(occ7, sort(1))`

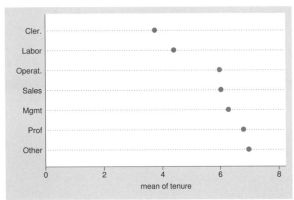

Here we sort the variables on the mean of **tenure**, yielding markers with means in ascending order. The `sort(1)` option sorts the markers according to the mean of the first y variable, the mean of **tenure**.

Uses nlsw.dta & scheme vg_brite

`graph dot tenure, over(occ7, sort(1) descending)`

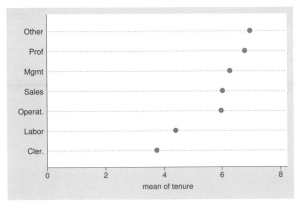

Adding the `descending` option yields markers in descending order, going from highest mean **tenure** to lowest mean **tenure**.

Uses nlsw.dta & scheme vg_brite

`graph dot tenure prev_exp, over(occ7, sort(2))`

Adding a second y variable and changing `sort(1)` to `sort(2)` sorts the markers according to the second y variable, the mean of `prev_exp`.
Uses nlsw.dta & scheme vg_brite

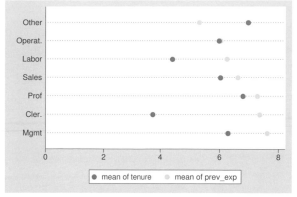

`graph dot tenure prev_exp, over(occ7, sort(1)) over(collgrad)`

We can use the `sort()` option when there are more `over()` variables. Here the markers are ordered according to the mean of `tenure` within each level of `collgrad`.
Uses nlsw.dta & scheme vg_brite

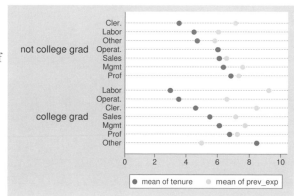

`graph dot tenure prev_exp, over(occ7, sort(1)) over(collgrad, descending)`

We add the `descending` option to the second `over()` option, and the levels of `collgrad` are now shown with college graduates appearing first.
Uses nlsw.dta & scheme vg_brite

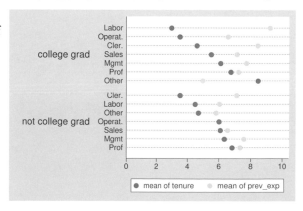

Introduction

Yvars and over

Editor

Dot gaps

Twoway

Dot sorting

Matrix

Cat axis

Bar

Legend

Box

Y axis

Dot

Dotlook options

Pie

By

Options

Standard options

Styles

Appendix

7.4   Controlling the categorical axis

This section describes ways that we can label the categorical axis in dot plots. Dot plots, like bar and box plots, are different from other plots because their x axis is formed by categorical variables. (Stata calls the axis with the categorical variable(s) the x axis, even though it may be placed on the left axis.) This section describes options we can use to customize the categorical axis. For more details on these options, see [G-3] *cat_axis_label_options* and [G-3] *cat_axis_line_options*. We will start by seeing how we can change the labels used for the categorical axis. These examples use the vg_past scheme.

graph dot tenure, over(occ7) over(south)

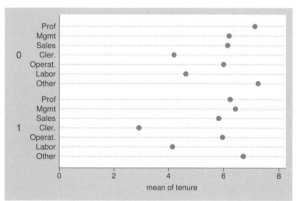

This is an example of a dot plot with two over() variables graphing the average tenure broken down by occupation and whether one lives in the South. The variable south is a dummy variable that does not have any value labels, so the x axis is not labeled well.
Uses nlsw.dta & scheme vg_past

graph dot wage, over(occ7) over(south, relabel(1 "N & W" 2 "South"))

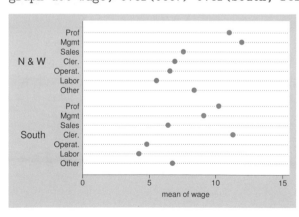

We can use the relabel() option to change the labels displayed for the levels of south, giving the x axis more meaningful labels. We wrote relabel(1 "N & W"), not relabel(0 "N & W"), because these numbers do not represent the actual levels of south but the ordinal position of the levels, i.e., first and second.

See the next graph.
Uses nlsw.dta & scheme vg_past

```
graph dot wage, over(occ7) over(south)
```

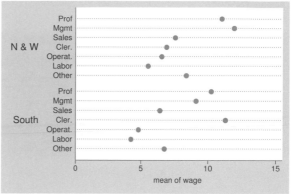

Using the Graph Editor, double-click on the label for `south` (e.g., 0 or 1) and click on the **Edit or add individual ticks and labels** button. Select the tick labeled 1 and click on **Edit**, change the **Label** to `South` and click on **OK**. Next select the tick labeled 0 and click on **Edit**, change the **Label** to `N & W`, and click on **OK**. *Uses nlsw.dta & scheme vg_past*

```
graph dot prev_exp tenure ttl_exp, ascategory
```

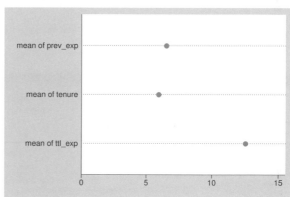

This **graph dot** command has multiple *y* variables but uses the `ascategory` option to plot the different *y* variables as if they were categorical variables. The dots for the different *y* variables are plotted on different lines using the same symbol, and each line is labeled on the *x* axis rather than using a legend. The default labels on the *x* axis are not bad, but we might want to change them. *Uses nlsw.dta & scheme vg_past*

```
graph dot prev_exp tenure ttl_exp, ascategory
   yvaroptions(relabel(1 "Previous" 2 "Current" 3 "Total"))
```

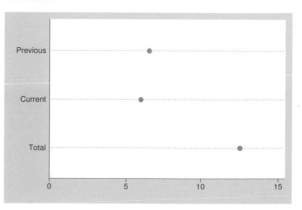

If we had an **over()** option, we could use the **relabel()** option to change the labels on the *x* axis. However, because we have multiple *y* variables that we have treated as categories, we use the `yvaroptions(relabel())` option to modify the labels on the *x* axis. See the next graph. *Uses nlsw.dta & scheme vg_past*

Introduction   Editor   Twoway   Matrix   Bar   Box   Dot   Pie   Options   Standard options   Styles   Appendix

Yvars and over   Dot gaps   Dot sorting   Cat axis   Legend   Y axis   Dotlook options   By

```
graph dot prev_exp tenure ttl_exp, ascategory
```

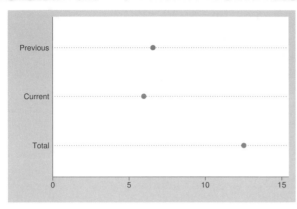

We can customize the graph as shown in the previous example using the Graph Editor. Double-click on the vertical axis (e.g., mean of `prev_exp`) and then click on **Edit or add individual ticks and labels**. Select *mean of ttl_exp* and click on **Edit** and change the **Label** to `Total`. We can likewise change the labels for the other two labels.

Uses nlsw.dta & scheme vg_past

```
graph dot prev_exp tenure ttl_exp, ascategory xalternate
    yvaroptions(relabel(1 "Previous" 2 "Current" 3 "Total"))
```

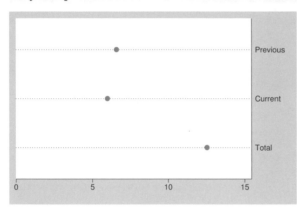

We add the `xalternate` option, which moves the labels for the x axis to the opposite side, here, from the left to the right.

Click the Grid Edit tool and, pointing to the vertical axis, click and hold and drag the vertical axis to the rightmost part of the graph. Then select the Pointer tool, double-click on the vertical axis, select **Advanced**, and change the **Position** to `Right`.

Uses nlsw.dta & scheme vg_past

```
graph dot wage, over(occ7)
    l1title("Occupations recoded into 7 categories")
```

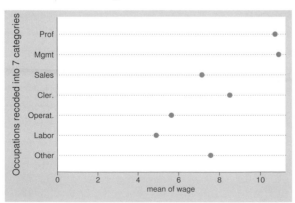

In this example, the categorical axis represents the occupation after recoding it into 7 categories. We can use the `l1title()` option to add a title to the left of the graph labeling this axis; see Standard options: Titles (395) for more details.

In the Object Browser, double-click on `left 1` (which is below **positional titles**) and enter a title under **text**.

Uses nlsw.dta & scheme vg_past

7.5   Controlling legends

This section discusses the use of legends for dot plots, emphasizing the features that are unique to dot plots. The section Options : Legend (361) goes into great detail about legends, as does [G-3] *legend_options*. We can use legends for multiple y variables or when the first over() variable is treated as a y variable via the asyvars option. See Dot : Yvars and over (263) for more information about using multiple y variables and more examples of treating the first over() variable as a y variable. These examples use the vg_rose scheme.

```
graph dot prev_exp tenure ttl_exp, over(occ7)
```

Consider this dot plot of three different variables. These variables are shown with different markers, and a legend is used to identify the variables.
Uses nlsw.dta & scheme vg_rose

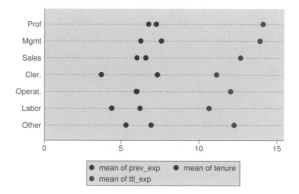

```
graph dot wage, over(collgrad) over(occ7) asyvars
```

This is another example of how a legend can arise in a Stata dot plot if the over() variable is used with the asyvars option. Stata treats the levels of the first over() variable as if they were really multiple y variables.
Uses nlsw.dta & scheme vg_rose

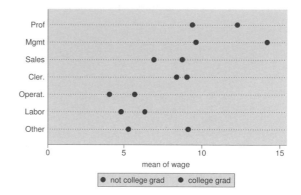

Unless otherwise mentioned, the legend options described below work the same whether we derived the legend from multiple y variables or from an `over()` variable that was combined with the `asyvars` option.

`graph dot prev_exp tenure ttl_exp, over(occ7) nolabel`

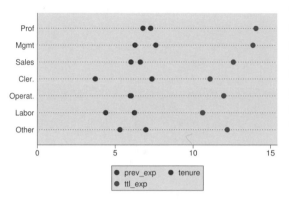

The `nolabel` option works only when we have multiple y variables. When we use this option, the variable names (not the derived variable labels) are used in the legend. For example, instead of showing the derived variable label *mean of prev_exp*, this option shows the variable name *prev_exp*.

Uses nlsw.dta & scheme vg_rose

`graph dot prev_exp tenure ttl_exp, over(occ7)`
`    legend(label(1 "Previous") label(2 "Current") label(3 "Total")`
`    title("Work Experience"))`

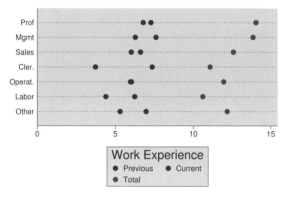

We use the `legend(label())` option to change the labels for the variables in the legend and the `title()` option to add a title to the legend.

📈 Click on *mean of prev_exp* and, in the Contextual Toolbar, change the **Text** to `Previous`. Likewise change the labels for *Current* and *Total*. Then, in the Object Browser, double-click on **Legend** and, in the *Titles* tab, change the **Title** to `Work Experience`.

Uses nlsw.dta & scheme vg_rose

```
graph dot prev_exp tenure ttl_exp, over(occ7)
   legend(rows(1))
```

Using the `legend(rows(1))` option makes the legend display as one row.

📖 In the Object Browser, double-click on **legend**. Change **Rows/Columns** to Rows and change **Rows** to 1.

Uses nlsw.dta & scheme vg_rose

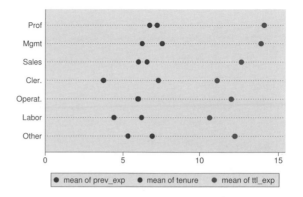

```
graph dot prev_exp tenure ttl_exp, over(occ7)
   legend(cols(1) position(9))
```

Here the legend is one column and moved to the left by using the `legend(cols(1) position(9))` option.

📖 Select the Grid Edit 🔲 tool and then, in the Object Browser, click on **legend** to select it. Click and hold the legend and drag it to the left of the graph. Then, in the Object Browser, double-click **legend**; in the *Organization* tab, change the **Columns** to 1; and, in the *Advanced* tab, change the **Alignment** to Center.

Uses nlsw.dta & scheme vg_rose

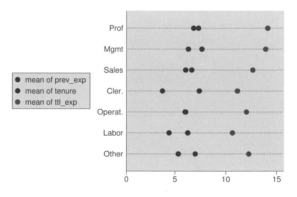

7.6   Controlling the y axis

This section describes options to customize the *y* axis with dot plots. To be precise, when Stata refers to the *y* axis on a dot plot, it refers to the axis with the continuous variable, which is placed on the bottom (where the *x* axis would traditionally be placed). This section emphasizes the features that are particularly relevant to dot plots. For more details, see Options: Axis titles (327), Options: Axis labels (330), and Options: Axis scales (339). Also see [G-3] *axis_title_options*, [G-3] *axis_label_options*, and [G-3] *axis_scale_options*. These examples use the **vg_teal** scheme.

Introduction Editor Twoway Matrix Bar Box Dot Pie Options Standard options Styles Appendix

Yvars and over Dot gaps Dot sorting Cat axis Legend Y axis Dotlook options By

```
graph dot hours, over(occ7) ytitle(Hours Worked Per Week)
```

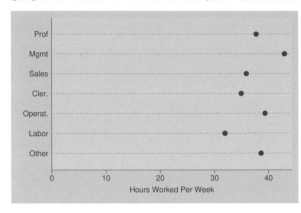

Consider this graph showing the mean hourly wage broken down by occupation. We use the `ytitle()` option to add a title to the y axis. See Options : Axis titles (327) and [G-3] *axis_title_options* for more details.

📖 Click on *Mean of Hours* and, in the Contextual Toolbar, change the **Text** to Hours Worked Per Week.

Uses nlsw.dta & scheme vg_teal

```
graph dot hours, over(occ7)
    ytitle(Hours Worked Per Week, fcolor(eggshell) box bexpand)
```

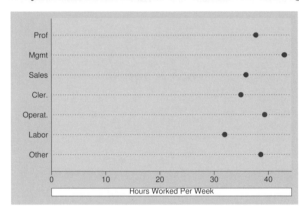

Because the title is considered a textbox, you can use textbox options as illustrated here to control the look of the title. See Options : Textboxes (379) for more examples of how to use textbox options to control the display of text.

📖 Double-click on the y title and, in the *Box* tab, check **Place box around text** and change the **Fill color** to Eggshell. Then, in the *Format* tab, check **Expand area to fill cell**.

Uses nlsw.dta & scheme vg_teal

```
graph dot hours, over(occ7)
    yline(40, lwidth(thin) lcolor(navy) lpattern(dash))
```

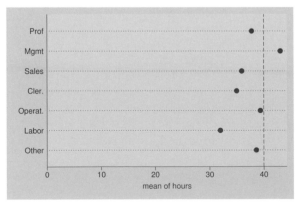

This example uses the `yline()` option to add a thin, navy, dashed line to the graph where the hours worked equals 40.

📖 In the Object Browser, double-click on **dot region**. In the *Reference lines* tab, click on **Add vertical reference line**, enter an **X axis value** of 40, and click on **OK**. Then double-click on the line and change the **Color** to Navy, the **Width** to Thin, and the **Pattern** to Dash.

Uses nlsw.dta & scheme vg_teal

```
graph dot hours, over(occ7) ylabel(30(5)45)
```

We use the `ylabel()` option to label the *y* axis from 30 to 45, incrementing by 5. See Options: Axis labels (330) and [G-3] **axis_label_options** for more details. The *y* axis still begins at 0, but see the next example for how we can override this.

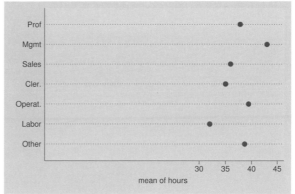

📊 Double-click on a label for `hours` (e.g., 10). In the *Axis properties* dialog box, select **Range/Delta** and enter 30 for the **Minimum value**, 45 for the **Maximum value**, and 5 for the **Delta**.

Uses nlsw.dta & scheme vg_teal

```
graph dot hours, over(occ7) ylabel(30(5)45) exclude0
```

When we add the `exclude0` option, the dot plot does not automatically begin at 0. Here it starts at 30 because that is the value we specified as the starting point on the `ylabel()` option.

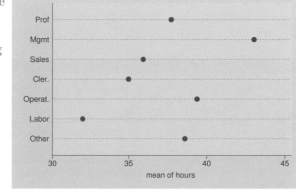

📊 In the Object Browser, double-click on **dot region** and uncheck **Include zero in scale**.

Uses nlsw.dta & scheme vg_teal

```
graph dot hours, over(occ7) yscale(off)
```

We can use the `yscale(off)` option to turn off the *y* axis. See Options: Axis scales (339) and [G-3] **axis_scale_options** for more details. Disregard any references to `xscale()` because this option is not valid when using `graph dot`.

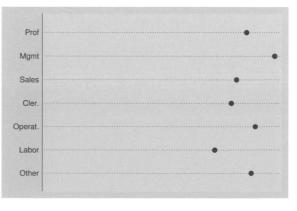

📊 In the Object Browser, right-click on **scaleaxis** and select **Hide**. You can redisplay the *y* axis by right-clicking on **scaleaxis** and selecting **Show**.

Uses nlsw.dta & scheme vg_teal

Introduction | Editor | Twoway | Matrix | Bar | Box | Dot | Pie | Options | Standard options | Styles | Appendix

Yvars and over | Dot gaps | Dot sorting | Cat axis | Legend | Y axis | Dotlook options | By

```
graph dot hours, over(occ7) yalternate
```

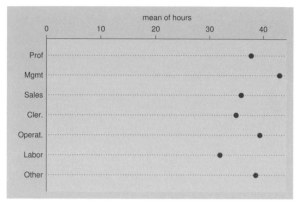

The `yalternate` option puts the y axis on the opposite side, here, on the top of the graph.

📊 Select the Grid Edit 🔲 tool and click and hold the y axis (bottom axis) and drag it to the top of the graph. Then select the Pointer 🔺 tool, double-click on the y axis (top axis), click on **Advanced**, and change **Position** to Top.

Uses nlsw.dta & scheme vg_teal

```
graph dot hours, over(occ7) yreverse
```

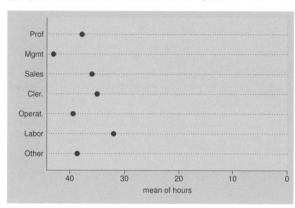

We can reverse the direction of the y axis with the `yreverse` option.

📊 In the Object Browser, double-click on **dot region** and check **Reverse scale**.

Uses nlsw.dta & scheme vg_teal

7.7   Changing the look of dot rulers

This section shows how we can control the look of the lines in our dot plots. We will see how we can control the space between the lines, the color of the lines, and other characteristics of the line. For more information, see the *linelook_options* table in [G-2] **graph dot**. These graphs are shown using the `vg_s2c` scheme.

```
graph dot prev_exp tenure, over(occ7)
```

Consider this dot plot showing previous and current work experience broken down by occupation. Each dot plot has a series of small dots that forms a line on which the symbols are plotted. *Uses nlsw.dta & scheme vg_s2c*

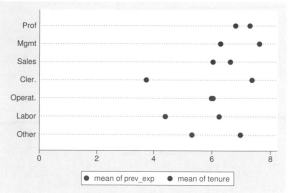

```
graph dot prev_exp tenure, over(occ7)
    dots(msymbol(Oh) msize(medium) mcolor(dkgreen))
```

With the dots() option, the small dots are displayed as medium-sized, dark green, hollow circles. See Styles:Symbols (427), Styles:Markersize (425), and Styles:Colors (412) for more information.

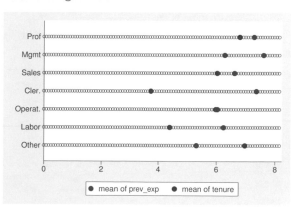

Double-click on, for example, the line associated with *Prof*. In the *Dot line properties* box, change the **Symbol** to Hollow circle, the **Size** to Medium, and the **Color** to Dark Green. *Uses nlsw.dta & scheme vg_s2c*

```
graph dot prev_exp tenure, over(occ7) linetype(line)
    lines(lwidth(thick) lcolor(erose))
```

The linetype(line) option displays lines instead of dots, and the lines() option makes the line width thick and the line color rose. You could also add the lpattern() option to control the line pattern. See Styles:Linewidth (422), Styles:Colors (412), and Styles:Linepatterns (420) for more information.

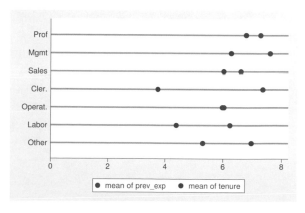

Double-click on, for example, the line associated with *Prof*. In the *Dot line properties* box, change the **Line type** to Line, the **Width** to Thick, and the **Color** to erose. *Uses nlsw.dta & scheme vg_s2c*

Introduction Editor Twoway Matrix Bar Box Dot Pie Options Standard options Styles Appendix

Yvars and over Dot gaps Dot sorting Cat axis Legend Y axis Dotlook options By

```
graph dot prev_exp tenure, over(occ7) linetype(rectangle)
    rwidth(3) rectangles(fcolor(erose) lcolor(maroon))
```

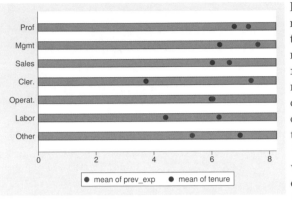

Here we change the `linetype()` to a rectangle. The `rwidth(3)` option sets the rectangle width to three times its normal width. We also use the `rectangle()` option to customize the rectangle, using the `fcolor()` (fill color) and `lcolor()` (line color) options to make the rectangle rose on the inside with a maroon outline.

These modifications can be made with the steps from the previous two examples.

Uses nlsw.dta & scheme vg_s2c

Now I show options that allow you to control the markers and whether the markers are displayed on the same line.

```
graph dot prev_exp tenure, over(occ7)
    marker(1, msymbol(D) mcolor(teal) msize(large))
```

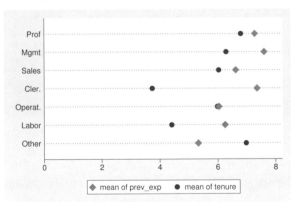

Here we use the `marker()` option to control the marker used for the first y variable, making it a large teal-colored diamond. See Options: Markers (307) for more details on how we can control markers.

Click on any marker for `prev_exp` and, in the Contextual Toolbar, change the **Color** to Teal, the **Size** to Large, and the **Symbol** to Diamond.

Uses nlsw.dta & scheme vg_s2c

```
graph dot prev_exp tenure, over(occ7)
   marker(1, msymbol(d) mfcolor(teal) mlcolor(dkgreen) mlwidth(thick))
   marker(2, msymbol(S) mfcolor(ltblue) mlcolor(blue) mlwidth(thick))
```

In this example, we use two `marker()` options, so we can control both markers. The first marker is now a diamond with a teal fill and a thick, dark green outline. The second marker is a square, light blue on the inside with a thick blue outline. The section Options : Markers (307) has more details on controlling markers.

See the next graph.

Uses nlsw.dta & scheme vg_s2c

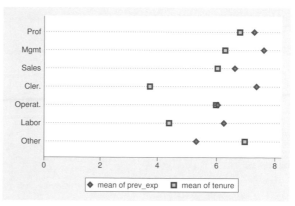

```
graph dot prev_exp tenure, over(occ7)
```

Using the Graph Editor, double-click on a marker for `prev_exp`. In the *Marker properties* dialog box, change the **Symbol** to Diamond and the **Color** to Teal; check **Different outline color**; change the **Outline color** to Dark green and the **Outline width** to Thick; and click on **OK**. Then double-click on a marker for *tenure*. In the *Marker properties* dialog box, change the **Symbol** to Square and the **Color** to Light blue; check **Different outline color**; change the **Outline color** to Blue and the **Outline width** to Thick; and click on **OK**.

Uses nlsw.dta & scheme vg_s2c

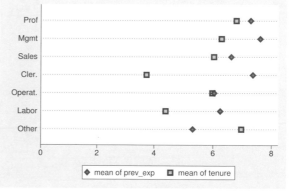

7.8 Graphing by groups

This section discusses the use of the `by()` option in combination with `graph dot`. Normally, you would use the `over()` option instead of the `by()` option, but sometimes the `by()` option is either necessary or more advantageous. For example, a `by()` option is useful if you exceed the maximum number of `over()` options (three if you have one *y* variable or two if you have multiple *y* variables). Thus the `by()` option allows you to break down your data by more categorical variables. `by()` also provides more flexibility in the placement of the separate panels. For more information about the `by()` option, see Options : By (346), and for more information about the `over()` option, see Dot : Yvars and over (263). The examples in this section use the `vg_s1m` scheme.

Introduction Editor Twoway Matrix Bar Box Dot Pie Options Standard options Styles Appendix

Yvars and over Dot gaps Dot sorting Cat axis Legend Y axis Dotlook options By

`graph dot wage, over(collgrad) over(occ5) over(urban2)`

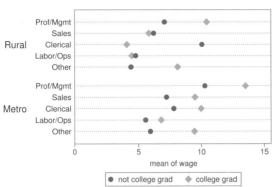

Consider this dot graph breaking down wages by three categorical variables. If we wanted to break this down by another categorical variable, we could not use another `over()` option because we can have only a maximum of three `over()` options with one *y* variable.
Uses nlsw.dta & scheme vg_s1m

`graph dot wage, over(collgrad) over(occ5) over(urban2) by(union)`

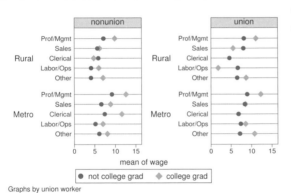

To break down `wage` further by `union`, we can use the by(union) option.
Uses nlsw.dta & scheme vg_s1m

`graph dot tenure ttl_exp, over(occ5) over(urban2)`

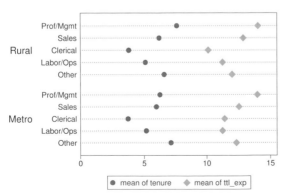

Consider this dot graph with multiple *y* variables, breaking down them by two categorical variables using two `over()` options. When we have multiple *y* variables, we can have a maximum of two `over()` options.
Uses nlsw.dta & scheme vg_s1m

```
graph dot tenure ttl_exp, over(occ5) over(urban2)
    by(union)
```

To further break down these results by
union membership, we can add the
by(union) option. Although this
example shows only one variable in the
by() option, we can specify multiple
variables.
Uses nlsw.dta & scheme vg_s1m

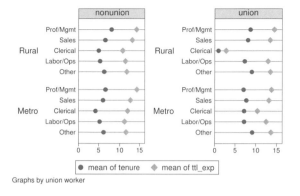

```
graph dot ttl_exp tenure, over(married) over(urban2)
    by(union, missing)
```

We can use the missing option to
include a panel for the missing values of
union. Note that we changed the first
over() variable to be over(married)
to make an example that was more
readable.
Uses nlsw.dta & scheme vg_s1m

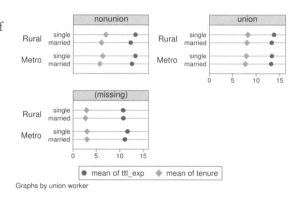

```
graph dot ttl_exp tenure, over(married) over(collgrad)
    by(union, total)
```

We can add the total option to include
a panel for all observations.
Uses nlsw.dta & scheme vg_s1m

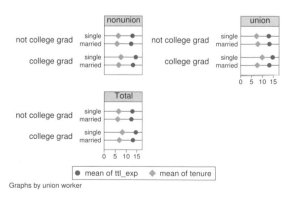

Introduction Editor Twoway Matrix Bar Box Dot Pie Options Standard options Styles Appendix

Yvars and over Dot gaps Dot sorting Cat axis Legend Y axis Dotlook options By

```
graph dot ttl_exp tenure, over(married) over(collgrad)
  by(union, total cols(1))
```

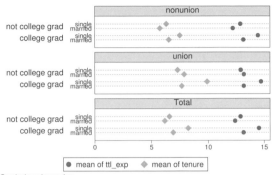

Graphs by union worker

We can use the `cols(1)` option to show the graphs in one column.

▣ Double-click on **Graph** in the Object Browser and, in the *Organization* tab, change the **Columns** to **1**.

Uses nlsw.dta & scheme vg_s1m

8 Pie charts

This chapter discusses the use of pie charts in Stata. The first section illustrates the different ways we can create pie charts in Stata, followed by a section showing how we can sort the slices in our pie charts. Next the chapter shows how we can customize the display of individual slices, as well as control the colors of the pie chart. The next sections demonstrate different ways we can label the pie slices and how we can control the legends for pie charts. Finally, the chapter discusses how to use the by() option.

8.1 Types of pie charts

This section describes different ways to produce pie charts with Stata. Stata allows you to produce pie charts of multiple y variables, with each y variable corresponding to a slice. We can also create a pie chart based on one y variable broken down by one over() variable. Finally, we can create a pie chart with no y variables broken down by an over() variable, which counts the number of observations by each level of the over() variable. For more details, see [G-2] **graph pie**. This section uses the **vg_s1c** scheme.

```
graph pie poplt5 pop5_17 pop18_64 pop65p
```

Here we supply multiple y variables, and each y variable corresponds to a slice in the pie. The first y variable is the population in the state that is younger than 5 years old, the next y variable is the population 5–17 years old, the next y variable is 18–64 years old, and the last y variable is 65 years and older. The entire pie would correspond to the sum of all these variables across all states. The first slice then corresponds to the percentage of the total population that is younger than 5 years old.

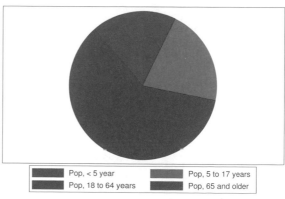

Uses allstates.dta & scheme vg_s1c

`graph pie pop, over(division)`

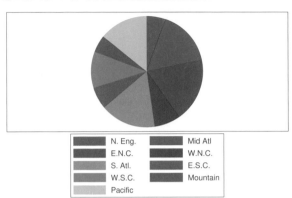

In this example, we supply a *y* variable and an `over()` option. The *y* variable corresponds to the population of the state, the entire pie corresponds to the entire population, and each slice corresponds to the percentage of the population for each level of `division`.
Uses allstates.dta & scheme vg_s1c

`graph pie, over(occ7)`

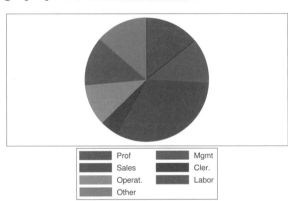

We now switch to the **nlsw** data file. Here we use an `over()` option but no *y* variable (in a sense, the observation itself serves as the *y* variable). This pie chart is much like a visual frequency distribution of **occ7**, where the size of each slice corresponds to the proportion of women in each occupation.
Uses nlsw.dta & scheme vg_s1c

`graph pie, over(union) missing`

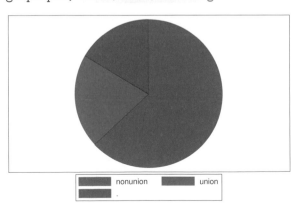

This example shows the proportion of women in union and nonunion jobs. With the `missing` option, another pie slice is added for the observations in which **union** is missing.
Uses nlsw.dta & scheme vg_s1c

8.2 Sorting pie slices

This section describes how we can sort and arrange slices in pie charts. For more details, see [G-2] **graph pie**. This section uses the `vg_lgndc` scheme, which places the legend at the left in one column.

`graph pie, over(occ7)`

Consider this pie chart showing the number of women who work in seven different occupations. The slices are ordered according to the levels of `occ7` from 1 to 7, rotating clockwise, starting with the first slice, which is positioned at 90 degrees.

Uses nlsw.dta & scheme vg_lgndc

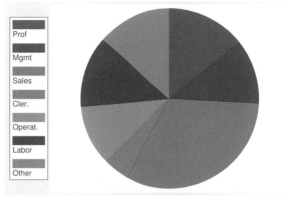

`graph pie, over(occ7) noclockwise`

With the `noclockwise` option, we can display the slices in counterclockwise order.

Uses nlsw.dta & scheme vg_lgndc

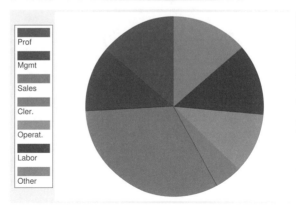

```
graph pie, over(occ7) angle0(0)
```

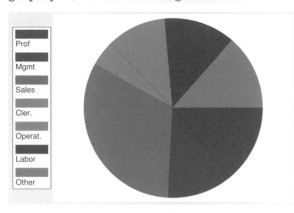

The angle0() option sets the angle of the line that begins the first pie slice. Here we make the first pie slice begin at 0 degrees.

Uses nlsw.dta & scheme vg_lgndc

```
graph pie, over(occ7) sort
```

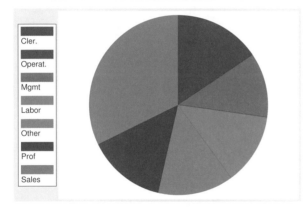

The sort option sorts the slices according to their size, from smallest to largest.

Uses nlsw.dta & scheme vg_lgndc

```
graph pie, over(occ7) sort descending
```

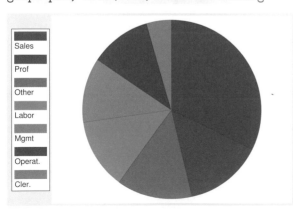

Adding the descending option to the sort option orders the slices from largest to smallest.

Uses nlsw.dta & scheme vg_lgndc

```
graph pie, over(occ7) sort(occ7alpha)
```

Say that we wanted to sort the slices (alphabetically) by occupation name. We have created a new variable, `occ7alpha`, which is a recoded version of `occ7`. It is recoded such that, as `occ7alpha` goes from 1 to 7, the occupations are alphabetized (except for *Other*, which is placed last). We add `sort(occ7alpha)`, and the slices are ordered alphabetically.
Uses nlsw.dta & scheme vg_lgndc

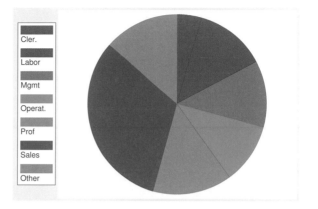

8.3   Changing the look and color and exploding pie slices

This section describes how to change the color of pie slices, explode pie slices, and control the characteristics of lines surrounding the pie slices. For more details, see [G-2] **graph pie**. This section uses the **vg_rose** scheme.

```
graph pie, over(occ7)
```

Consider this pie chart showing the number of women who work in these seven different occupations. The slices are colored using the colors indicated by the scheme. None of the slices are exploded, and no lines surround the slices.
Uses nlsw.dta & scheme vg_rose

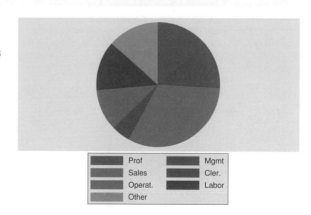

```
graph pie, over(occ7) pie(3, explode)
```

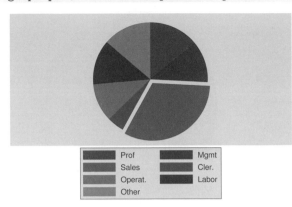

In this example, we use the `pie()` option to explode the third pie slice, calling attention to this slice.

📊 Click on the slice for *Sales* and, in the Contextual Toolbar, change **Explode** to Medium.

Uses nlsw.dta & scheme vg_rose

```
graph pie, over(occ7) pie(3, explode(vlarge) color(cyan))
```

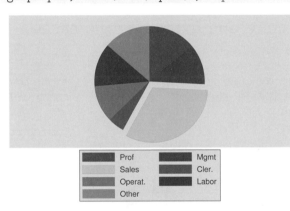

Here we specify `explode(vlarge)` to increase the distance this slice is exploded. We also make the third slice cyan to make it more noticeable. See **Styles : Colors** (412) for other available colors.

📊 Assuming that we are still pointing to this slice, in the Contextual Toolbar, change **Explode** to v Large and **Color** to Cyan.

Uses nlsw.dta & scheme vg_rose

```
graph pie, over(occ7) pie(3, explode(vlarge) color(cyan))
    pie(1, explode(small) color(gold))
```

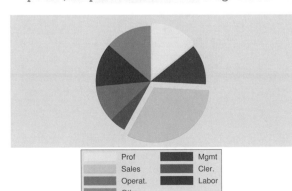

We can use the `pie()` option repeatedly. Here we change the color and explode slices 1 and 3.

📊 Continuing from the previous graph, point to the *Prof* slice and, in the Contextual Toolbar, change **Explode** to Small and **Color** to Gold.

Uses nlsw.dta & scheme vg_rose

```
graph pie, over(occ7) line(lcolor(sienna) lwidth(thick))
```

We can use the `line()` option to
change the characteristics of the lines
surrounding the pie slices. Here we add
the `lcolor()` (line color) and `lwidth()`
(line width) options to make the line
sienna and thick.

Uses nlsw.dta & scheme vg_rose

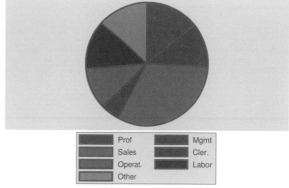

8.4   Slice labels

This section describes how we can label the pie slices. For more details, see [G-2] **graph
pie**. This section uses the `economist` scheme.

```
graph pie, over(occ7) plabel(_all sum)
```

Consider this pie chart showing the
number of women who work in seven
different occupations. Here we use
the `plabel()` (pie label) option to
label all slices with the sum, i.e., the
frequency of women who work in each
occupation.

Uses nlsw.dta & scheme economist

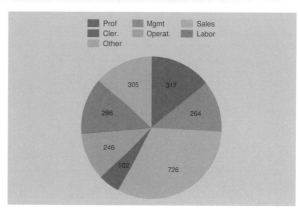

`graph pie, over(occ7) plabel(_all percent)`

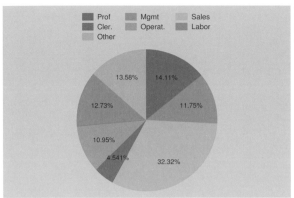

By using the `percent` option, we can show the percentage of women who work in each occupation.

Uses nlsw.dta & scheme economist

`graph pie, over(occ7) plabel(_all name)`

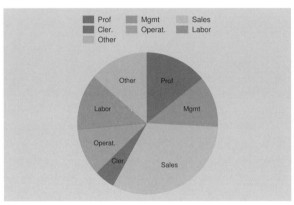

The `name` option adds a label, here, the name of the occupation.

Uses nlsw.dta & scheme economist

`graph pie, over(occ7) plabel(_all name) legend(off)`

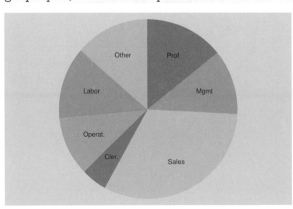

When we use the `name` option, the legend is not necessary, and we can suppress it by using the `legend(off)` option.

In the Object Browser, right-click on **legend** and select **Hide**.

Uses nlsw.dta & scheme economist

```
graph pie, over(occ7) plabel(1 "Prof=14.11%")
   plabel(3 "Sales=32.32%")
```

We can also use the `plabel()` option to put text into all slices or individual slices. Here we add text to the first and third slices.

Uses nlsw.dta & scheme economist

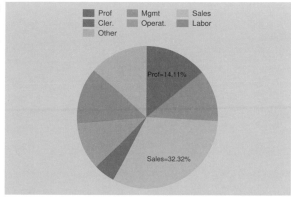

```
graph pie, over(occ7) plabel(_all percent, format("%2.0f"))
```

When we use `plabel()` to label slices with a `sum` or `percent`, we can use the `format()` option to control the format of the numeric values displayed. Here we display the percentages as whole numbers.

Uses nlsw.dta & scheme economist

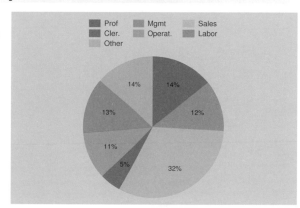

```
graph pie, over(occ7) plabel(_all percent, gap(-5))
```

We can use the `gap()` option to adjust the position of the label with respect to the center of the pie. A positive number pushes the label away from the center of the pie, and a negative value pushes the label closer to the center of the pie.

Uses nlsw.dta & scheme economist

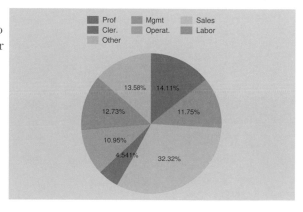

```
graph pie, over(occ7) plabel(_all percent, size(large) color(maroon))
```

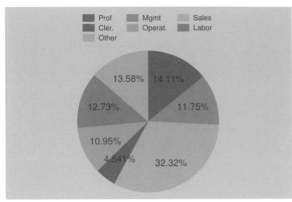

We can use textbox options to modify the display of the text labeling the pie slices. Here we increase the size of the text and change its color to maroon. See Options : Textboxes (379) for more options you can use.

📈 Click on any of the labels and, in the Contextual Toolbar, change the **Size** to Large and the **Color** to Maroon.

Uses nlsw.dta & scheme economist

```
graph pie, over(occ7) plabel(_all name, gap(-5))
     plabel(1 "32%", gap(5)) legend(off)
```

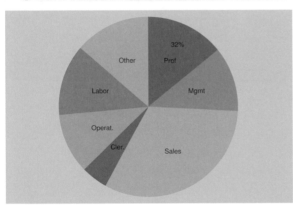

We can include multiple plabel() options. In this example, the first plabel() option assigns the occupation names to all the slices and moves the names 5 units inward. The second plabel() option assigns text to the second slice and displays it 5 more units from the center. Because the legend was not needed, we suppressed it with the legend(off) option.

Uses nlsw.dta & scheme economist

```
graph pie, over(occ7) plabel(_all name, gap(-5))
   plabel(_all percent, gap(5) format("%2.0f"))
   legend(off)
```

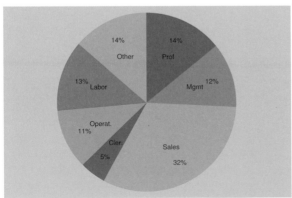

Here we use the plabel() option twice to label the slices with the occupation name and with the percentage. We use the gap() option to move the names closer to the center by 5 units and move the percentage 5 extra units from the center.

Uses nlsw.dta & scheme economist

```
graph pie, over(occ7) ptext(0 30 "This is some text")
```

We can use the `ptext()` (pie text)
option to add text to the pie chart. We
use polar coordinates to determine the
location of the text by specifying the
angle and distance from the center.
Here the angle is 0, and the distance
from the center is 30.

🖾 Select the Add Text **T** tool and
click where the text should
begin. Enter the desired text and click
on **OK**.

Uses nlsw.dta & scheme economist

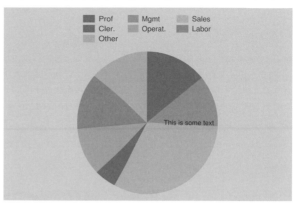

```
graph pie, over(occ7) ptext(-10 10 "This is some text")
```

Here we choose an angle of −10
(putting it 10 degrees below 0) and a
distance of 10 units from the center.
The angle determines only the position
of the text but not its actual angle of
display, which is controlled with the
`orientation()` option. See the next
example for more details.

🖾 Select the Add Text **T** tool and
click where the text should
begin. Enter the desired text and click
on **OK**.

Uses nlsw.dta & scheme economist

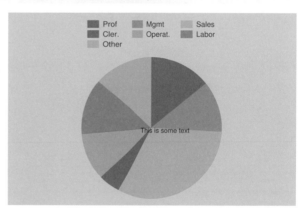

```
graph pie, over(occ7)
   ptext(-10 10 "This is some text", orientation(rvertical)
   placement(s))
```

Now we make the text reverse vertical
and start the text from the south of the
given coordinates. For more
information on these kinds of textbox
options, see Options : Textboxes (379).

🖾 Double-click on the text and, in the
Format tab, change the **Orientation**
to Rev. Vertical and the **Placement**
to South.

Uses nlsw.dta & scheme economist

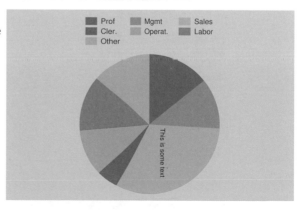

Introduction Editor Twoway Matrix Bar Box Dot Pie Options Standard options Styles Appendix

Types of pie charts Sorting Colors and exploding Labels Legend By

8.5   Controlling legends

This section illustrates some of the options that we can use to control the display of legends with pie charts. Although this section illustrates the use of legends, it emphasizes options that may be particularly useful with pie charts. See Options : Legend (361) for more details about legends; those details apply well to pie charts, even if the examples use other kinds of graphs. Also, see [G-3] *legend_options* for more details. This section uses the vg_brite scheme.

```
graph pie, over(occ7)
```

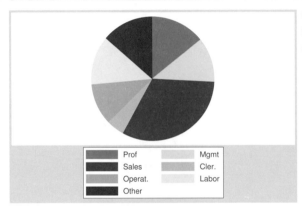

Consider this pie chart showing the frequencies of women in seven occupational categories.
Uses nlsw.dta & scheme vg_brite

```
graph pie, over(occ7) legend(label(1 "Professional"))
```

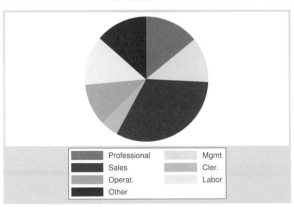

We use the legend(label()) option to change the label for the first occupation.

🖱 Click on the word *Prof* within the legend and, in the Contextual Toolbar, change the **Text** to Professional.
Uses nlsw.dta & scheme vg_brite

```
graph pie, over(occ7) legend(title(Occupation))
```

The `title()` option within the
`legend()` option adds a title to the
legend. We can also use `subtitle()`,
`note()`, and `caption()` options, much
as we would for adding titles to a
graph; see Standard options : Titles (395)
for more details.

In the Object Browser, double-click
on **Legend**, select the *Titles* tab, and
enter `Occupation` for the **Title**.
Uses nlsw.dta & scheme vg_brite

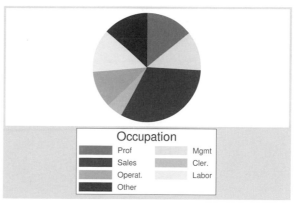

```
graph pie, over(occ7) legend(title(Occupation, position(6)))
```

We can use the `position()` option
within the `title()` option to control
the position of the title. Here we put
the title in the six o'clock position,
placing it at the bottom of the legend.

Click on the Grid Edit tool and
select *Occupation* and drag it to the
bottom of the legend.
Uses nlsw.dta & scheme vg_brite

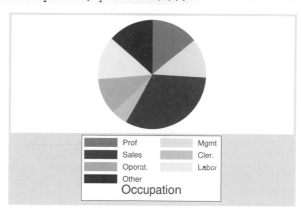

```
graph pie, over(occ7) legend(colfirst)
```

We can use the `legend(colfirst)`
option to order the items in the legend
by columns instead of rows.

In the Object Browser, double-click
on **legend**. Then change the **Key
sequence** to Down **first**.
Uses nlsw.dta & scheme vg_brite

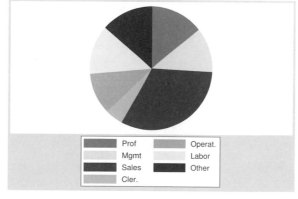

Introduction   Editor   Twoway   Matrix   Bar   Box   Dot   Pie   Options   Standard options   Styles   Appendix

Types of pie charts   Sorting   Colors and exploding   Labels   Legend   By

```
graph pie, over(occ7) legend(colfirst order(7 6 5 1 2 3 4) holes(1))
```

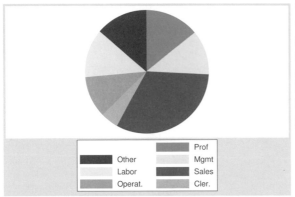

The pie wedges rotate clockwise; here we make the items within the legend rotate in a similar clockwise fashion, starting from the top right. The order() option puts the items in the legend in a clockwise order, and the holes(1) option leaves the first position empty.

Uses nlsw.dta & scheme vg_brite

```
graph pie, over(occ7) legend(position(9))
```

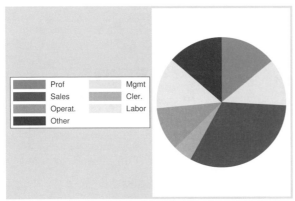

Here we put the legend at the nine o'clock position, placing it at the left of the chart.

📊 Click on the Grid Edit 🔧 tool and then select the entire legend by clicking on **legend** in the Object Browser. Then click on the legend and drag it to the far left of the graph. Finally, in the Object Browser, double-click on **legend** and, in the *Advanced* tab, change the **Alignment** to Center.

Uses nlsw.dta & scheme vg_brite

```
graph pie, over(occ7) legend(position(9) cols(1) stack)
```

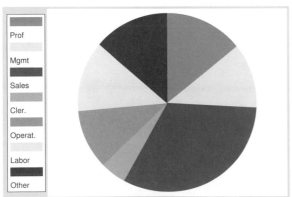

Adding the cols(1) and stack options makes the legend display in one column with the symbol stacked above the descriptive text.

📊 In the Object Browser, double-click on **legend**. Change the **Columns** to 1 and the **Stack symbols and text** to Yes.

Uses nlsw.dta & scheme vg_brite

```
graph pie, over(occ7)
```

Here we use the **vg_lgndc** scheme. This
scheme places the legend to the left in
one column with the symbol stacked
above the description.
Uses nlsw.dta & scheme vg_lgndc

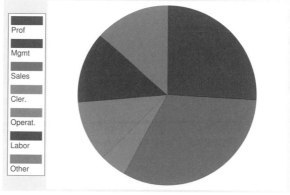

8.6   Graphing by groups

This section describes the use of the by() option with pie charts, focusing on features
that are specifically relevant to pie charts. For more details, see Options: By (346) and
[G-3] ***by_option***.

```
graph pie, over(occ7)
```

Here we see a basic pie chart showing
the distribution of occupations.
Uses nlsw.dta & scheme vg_s2c

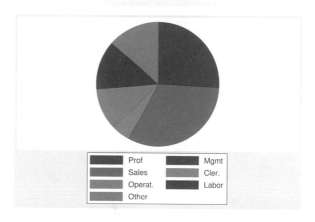

```
graph pie, over(occ7) by(union)
```

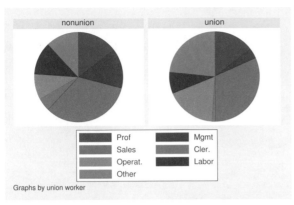

In this graph, the occupations are broken down by whether one belongs to a union.

Uses nlsw.dta & scheme vg_s2c

```
graph pie, over(occ7) by(union) pie(2, explode)
```

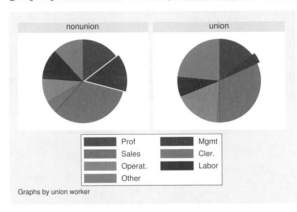

If we add the `pie(2, explode)` option, the second slice is exploded in both graphs.

Double-click on the second slice (from either graph) and check **Explode Slice**.

Uses nlsw.dta & scheme vg_s2c

```
graph pie, over(occ7) by(union) sort
```

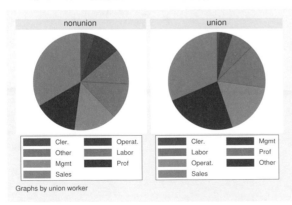

Here we sort the slices from least frequent to most frequent. Separate legends are shown for each chart. This is because we can order the slices differently in the two different graphs when sorted. Thus, when we use the `sort` option for pie charts, Stata shows two separate legends to assure proper labeling of the slices.

Uses nlsw.dta & scheme vg_s2c

```
graph pie, over(occ7) by(union, legend(off))
    plabel(_all name)
```

In this example, we add the `plabel()`
option to label the inside of each slice
with the name of the slice, so the
legend is no longer needed. We
suppress the legend with the
`legend(off)` option, which is placed
within the `by()` option because it, in a
way, determines the placement of the
legend by turning it off.

Uses nlsw.dta & scheme vg_s2c

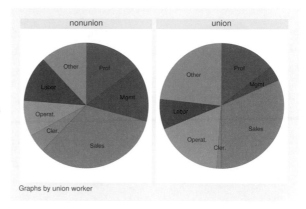

```
graph pie, over(occ7) by(union, legend(pos(9))) legend(cols(1) stack)
```

Here we place the legend to the left by
using the `legend(pos(9))` option. This
option is contained within the `by()`
option because it alters the position of
the legend. We also make the legend
one column with the legend symbols
and labels stacked with the
`legend(cols(1) stack)` option. This
option is outside the `by()` option
because it does not determine the
position of the legend.

See the second to last example in
the previous section.

Uses nlsw.dta & scheme vg_s2c

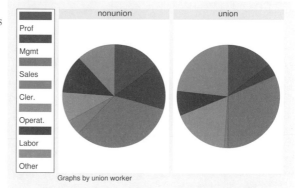

```
graph pie, over(occ7) by(union) legend(pos(3) cols(1) stack) sort
```

This example is similar to the previous
one, but we have added the `sort`
option. When we add the `sort` option,
we need to move the `pos()` option from
within the `by()` option to outside the
`by()` option. This is an exception to
the general rule that legend options
that control the position of the legend
are placed within the `by()` option. Here
we get the legends that we desire, each
to the right of the pie.

Uses nlsw.dta & scheme vg_s2c

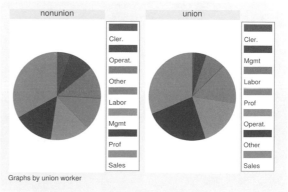

```
graph pie, over(occ7) by(urban3, legend(at(4)))
```

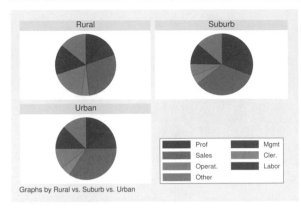

Graphs by Rural vs. Suburb vs. Urban

Here we break down occupation by a three-level variable, leaving a fourth position open. We can specify the `legend(at(4))` option within the `by()` option to place the legend in the space in the fourth position.

⊞ Select the Grid Edit ⊞ tool and then, in the Object Browser, click on **legend**. Drag the legend to the center of the plot region. Then select the Pointer ↖ and drag the legend to the desired location.

Uses nlsw.dta & scheme vg_s2c

Introduction Editor Twoway Matrix Bar Box Dot Pie Options Standard options Styles Appendix

Markers Marker labels Connecting Axis titles Axis labels Axis scales Axis selection By Legend Adding text Textboxes Text display

9 Options available for most graphs

This chapter discusses options that are used in many, but not all, kinds of graphs in Stata, as compared with the Standard options (395) chapter, which covers options that are standard in all Stata graphs. This chapter goes into greater detail about how to use these options to customize your graphs. As you can see from the *Visual Table of Contents* at the right, this chapter covers markers, how to connect points and markers, axis titles, axis labels, axis scales, axis selection, the by() option, legends, added text, textboxes, and options for controlling the display of text. For further details, the examples will frequently refer to Styles (411) and [G-2] **graph**.

9.1 Changing the look of markers

This section looks at options for controlling markers. Although the examples in this section focus on twoway scatter, these options apply to any graph where you have markers and can control them. This section will show how to change the marker symbol, size, and color (both fill and outline color). For more information, see [G-3] ***marker_options***. The following graphs use the vg_s2c scheme.

`twoway scatter ownhome borninstate`

Consider this scatterplot showing the relationship between the percentage of people in a state who own their home and the percentage of people born in their state of residence. The markers in this plot are filled circles.

Uses allstates.dta & scheme vg_s2c

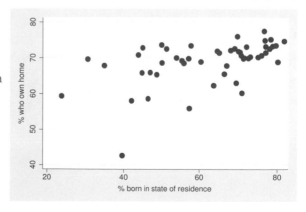

`twoway scatter ownhome borninstate, msymbol(S)`

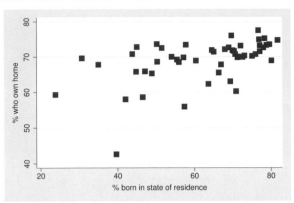

We can control the shape of the marker with the `msymbol()` (marker symbol) option. Here we make the symbols large squares.

 Click on any marker and, in the Contextual Toolbar, change the **Symbol** to `Square`.

Uses allstates.dta & scheme vg_s2c

`twoway scatter ownhome borninstate, msymbol(s)`

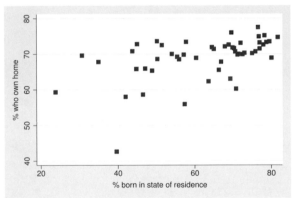

Specifying `msymbol(s)`, which uses a lowercase `s`, displays smaller squares.

 Click on any marker and, in the Contextual Toolbar, change the **Symbol** to `Small square`.

Uses allstates.dta & scheme vg_s2c

`twoway scatter ownhome borninstate, msymbol(sh)`

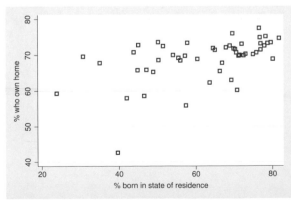

We can append an `h` (i.e., `msymbol(sh)`) to yield small, hollow squares. In addition to choosing `S` for larger squares and `s` for small squares, we can specify `D` (large diamond), `T` (large triangle), and `O` (large circles). We can specify lowercase letters to get smaller versions of these symbols and append `h` for hollow versions.

 Click on any marker and, in the Contextual Toolbar, change the **Symbol** to `Hollow small square`.

Uses allstates.dta & scheme vg_s2c

`twoway scatter ownhome borninstate, msymbol(X)`

We can also specify `msymbol(X)` to use a large X for the markers. A lowercase x could be used for smaller markers. We cannot append an `h` because we cannot make a hollow X.

📖 Click on any marker and, in the Contextual Toolbar, change the **Symbol** to `Large X`.

Uses allstates.dta & scheme vg_s2c

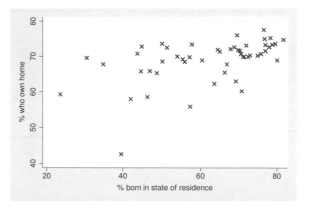

`twoway scatter ownhome borninstate, msymbol(+)`

Specifying `msymbol(+)` yields a plus sign for the markers. We cannot make the plus signs hollow or make a smaller version.

📖 Click on any marker and, in the Contextual Toolbar, change the **Symbol** to `Plus`.

Uses allstates.dta & scheme vg_s2c

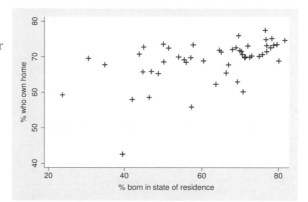

`twoway scatter heatdd cooldd, msymbol(p)`

Here we switch to the `citytemp` data file to illustrate the use of the `msymbol(p)` option to plot small points. Although each point is hard to see because it is so small, we can see the overall pattern of the data because of the many points and the strong trend in the data. See **Styles : Symbols** (427) for more information about symbols.

📖 Click on any marker and, in the Contextual Toolbar, change the **Symbol** to `Point`.

Uses citytemp.dta & scheme vg_s2c

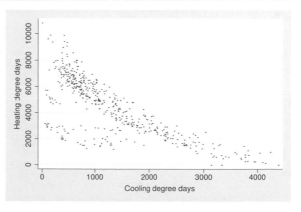

`twoway scatter ownhome propval100 borninstate`

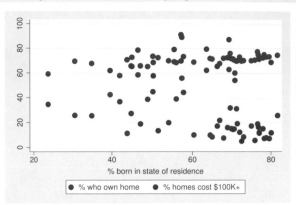

Aside from aesthetics, choosing different marker symbols is useful to differentiate multiple markers displayed in the same plot. In this example, we plot two *y* variables, and Stata displays both as solid circles, differing in color.
Uses allstates.dta & scheme vg_s2c

`twoway scatter ownhome propval100 borninstate,`
`   msymbol(t Oh)`

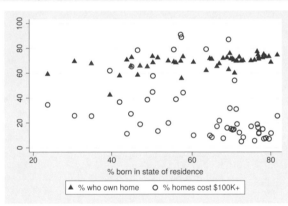

To further differentiate the symbols, we add the `msymbol(t Oh)` option to control both markers. Here we make the first marker a small triangle and the second a larger hollow circle.

Click on the blue marker (i.e., `ownhome`) and, in the Contextual Toolbar, change the **Symbol** to `Triangle`. Click on the red marker (i.e., `propval100`) and change the **Symbol** to `Hollow circle`.
Uses allstates.dta & scheme vg_s2c

`twoway scatter ownhome propval100 borninstate,`
`   msymbol(. Oh)`

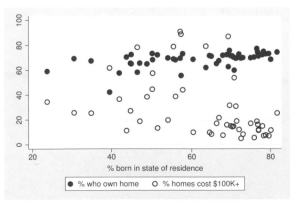

By using the `msymbol(. Oh)` option, we can leave the first symbol unchanged (as indicated by the dot) and change the second symbol to a hollow circle. We might think that the dot indicates a small point but that is indicated by the `p` option.

Click on the red marker (i.e., `propval100`) and, in the Contextual Toolbar, change the **Symbol** to `Hollow circle`.
Uses allstates.dta & scheme vg_s2c

```
twoway scatter ownhome borninstate, mlabel(stateab) mlabpos(center)
```

One last marker symbol is `i` for invisible, allowing us to hide the marker symbol. Here we use the `mlabel(stateab)` (marker label) option to display a marker label with the state abbreviation for each observation and the `mlabpos(center)` (marker label position) option to center the marker label. However, the marker symbol (the circle) and the marker label (the abbreviation) are on top of each other. *Uses allstates.dta & scheme vg_s2c*

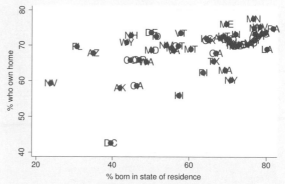

```
twoway scatter ownhome borninstate, mlabel(stateab) mlabpos(center)
    msymbol(i)
```

If we use `msymbol(i)` to make the marker symbol invisible, the marker label (the state abbreviation) can be displayed without being obscured by the marker symbol. See Styles : Symbols (427) for more information about symbols.

📈 See the next graph.
Uses allstates.dta & scheme vg_s2c

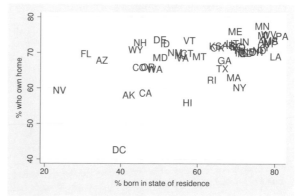

```
twoway scatter ownhome borninstate, mlabel(stateab)
```

📈 As an alternative to the previous graph command, you can make the same changes using the Graph Editor. Click on any marker (i.e., any dot) and, in the Contextual Toolbar, change the **Symbol** to none. Next click on any label (e.g., NV) and change the **Position** to Middle.
Uses allstates.dta & scheme vg_s2c

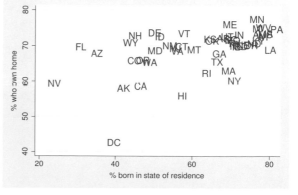

Introduction Editor Twoway Matrix Bar Box Dot Pie Options Standard options Styles Appendix

Markers Marker labels Connecting Axis titles Axis labels Axis scales Axis selection By Legend Adding text Textboxes Text display

So far, we have seen that we can use the `msymbol()` option to control the marker symbol and, to a certain extent, to control the marker size (e.g., using `O` yields large circles and using `o` yields small circles). As the following examples show, we can use the `msize()` option to exert more flexible control over the size of the markers. The following examples will use the `vg_s1m` scheme.

`twoway scatter ownhome borninstate, msymbol(+) msize(small)`

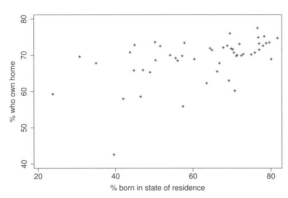

Here we use the `msize()` (marker size) option to control the size of the marker symbol, making the marker symbol small. Values we could have chosen include `vtiny`, `tiny`, `vsmall`, `small`, `medsmall`, `medium`, `medlarge`, `large`, `vlarge`, `huge`, `vhuge`, and `ehuge`.

📊 Click on a marker and, in the Contextual Toolbar, change the **Size** to `Small`.

Uses allstates.dta & scheme vg_s1m

`twoway scatter ownhome borninstate, msymbol(Oh) msize(*2)`

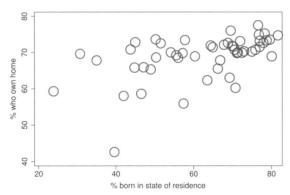

We can specify the sizes as multiples of the original size of the marker. In this example, we make the markers twice their original size by specifying `msize(*2)`. Specifying a value less than one reduces the marker size; e.g., `msize(*.5)` would make the marker half its normal size. See Styles : Markersize (425) for more details.

📊 Double-click on a marker and, next to **Size**, type in `*2`.

Uses allstates.dta & scheme vg_s1m

```
twoway scatter ownhome borninstate [aweight=propval100],
    msymbol(oh)
```

Stata even allows us to size the symbols
on the values of another variable in
your data file. This allows us, in a
sense, to graph three variables at once.
Here we look at the relationship
between `borninstate` and `ownhome` and
then size the markers on `propval100`
by using `[aweight=propval100]`,
weighting the markers by `propval100`.
Uses allstates.dta & scheme vg_s1m

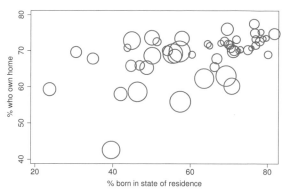

```
twoway scatter ownhome borninstate [aweight=propval100],
    msymbol(oh) msize(small)
```

Even if we weight the size of the
markers by using `aweight`, we can still
control the general size of the markers.
Here we make all markers smaller by
using the `msize(small)` option. The
markers are smaller than they were
previously but are still sized according
to the value of `propval100`.

Click on a marker and, in the
Contextual Toolbar, change the **Size** to
Small.
Uses allstates.dta & scheme vg_s1m

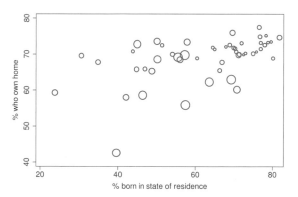

```
twoway scatter ownhome borninstate [aweight=propval100],
    msymbol(oh) msize(large) mlabel(stateab)
```

We might even try to graph a fourth
variable in the plot by using the
`mlabel()` (marker label) option. Here
we try to use the `mlabel(stateab)`
option to label each marker with the
abbreviation of the state. However,
when we add the `mlabel()` option, the
weights no longer affect the size of the
markers. See the following example for
a solution to this.
Uses allstates.dta & scheme vg_s1m

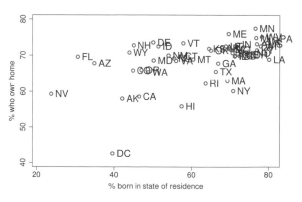

```
twoway (scatter ownhome borninstate [aweight=propval100],
    msymbol(oh) msize(large))
    (scatter ownhome borninstate, mlabel(stateab) msymbol(i) mlabpos(center))
```

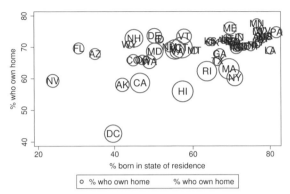

We can solve the problem from the previous example by overlaying a scatterplot that has the symbols weighted by `propval100` with a scatterplot that shows just the marker labels. The second scatterplot uses the `mlabel(stateab) msymbol(i) mlabpos(center)` options to label the markers with the state abbreviation. See Options : Marker labels (320) for more details.

Uses allstates.dta & scheme vg_s1m

Stata also allows us to control the color of the markers. We can control the overall color of the marker: create a solid color or make the inner part of the marker one color (called a fill color) and the outline of the marker a different color. We can also vary the thickness of the outline of the marker. These next examples use the `vg_rose` scheme.

```
twoway scatter ownhome borninstate, mcolor(navy)
```

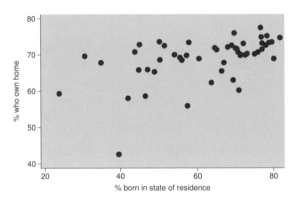

We can use the `mcolor()` (marker color) option to control the color of the markers. Here we make the markers navy blue by using the `mcolor(navy)` option. See Styles : Colors (412) for more information about specifying colors.

📈 Click on a marker and use the Contextual Toolbar to change the **Color** to Navy.

Uses allstates.dta & scheme vg_rose

```
twoway scatter ownhome borninstate, mfcolor(ltblue) mlcolor(navy)
```

We can separately control the fill color (inside color) and outline color with the `mfcolor()` (marker fill color) and `mlcolor()` (marker line color) options, respectively. Here we make the fill color light blue by specifying `mfcolor(ltblue)` and the line color navy by specifying `mlcolor(navy)`.

🖼 Double-click on a marker and change the **Color** to `Light Blue`. Then check **Different Outline Color** and change the **Outline color** to `Navy`.

Uses allstates.dta & scheme vg_rose

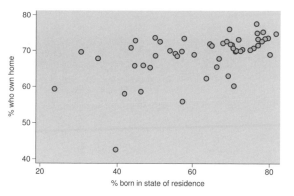

```
twoway scatter ownhome borninstate, mlcolor(black)
```

We can change the line color surrounding the marker with the `mlcolor()` option. Here we specify `mlcolor(black)` to make the line surrounding the markers black.

🖼 Double-click on a marker. Check **Different Outline Color** and change the **Outline color** to `Black`.

Uses allstates.dta & scheme vg_rose

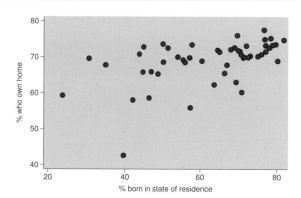

```
twoway scatter ownhome borninstate, mfcolor(ltblue)
```

We can also separately control the fill color using the `mfcolor()` option. Here we choose `mfcolor(ltblue)` to make the fill color light blue.

🖼 Double-click on a marker and check **Different Outline Color**. Then change the **Color** to `Light Blue`.

Uses allstates.dta & scheme vg_rose

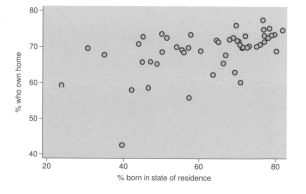

`twoway scatter ownhome borninstate, mfcolor(ltblue)`
`    mlwidth(thick)`

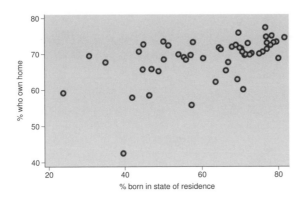

Adding the `mlwidth(thick)` (marker line width) option makes the line around the marker thick. We can also indicate the thickness as a multiple of the original thickness; e.g., `mlwidth(*3)` indicates the line should be three times as thick as it would normally be. See Styles : Linewidth (422) for more details.

Double-click on a marker and change the **Outline width** to `Thick`.
Uses allstates.dta & scheme vg_rose

`twoway scatter ownhome borninstate, mlwidth(medthick)`

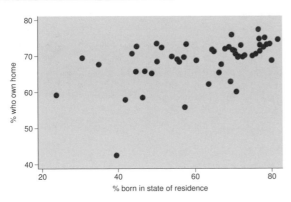

If we do not specify a different color for the line that outlines the marker (e.g., via the `mlcolor()` option), we may not see any effect in specifying the `mlwidth()` option. This is because the color of the line surrounding the marker is the same as the fill color, so we cannot see the effect of modifying the width of the line surrounding the marker, as illustrated here.
Uses allstates.dta & scheme vg_rose

So far, these examples have focused on controlling the individual elements of markers: the symbol, color, size, fill color, line color, and so forth. There is another way to change the appearance of a marker: by specifying a marker style. The marker style controls all these attributes at once, and sometimes it can be more efficient to use a marker style to control the elements individually, as we will see in the following examples. The next examples use the `vg_s2m` scheme.

`twoway scatter ownhome borninstate`

The marker styles are named/numbered **p1–p15**. The markers here are displayed using the **p1** style because we are plotting only one *y* variable and have not specified a marker style.
Uses allstates.dta & scheme vg_s2m

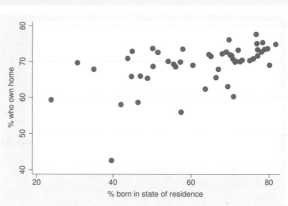

`twoway scatter ownhome borninstate, mstyle(p1)`

Here we explicitly select the default marker style by using the **mstyle(p1)** (marker style) option, and the markers look identical to the previous graph.
Uses allstates.dta & scheme vg_s2m

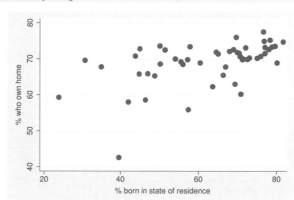

`twoway scatter ownhome borninstate, mstyle(p2)`

Here we use **mstyle(p2)** to explicitly select the **p2** style for displaying the markers, and now the markers are different in size, shape, and color. The markers are now larger diamonds that are a middle-level gray color.
Uses allstates.dta & scheme vg_s2m

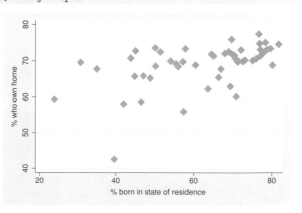

Introduction Editor Twoway Matrix Bar Box Dot Pie Options Standard options Styles Appendix

Markers Marker labels Connecting Axis titles Axis labels Axis scales Axis selection By Legend Adding text Textboxes Text display

`twoway scatter ownhome propval100 borninstate`

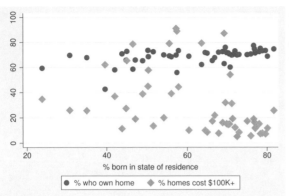

Now, when we plot two variables, the first variable is plotted with the `p1` style and the second variable is plotted with the `p2` style. We would have gotten the same result if we had specified the option `mstyle(p1 p2)`.
Uses allstates.dta & scheme vg_s2m

`twoway scatter ownhome propval100 borninstate, mstyle(p1 p10)`

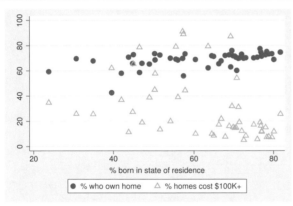

Here we use the `mstyle(p1 p10)` option to request that the first variable be plotted with the `p1` style and the second be plotted with the `p10` style. A style is just a starting point, and we can use more options to modify the markers to suit our taste.
Uses allstates.dta & scheme vg_s2m

`twoway scatter ownhome propval100 borninstate, mstyle(p1 p10)`
`  msize(. medium)`

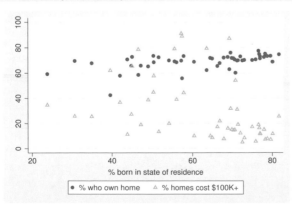

Say that in the previous graph you wanted medium-sized triangles. We can add the `msize(. medium)` option to control the size of the second marker, leaving the first unchanged. So, even though a style chooses a number of characteristics for the markers, we can override them.
 Click on any of the triangular markers and, in the Contextual Toolbar, change the **Size** to `Medium`.
Uses allstates.dta & scheme vg_s2m

```
twoway scatter ownhome propval100 borninstate, mstyle(p1 p1)
   mfcolor(. white)
```

In this example, we use the p1 style for both the first and second markers, creating small, dark gray, filled circles. Adding the mfcolor(. white) option makes the fill color as-is for the first variable and white for the second variable.

▨ Double-click on the second marker (e.g., in the legend) and change the **Color** to White and check **Different outline color**.

Uses allstates.dta & scheme vg_s2m

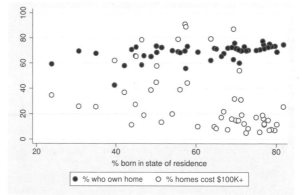

```
twoway scatter ownhome propval100 borninstate, msymbol(. Sh)
   scheme(vg_samem)
```

Another strategy for controlling the marker symbols is choosing or creating a scheme. The vg_samem scheme makes all markers solid, dark gray circles, allowing you to customize them all from a common base. Adding the msymbol(. Sh) option makes the second symbol hollow squares.

▨ Click on the second marker (e.g., in the legend) and, in the Contextual Toolbar, change the **Symbol** to Hollow square.

Uses allstates.dta & scheme vg_samem

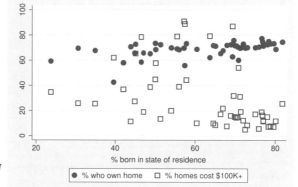

```
twoway scatter ownhome propval100 borninstate, scheme(vg_outm)
```

Say that you wanted the markers to be displayed as outlines filled with white. Rather than specifying the mfcolor() option, you could use the vg_outm scheme, as shown here. Even if you overlaid multiple commands, using this scheme would display the markers, by default, as white-filled outlines.

Uses allstates.dta & scheme vg_outm

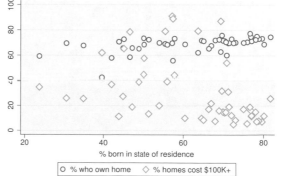

Introduction Editor Twoway Matrix Bar Box Dot Pie Options Standard options Styles Appendix

Markers Marker labels Connecting Axis titles Axis labels Axis scales Axis selection By Legend Adding text Textboxes Text display

9.2   Creating and controlling marker labels

This section looks at the details of using marker labels. We can use marker labels to identify the markers with `graph twoway`, but we can also use them with other types of graphs, such as `graph matrix` and `graph box`, affecting the outside values. We can even use marker labels instead of markers. For more information, see [G-3] *marker_label_options*. This section will use the `vg_s2c` scheme and the `allstates3` file, which keeps the states that are in the South, i.e., if `region` is equal to 3.

`twoway scatter ownhome borninstate, mlabel(stateab)`

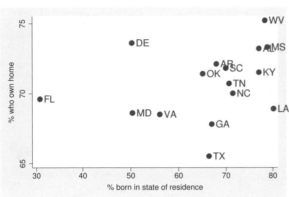

Consider this scatterplot showing the relationship between the percentage of people in a state who own their home and the percentage of people born in their state of residence. We might want to be able to identify some of the observations, and we can use the `mlabel()` (marker label) option to label the markers with the two-letter abbreviation of the state.

Uses allstates3.dta & scheme vg_s2c

`twoway scatter ownhome borninstate, mlabel(stateab) mlabpos(12)`

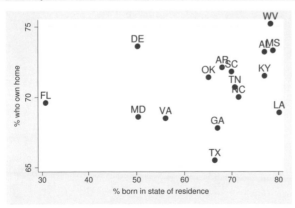

We can use the `mlabpos()` (marker label position) option to give the marker labels a different position. Here we place the marker labels in the twelve o'clock position (above the markers).

Click on one of the marker labels (e.g., FL) and, using the Contextual Toolbar, change the **Position** to 12 o'clock. The next two examples show how we can deal with the marker labels that overlap.

Uses allstates3.dta & scheme vg_s2c

`twoway scatter ownhome borninstate, mlabel(stateab) mlabvpos(pos)`

The `mlabvpos()` (marker label variable position) option allows us to assign a different marker label position for each observation using a variable in the data file. The variable `pos` has a value of 3, placing the markers in the three o'clock position, except for states AL, MS, and LA, where `pos` is 9, 12, and 6, respectively.

🔖 See the next graph.

Uses allstates3.dta & scheme vg_s2c

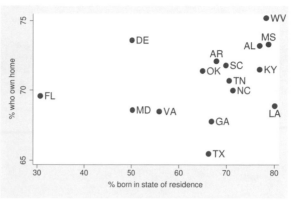

`twoway scatter ownhome borninstate, mlabel(stateab)`

🔖 The Graph Editor provides a much simpler way of changing the positions of the marker labels. Right-click on AL in the graph, choose **observation properties**, and change the **position** to 9 o'clock. Then right-click on MS, select **observation properties**, and change the **position** to 12 o'clock. Finally, right-click on LA, choose **observation properties**, and change the **position** to 6 o'clock.

Uses allstates3.dta & scheme vg_s2c

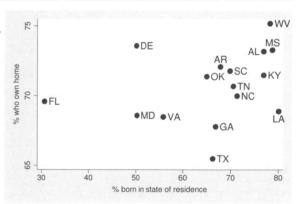

`twoway scatter ownhome borninstate, mlabel(stateab) mlabsize(vsmall)`

The `mlabsize()` (marker label size) option controls the size of the marker labels. Here we make the marker labels very small. Some other sizes you could choose include `small`, `medsmall`, `medium`, `medlarge`, `large`, and `vlarge`; see Styles:Textsize (428) for more options.

🔖 Click on a marker label (e.g., DE) and, in the Contextual Toolbar, change the **Size** to v Small.

Uses allstates3.dta & scheme vg_s2c

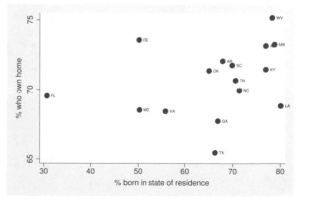

`twoway scatter ownhome borninstate, mlabel(stateab) mlabsize(*.6)`

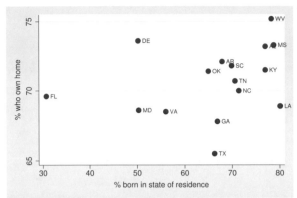

We can also specify the `mlabsize()` as a relative size, a multiple of the original size. Here the labels are .6 times their normal size.

Double-click on a marker label (e.g., DE) and, in the space after **Size**, type in the value `*.6`.

Uses allstates3.dta & scheme vg_s2c

`twoway scatter ownhome borninstate, mlabel(stateab) mlabangle(45)`

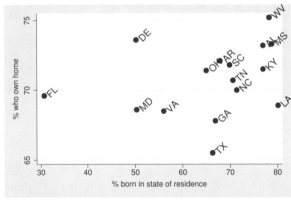

The `mlabangle()` (marker label angle) option controls the angle of the marker label. For example, 0 degrees indicates horizontal text, 90 degrees vertical text, 180 degrees reverse horizontal text, and 270 degrees reverse vertical text. We can also specify negative degrees (for example, −90 degrees is the same as 270 degrees). See Styles : Angles (411) for more details.

Double-click on a marker label (e.g., DE) and change the **Angle** to 45.

Uses allstates3.dta & scheme vg_s2c

`twoway scatter ownhome borninstate, mlabel(stateab) mlabcolor(red)`

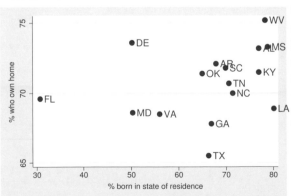

The `mlabcolor()` (marker label color) option controls the color of the marker labels. Here we make the marker labels red. See Styles : Colors (412) for more details.

Click on a marker label (e.g., DE) and, in the Contextual Toolbar, change the **Color** to Red.

Uses allstates3.dta & scheme vg_s2c

```
twoway scatter ownhome borninstate, mlabel(stateab) mlabgap(*3)
```

The `mlabgap()` (marker label gap)
option controls the gap between the
marker and the marker label. Here we
make the gap three times the size that
it would normally be. We can also
specify a value less than 1 to place the
marker label closer to the marker.
📖 Double-click on a marker label (e.g.,
DE) and change the **Gap** to *3.
Uses allstates3.dta & scheme vg_s2c

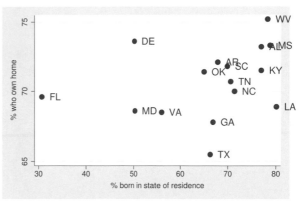

9.3   Connecting points and markers

Stata supports a variety of methods for connecting points by using different values for
the *connectstyle*. These include `l` to connect with a straight line, `L` to connect with a straight
line only if the current x value is greater than the prior x value, `J` for stairstep, `stepstair`
for step then stair, and `i` for invisible connections. For the next few examples, we will use
the `spjanfeb2001` data file, keeping just the data for January and February of 2001. These
examples of connect styles do not demonstrate how you would normally use these styles
but illustrate the different ways we can connect points. See [G-4] ***connectstyle*** for more
information. This section uses the `vg_blue` scheme.

```
twoway scatter close tradeday
```

Consider this graph showing the closing
price of the S&P 500 index for January
and February of 2001 by `tradeday`, the
trading day numbered from 1 to 40.
The graph would be easier to follow if
the points were connected.
*Uses spjanfeb2001.dta & scheme
vg_blue*

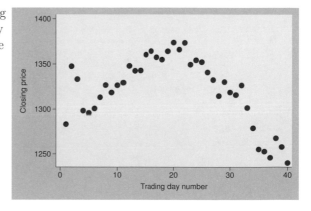

Introduction   Editor   Twoway   Matrix   Bar   Box   Dot   Pie   Options   Standard options   Styles   Appendix

Markers   Marker labels   Connecting   Axis titles   Axis labels   Axis scales   Axis selection   By   Legend   Adding text   Textboxes   Text display

`twoway connected close tradeday`

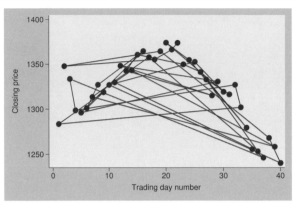

We can change `scatter` to `connected` to connect the points. This, however, does not lead to the kind of graph we really wanted to create, because the observations are connected in the order in which they appear in the data file. We would obtain the desired graph if the observations were sorted according to `tradeday` or if we used the `sort` option as shown in the next example. *Uses spjanfeb2001.dta & scheme vg_blue*

`twoway connected close tradeday, sort`

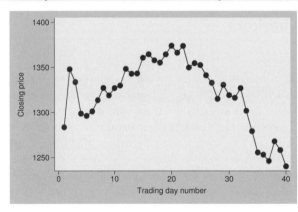

Here, by adding the `sort` option, the observations are connected after sorting them by `tradeday`. We could have first typed `sort tradeday`, and all subsequent graphs would have been ordered on `tradeday`, even without the `sort` option.

Double-click on any of the observations to bring up the **Connected properties** dialog and then click on the **Sort** button. Also see the next example. *Uses spjanfeb2001.dta & scheme vg_blue*

`twoway scatter close tradeday`

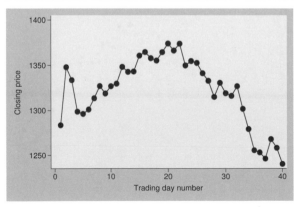

To connect the points in the proper order with the Graph Editor, start with a scatterplot. Click on any marker and, in the Contextual Toolbar, change the **Plottype** to `Connected`. Then double-click on any marker and click the **Sort** button. *Uses spjanfeb2001.dta & scheme vg_blue*

`twoway connected close tradeday, connect(J) sort`

This connection method, obtained by
using the `connect(J)` option, would
normally be used in a graph showing a
survival function over time.
📈 Double-click on any of the
observations and change the
Connecting method to Stairstep.
*Uses spjanfeb2001.dta & scheme
vg_blue*

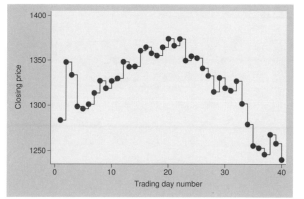

`twoway connected close tradeday, connect(stepstair) sort`

We can obtain a connection method
related to the
one above by using
the `connect(stepstair)` option.
📈 Double-click on any of the
observations and change the
Connecting method to
Stairstep (up first).
*Uses spjanfeb2001.dta & scheme
vg_blue*

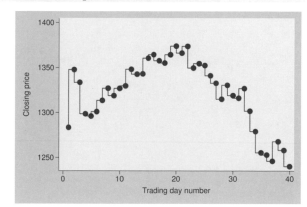

`twoway connected close dom, sort`

Say that we wanted to show the closing
price as a function of the day of the
month for the two months for which we
have data. Here we have the variable
dom (day of the month) on the x axis.
With the `sort` option, the data are
shown as one continuous line, as
opposed to having one line for January
and a second line for February.
*Uses spjanfeb2001.dta & scheme
vg_blue*

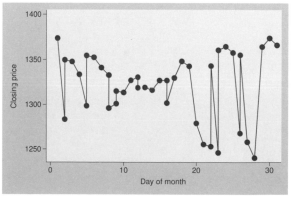

`twoway connected close dom, sort(tradeday)`

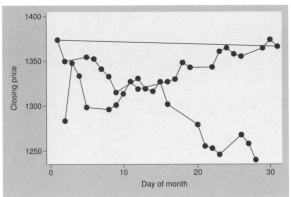

The `sort(tradeday)` option sorts the observations by `tradeday`. This graph is almost what we want, but the observation for January 31 is connected to the observation for February 1. *Uses spjanfeb2001.dta & scheme vg_blue*

`twoway connected close dom, connect(L) sort(tradeday)`

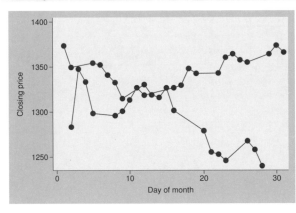

This graph is what we wanted to create. The `connect(L)` option avoids the line connecting January 31 and February 1 because it connects points only as long as `dom` is increasing. When `dom` decreases from 31 to 1, the `connect(L)` option does not connect those two points. See Styles : Connect (416) for more details on `connect()` options.

📊 Double-click on any observation and change the **Connecting method** to Ascending. *Uses spjanfeb2001.dta & scheme vg_blue*

`twoway connected close tradeday, sort`
      `lcolor(green) lwidth(thick) lpattern(dash)`

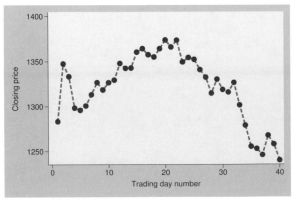

The `connect()` option determines how the markers are connected but not the color, width, or pattern of the line. Here we use the `lcolor()` (line color), `lwidth()` (line width), and `lpattern()` (line pattern) options to make the line green, thick, and dashed. See Styles : Colors (412), Styles : Linewidth (422), and Styles : Linepatterns (420) for more information.

📊 See the next graph. *Uses spjanfeb2001.dta & scheme vg_blue*

```
twoway connected close tradeday, connect(l) sort
```

📊 You can make the customizations shown in the previous example with the Graph Editor. Double-click on the connecting line and, in the *Connected Properties* dialog, change the **Color** to Green, the **Width** to Thick, and the **Pattern** to Dash.
Uses spjanfeb2001.dta & scheme vg_blue

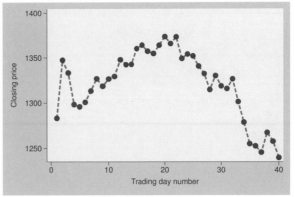

9.4   Setting and controlling axis titles

This section provides more details about the use of axis-title options for providing titles for axes. For more information, see [G-3] ***axis_title_options***. This section uses the vg_past scheme.

```
twoway scatter ownhome propval100
```

Consider this graph of the percentage of homeowners by the percentage of homes that cost over $100,000. The titles of the x and y axes are the names of the variables, unless the variables are labeled; i.e., the default title is the variable label. Here the axes are labeled with the variable labels.
Uses allstatesdc.dta & scheme vg_past

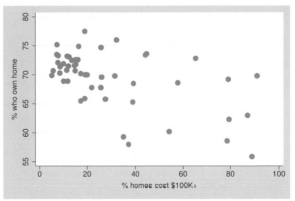

Introduction   Editor   Twoway   Matrix   Bar   Box   Dot   Pie   Options   Standard options   Styles   Apperdix

Markers   Marker labels   Connecting   Axis titles   Axis labels   Axis scales   Axis selection   By   Legend   Adding text   Textboxes   Text display

```
twoway scatter ownhome propval100,
    ytitle("Percent of households that own their homes")
    xtitle("Percent of homes over $100,000")
```

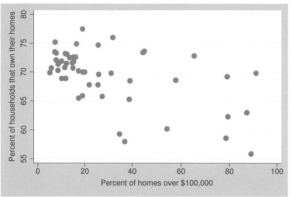

We can use the `ytitle()` and `xtitle()` options to supply our own titles.

Click on the y-axis title and, in the Contextual Toolbar, replace the **Text** with the desired title. Likewise, click on the x-axis title and change the text of that title.

Uses allstatesdc.dta & scheme vg_past

```
twoway scatter ownhome propval100,
    ytitle("Percent of households that own their homes")
    xtitle("Percent of homes over $100,000", size(small) color(navy))
```

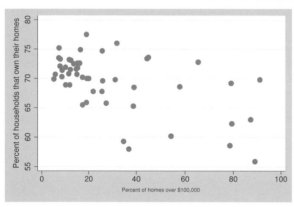

Because an axis title is considered a textbox, we can use textbox options to control the look of the axis title. See **Options : Textboxes** (379) for more examples.

Click on the x-axis title and use the Contextual Toolbar to change the **Size** to `Small` and **Color** to Navy.

Uses allstatesdc.dta & scheme vg_past

```
twoway scatter ownhome propval100,
    ytitle("Percent of households" "that own their homes")
    xtitle("Percent of homes" "over $100,000")
```

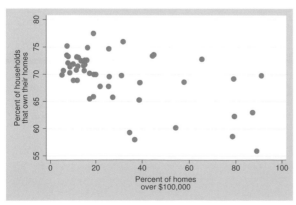

Here we supply the same titles but divide them into two separate, quoted strings, which are then displayed on separate lines.

Use quotation marks in the same way when typing the title into the Graph Editor to make a title appear across two lines.

Uses allstatesdc.dta & scheme vg_past

```
twoway scatter ownhome propval100,
   ytitle("1990 Census Data", suffix)
   xtitle("In 1990 dollars", prefix)
```

We can use the `prefix` and `suffix` options to add information before or after the existing title, respectively.

🔖 Click on a title and, in the Contextual Toolbar, add the desired text to the title.

Uses allstatesdc.dta & scheme vg_past

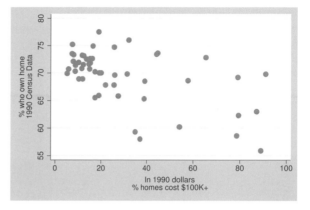

```
twoway (scatter rent700 ownhome )
   (scatter propval100 ownhome, yaxis(2))
```

Consider this overlaid twoway graph. The two y variables are both scaled in percentages, but they have different ranges. We use the `yaxis(2)` option on the second `scatter` command to place that axis on the second y axis, which is then placed on the right axis.

Uses allstatesdc.dta & scheme vg_past

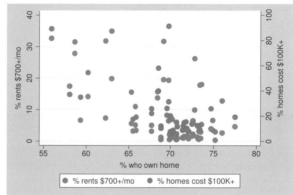

```
twoway (scatter rent700 ownhome) (scatter propval100 ownhome, yaxis(2)),
   ytitle("Percent rents over $700", axis(1))
   ytitle("Percent homes over $100,000", axis(2))
```

Now that we have two y axes, the `ytitle()` option would change the y title for the first y axis, unless we specify otherwise. Here we supply a `ytitle()` option with the `axis(1)` option to indicate that the title belongs to the first y axis, and a second `ytitle()` option using the `axis(2)` option to indicate that the second title belongs to the second y axis.

🔖 Click on the title and, in the Contextual Tool, add the desired text to the title.

Uses allstatesdc.dta & scheme vg_past

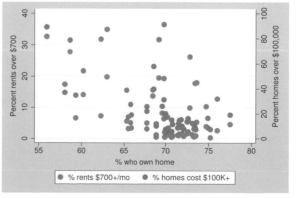

9.5   Setting and controlling axis labels

This section describes more details about axis labels, including major and minor (numeric) labels, major and minor tick marks, and grid lines. This section also shows how to control the appearance of these objects (e.g., size, color, thickness, or angle). For more information, see [G-3] ***axis_label_options***. This section uses the vg_s1c scheme.

`twoway scatter propval100 faminc`

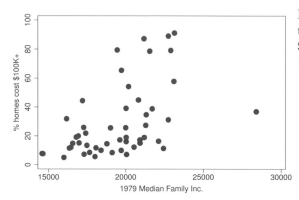

Let's start with a basic graph showing the percentage of homes costing over $100,000 by the median family income. *Uses allstatesdc.dta & scheme vg_s1c*

`twoway scatter propval100 faminc, xlabel(#10) ylabel(#10)`

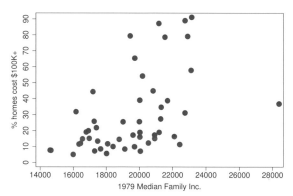

By using the xlabel(#10) and ylabel(#10) options, we ask for about 10 values to be labeled on each axis. Stata chose to use 10 values for the *y* axis, labeling it from 0 to 90, incrementing by 10, and 8 values for the *x* axis going from 14,000 to 28,000, incrementing by 2,000. Sometimes Stata follows the suggestion exactly, and sometimes it chooses a different number of values to make more logical labels.

See the next graph.

Uses allstatesdc.dta & scheme vg_s1c

`twoway scatter propval100 faminc`

In the Graph Editor, double-click on the *y*-axis label (e.g., double-click on 60) to modify it. Within **Axis rule** select **Suggest # of ticks** and select **10 Ticks**. Likewise, double-click on an *x*-axis label (e.g., 20000) and, within **Axis rule**, select **Suggest # of ticks** and select **10 Ticks**.

Uses allstatesdc.dta & scheme vg_s1c

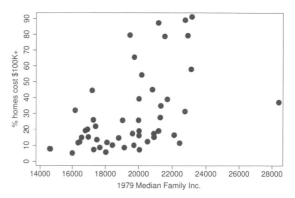

`twoway scatter propval100 faminc, ylabel(0(10)100)`

We can change the major labels for the *y* variable to range from 0 to 100, incrementing by 10, by using the `ylabel(0(10)100)` option.

Double-click on the *y*-axis label. Within **Axis rule** select **Range/Delta** and supply a **Minimum value** of 0, a **Maximum value** of 100, and a **Delta** of 10.

Uses allstatesdc.dta & scheme vg_s1c

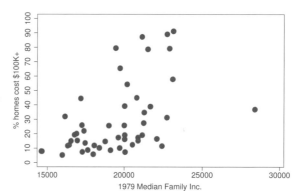

`twoway scatter propval100 faminc, xlabel(minmax) ylabel(none)`

Here we use the `xlabel(minmax)` option to label the *x* axis with only the minimum and maximum and use `ylabel(none)` so that the *y* axis will have no major labels or ticks.

Double-click on the *x*-axis label and, within **Axis rule**, select **Min Max**. Then double-click on the *y*-axis label and, within **Axis rule**, select **None**.

Uses allstatesdc.dta & scheme vg_s1c

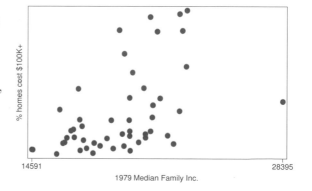

`twoway scatter propval100 faminc, ymlabel(10(20)90)`

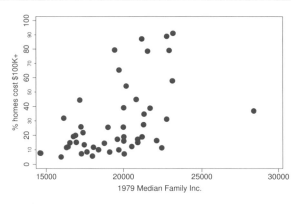

1979 Median Family Inc.

The default graph had major labels for the y axis at 0, 20, 40, 60, 80, and 100. We add minor labels for the y variable at 10, 30, 50, 70, and 90 by using the `ymlabel(10(20)90)` option. The m in `ymlabel()` stands for minor.

Double-click on the y-axis label, select the *Minor* tab, within **Axis rule** select **Range/Delta**, and supply a **Minimum value** of 10, a **Maximum value** of 90, and a **Delta** of 20.

Uses allstatesdc.dta & scheme vg_s1c

`twoway scatter propval100 faminc, ytick(10(10)90)`

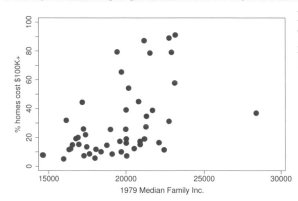

1979 Median Family Inc.

By using the `ytick(10(10)90)` option, we change the major ticks to range from 10 to 90, incrementing by 10.

Uses allstatesdc.dta & scheme vg_s1c

`twoway scatter propval100 faminc, ymtick(10(20)90)`

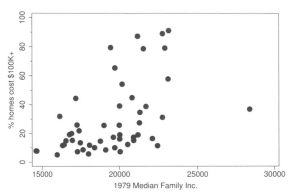

1979 Median Family Inc.

The `ymtick()` option adds minor ticks to the graph. Here we add minor ticks at 10, 30, 50, 70, and 90. The m in `ymtick()` stands for minor.

Uses allstatesdc.dta & scheme vg_s1c

`twoway scatter propval100 faminc, ymtick(##10)`

The default graph had major labels for
the y axis at 0, 20, 40, 60, 80, and 100.
We can place 9 minor ticks between the
major ticks with the `ymtick(##10)`
option. The value of 10 includes the 9
minor ticks plus the 10th major tick.
Uses allstatesdc.dta & scheme vg_s1c

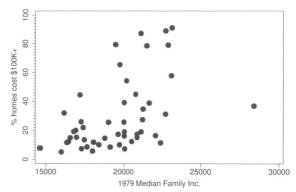

`twoway scatter propval100 faminc, ylabel(0(10)100, noticks)`

We specify the `noticks` option to label
the y axis using values ranging from 0
to 100, incrementing by 10, but
suppressing the display of ticks.

📊 Double-click on the y-axis label.
Within **Global properties**, click on
the **Tick properties** button and
uncheck the box that says **Show ticks**.

Uses allstatesdc.dta & scheme vg_s1c

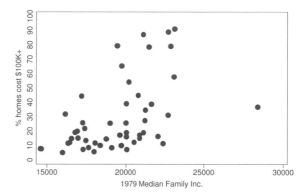

`twoway scatter propval100 faminc, ylabel(, nolabel)`

We suppress the labels by using the
`nolabel` option, thus showing only the
ticks.

📊 Double-click on the y-axis label.
Within **Global properties** click on the
Label properties button and uncheck
the box that says **Show labels**.
Uses allstatesdc.dta & scheme vg_s1c

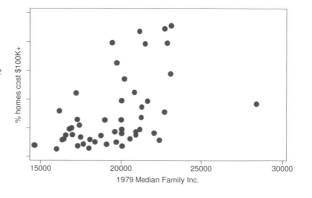

Introduction   Editor   Twoway   Matrix   Bar   Box   Dot   Pie   Options   Standard options   Styles   Appendix

Markers   Marker labels   Connecting   Axis titles   Axis labels   Axis scales   Axis selection   By   Legend   Adding text   Textboxes   Text display

`twoway scatter propval100 region`

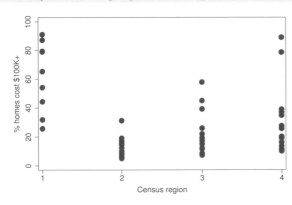

If a variable has meaningful value labels, we can display the value labels instead of the values. For example, we can look at `propval100` broken down by census region, but we do not know which regions correspond to the values 1–4.

Uses allstatesdc.dta & scheme vg_s1c

`twoway scatter propval100 region, xlabel(, valuelabels)`

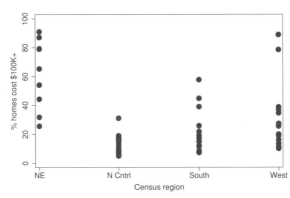

With the `xlabel(, valuelabels)` option, the value labels are displayed, making the graph much easier to understand.

Double-click the *x*-axis label. Within **Global properties**, click on the **Label properties** button and check the box that says **Use value labels**.

Uses allstatesdc.dta & scheme vg_s1c

`twoway scatter propval100 region,`
`   xlabel(1 "NorthEast" 2 "NorthCentral" 3 "Southern" 4 "Western")`

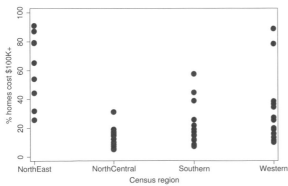

We can use the `xlabel()` option to add labels to the *x* axis.

Double-click on the *x*-axis label and click on the **Edit or add individual ticks and labels button**. If it is not already, select the first tick, click on the **Edit** button, and change the **Label** to NorthEast. Then choose the second tick, click on the **Edit** and change the **Label** to NorthCentral. We can edit the label for the third and fourth tick in the same way.

Uses allstatesdc.dta & scheme vg_s1c

`twoway scatter propval100 faminc, ylabel(, angle(0))`

By default, the values on the y axis are shown vertically (at a 90-degree angle), but we can use the `ylabel(, angle(0))` option to display the labels horizontally.

⧄ Click on a y-axis label (e.g., 100) and, in the Contextual Toolbar, change the **Label angle** to **Horizontal**.

Uses allstatesdc.dta & scheme vg_s1c

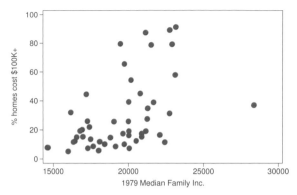

`twoway scatter propval100 faminc, xlabel(15000(1000)30000, angle(45))`

If we label an axis with a large number of values (and especially with wide values), the labels may crowd each other and overlap. Here we label the x axis from 15000 to 30000, incrementing by 1000. To avoid overlapping, we add the `angle(45)` option to show the labels at a 45-degree angle.

⧄ Click on an x-axis label (e.g., 15000) and, in the Contextual Toolbar, change the **Label angle** to **45 degrees**.

Uses allstatesdc.dta & scheme vg_s1c

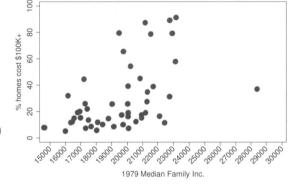

`twoway scatter propval100 faminc, xlabel(15000(1000)30000, alternate)`

We can also avoid overlapping the axis labels by adding the `alternate` option to `xlabel()`. The labels are now displayed in two rows of alternating text, so they are not crowded or overlapped.

⧄ Double-click on the x-axis label. Within **Global properties**, click on the **Label properties** button and place a check next to **Alternate spacing of adjacent labels**.

Uses allstatesdc.dta & scheme vg_s1c

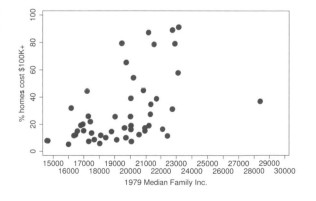

Introduction Editor Twoway Matrix Bar Box Dot Pie Options Standard options Styles Appendix

Markers Marker labels Connecting Axis titles Axis labels Axis scales Axis selection By Legend Adding text Textboxes Text display

`twoway scatter propval100 faminc, ylabel(0(5)90, labsize(vsmall))`

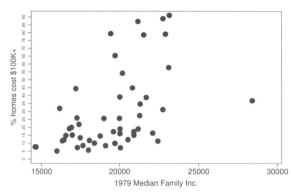

We can control the size of labels with the `labsize()` option. Here we label our y axis from 0 to 90, incrementing by 5, but we specify `labsize(vsmall)` so that smaller labels are used to minimize crowding and overlapping labels.

⊞ Click on the y-axis label and, in the Contextual Toolbar, change the **Label Size** to v Small.

Uses allstatesdc.dta & scheme vg_s1c

`twoway scatter propval100 faminc, ylabel(, labgap(large))`

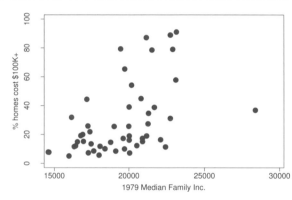

We can control the gap between the label and the tick with the `labgap()` option. In this example, we make the gap between the y labels and the y ticks `large`.

⊞ Double-click on the y-axis label. Within **Global properties**, click on the **Label properties** button and change the **Text gap** to Large.

Uses allstatesdc.dta & scheme vg_s1c

`twoway scatter propval100 faminc,`
`    ylabel(, tlength(large) tlwidth(thick) tposition(crossing))`

Here we use the `tlength()`, `tlwidth()`, and `tposition()` options to make the tick length large, the width thick, and the ticks cross the y axis.

⊞ Double-click on the y-axis label. Within **Global properties**, click on the **Tick properties** button and change the **Length** to Large, the **Width** to Thick, and the **Placement** to Crossing.

Uses allstatesdc.dta & scheme vg_s1c

```
twoway scatter propval100 faminc,
    ytick(#10, tposition(outside))
    ymtick(#5, tposition(inside))
```

In this example, we suggest 10 major ticks and 5 minor ticks, locating the major ticks on the outside of the plot and the minor ticks on the inside of the plot.

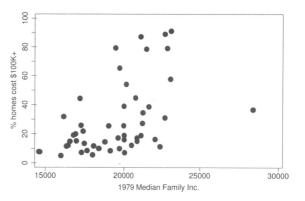

🖱 Double-click on the *y*-axis label. Within **Global properties**, click on the **Tick properties** button and change the **Placement** to `Outside`. Then, on the *Minor* tab, click on the **Tick properties** button and change the **Placement** to `Inside`.
Uses allstatesdc.dta & scheme vg_s1c

```
twoway scatter propval100 faminc, ylabel(, nogrid)
```

Here we suppress the grid corresponding to the *y* label by using the `nogrid` option. You can also use the `grid` and `nogrid` options with the `ymlabel()`, `ytick()`, and `ymtick()` options as well as the `xlabel()`, `xmlabel()`, `xtick()`, and `xmtick()` options.

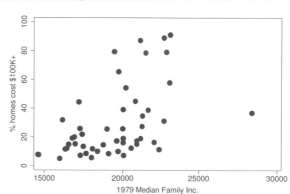

🖱 Click on the *y* label and then, in the Contextual Toolbar, select **Show grid** to toggle the display of the grid.
Uses allstatesdc.dta & scheme vg_s1c

```
twoway scatter propval100 faminc, ylabel(, grid) xlabel(, grid)
```

We display a grid for the values that correspond to the `ylabel()` and the `xlabel()` options with the `grid` option, as shown here. Depending on the scheme we choose, grids may be included or omitted by default.

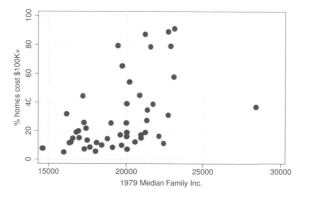

🖱 Click on the *y* label and then, in the Contextual Toolbar, select **Show grid** to toggle the display of the grid. We can do likewise for the *x* label.
Uses allstatesdc.dta & scheme vg_s1c

Introduction   Editor   Twoway   Matrix   Bar   Box   Dot   Pie   Options   Standard options   Styles   Appendix

Markers   Marker labels   Connecting   Axis titles   Axis labels   Axis scales   Axis selection   By   Legend   Adding text   Textboxes   Text display

```
twoway scatter propval100 faminc,
    ylabel(, grid glwidth(vthin) glcolor(gs10) glpattern(shortdash))
```

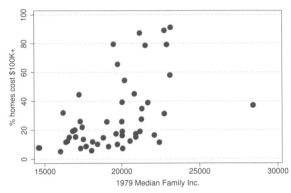

We can control the grid-line width, grid-line color, and grid-line pattern with the `glwidth()`, `glcolor()`, and `glpattern()` options. Here we make the grid line very thin, the color gray (`gs10`), and the pattern of the lines short dashes. See Styles : Linewidth (422), Styles : Colors (412), and Styles : Linepatterns (420) for more details.

See the next graph.

Uses allstatesdc.dta & scheme vg_s1c

```
twoway scatter propval100 faminc
```

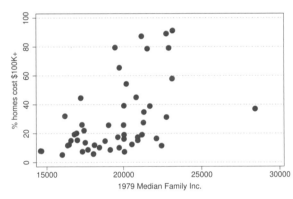

Using the Graph Editor, double-click on the *y* label and click on the **Grid Lines** button. Check **Show Grid** and change the **Width** to v Thin, the **Color** to Gray 10, and the **Pattern** to Short-dash.

Uses allstatesdc.dta & scheme vg_s1c

```
twoway scatter propval100 faminc,
    ylabel(0(20)100, grid glcolor(gs8) glpattern(solid))
    ymlabel(10(20)90, grid glcolor(gs11) glpattern(shortdash))
```

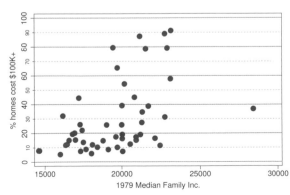

This example shows a solid, darker gray line for the major axis labels and a lighter gray, short, dashed line for the minor axis labels.

Double-click on the *y* label and click on the **Grid Lines** button. Change the **Color** to Gray 8 and the **Pattern** to Solid. From the *Minor* tab, click on **Grid Lines** and change the **Color** to Gray 11 and the **Pattern** to Short-dash.

Uses allstatesdc.dta & scheme vg_s1c

9.6   Controlling axis scales

   This section provides more details about axis-scale options, which allow us to control whether an axis is displayed, where it is displayed, the direction it is displayed, and the scale of the axis. For more information about these options, see [G-3] *axis_scale_options*. This section begins by using data on the S&P 500 from January 2, 2001, to December 31, 2001, stored in the file sp2001. For simplicity, we will use tradeday on the x axis, representing the trading day of the year. This section uses the vg_s2m scheme.

twoway rspike high low tradeday

Consider this rspike graph showing the high and low prices across 248 trading days.
Uses sp2001.dta & scheme vg_s2m

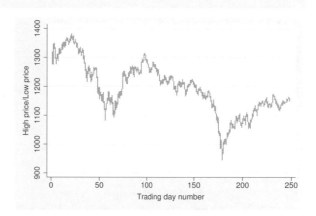

twoway rspike high low tradeday, xscale(off)

The xscale(off) option removes the display of the x axis entirely. Although it is not shown, you could do the same for the y axis by using the yscale(off) option. This option can be useful when combining multiple graphs on the same scale.

   Right-click on **xaxis1** in the Object Browser and select **Hide**. Right-clicking again and selecting **Show** will reshow the axis.
Uses sp2001.dta & scheme vg_s2m

`twoway rspike high low tradeday, `xscale(alt)

We shift the display of the x axis from the bottom of the graph to the top of the graph by using the `xscale(alt)` option. Likewise, you can shift the y axis from the left to the right by using the `yscale(alt)` option.

📖 See the next graph.

Uses sp2001.dta & scheme vg_s2m

`twoway rspike high low tradeday`

📖 Within the Graph Editor, select the Grid Edit 🔳 tool. Click and hold the x axis and drag it to the top of the graph. Then select the Pointer ➤ tool. Double-click on the x axis, select **Advanced**, and change the **Position** to Above.

Uses sp2001.dta & scheme vg_s2m

`twoway rspike high low tradeday, `xscale(reverse)

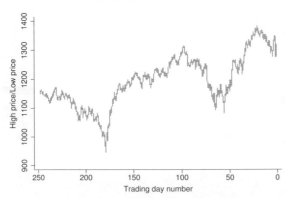

We reverse the scale of the x axis by specifying the `xscale(reverse)` option, as illustrated here. You can reverse the y axis by indicating the `yscale(reverse)` option.

📖 Double-click on the x axis. Click on the Scale button and check **Reverse scale to run from maximum to minimum**.

Uses sp2001.dta & scheme vg_s2m

twoway scatter educ popden, xscale(log)

We briefly return to the **allstates**
data file to show the **xscale(log)**
option, which indicates that the x axis
should be displayed on a log scale. The
labels for the x axis overlap each other.

📊 Double-click on the x axis, click on
the **Scale** button, and check **Use
logarithmic scale**.

Uses allstates.dta & scheme vg_s2m

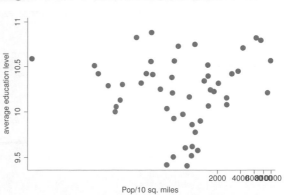

twoway scatter educ popden, xscale(log)
    xlabel(1 10 100 1000 10000)

Here we use the **xlabel()** option to
change the labels for the x axis using
the values 1, 10, 100, 1,000, and 10,000.
Note how these powers of 10 are more
equally spaced, reflecting the log scale
of the x axis.

📊 Double-click on an x-axis label and,
within **Axis rule**, select **None**. Then
click on **Edit or add individual ticks
and labels**. Click add and enter 1 for
the **Value** and **Label**. Repeat this for
10, 100, 1000, and 10000.

Uses allstates.dta & scheme vg_s2m

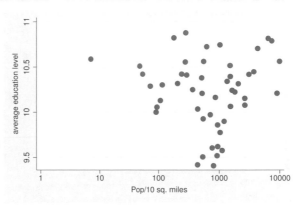

twoway rspike high low tradeday, xscale(lwidth(thick))

We now return to the **sp2001** data. You
can use the **xscale()** and **yscale()**
options to control the axis lines. Here
we make the x-axis line thick by
specifying **xscale(lwidth(thick))**.

📊 Double-click on the x axis, click on
Axis line, and change the **Width** to
Thick.

Uses sp2001.dta & scheme vg_s2m

Introduction   Editor   Twoway   Matrix   Bar   Box   Dot   Pie   Options   Standard options   Styles   Appendix

Markers   Marker labels   Connecting   Axis titles   Axis labels   Axis scales   Axis selection   By   Legend   Adding text   Textboxes   Text display

`twoway rspike high low tradeday, xscale(noline)`

We suppress the display of the x-axis line by using the `xscale(noline)` option.

📖 Double-click on the x axis, click on **Axis line**, and change the **Color** to **None**.

Uses sp2001.dta & scheme vg_s2m

`twoway rspike high low tradeday, yscale(range(700 1400))`

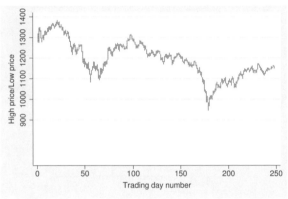

We can use the `yscale(range())` option to expand the scale of the y axis without needing to expand the labels for the axis (as the `ylabel()` option would). Here we have expanded the y axis to range from 700 to 1400. The next examples illustrate the usefulness of this option.

📖 Double-click on the y axis. Then click on **Scale**, check **Extend range of axis scale**, and change the **Lower limit** to 700.

Uses sp2001.dta & scheme vg_s2m

`twoway (rspike high low tradeday)`
`      (line volmil tradeday, sort yaxis(2))`

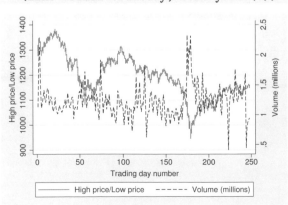

Consider that, in addition to the spike graph that shows the high and low values for a given trading day, we wish to see the volume for a given trading day. We can combine the plots into one graph, but this is difficult to read because the two plots overlap.

Uses sp2001.dta & scheme vg_s2m

```
twoway (rspike high low tradeday)
   (line volmil tradeday, sort yaxis(2)),
   yscale(range(700 1400) axis(1)) yscale(range(0 10) axis(2))
```

The `yscale(range(700 1400) axis(1))` option sets the range of the left y axis to occupy the upper third of the graph, whereas the `yscale(range(0 10) axis(2))` option sets the range of the right y axis to occupy the lower third of the graph.

🖱 Double-click on the left y axis. Click on **Scale**, check **Extend range of axis scale**, and change the **Lower limit** to 700. Repeat these steps for the right y axis setting the range from 0 to 10.
Uses sp2001.dta & scheme vg_s2m

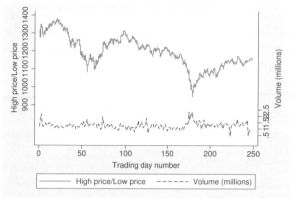

```
twoway (rspike high low tradeday)
   (line volmil tradeday, sort yaxis(2)),
   yscale(range(700 1400) axis(1)) yscale(range(0 10) axis(2))
   ylabel(1000 1200 1400, axis(1)) ylabel(0 1 2, axis(2))
```

We add the `ylabel(1000 1200 1400, axis(1))` and `ylabel(0 1 2, axis(2))` options to the previous example to make the labels for the y axes more readable.

🖱 Double-click on the left y axis, and, under **Axis rule**, select **Range/Delta**, and enter 1000 as the **Minimum**, 1400 and the **Maximum**, and 200 as the **Delta**. Repeat these steps for the right y axis using 0 as the **Minimum**, 2 as the **Maximum**, and 1 as the **Delta**.
Uses sp2001.dta & scheme vg_s2m

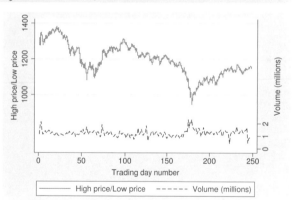

9.7   Selecting an axis

This section provides more details about how to select different axes and how to modify them. By default, any modifications you make to an axis are applied to the first axis, so you need to take extra action to modify other axes that you may create. For more information about these options, see [G-3] ***axis_choice_options***. This section uses the `vg_outc` scheme.

Although you can do almost anything to customize an axis by using the Graph Editor, you cannot, however, add an axis with the Graph Editor. So, be sure your graph has all the axes it needs by using the graph commands, and then feel free to move and change them by using the Graph Editor.

Markers   Marker labels   Connecting   Axis titles   Axis labels   Axis scales   Axis selection   By   Legend   Adding text   Textboxes   Text display

Introduction   Editor   Twoway   Matrix   Bar   Box   Dot   Pie   Options   Standard options   Styles   Appendix

`twoway scatter faminc educ`

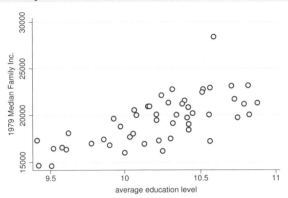

This graph shows the relationship between one x variable, `educ`, and one y variable, `faminc`.
Uses allstatesdc.dta & scheme vg_outc

`twoway (scatter faminc educ, xaxis(1) yaxis(1))`

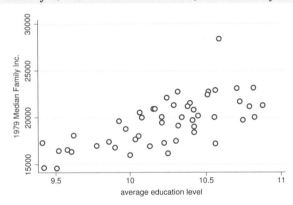

By default, the x variable is placed on the first x axis, and the y variable is placed on the first y axis. It is as though we had added the options `xaxis(1)` and `yaxis(1)`, as illustrated here. Note the parentheses emphasize that the options `xaxis(1)` and `yaxis(1)` belong to the `scatter` command and are not general options for the overall graph, which would appear after the parentheses.
Uses allstatesdc.dta & scheme vg_outc

`twoway (scatter faminc educ)`
`      (scatter workers2 educ)`

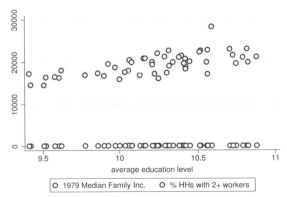

Now let's overlay a second scatterplot showing `workers2` by `educ`, which has the effect of adding a second variable to the y axis. Stata assumes that all variables are on the first (and thus the same) axis, unless we specify otherwise. As a result, this graph is hard to read because `faminc` is scaled differently from `workers2` but scaled on the same axis.
Uses allstatesdc.dta & scheme vg_outc

```
twoway (scatter faminc educ, yaxis(1))
     (scatter workers2 educ, yaxis(2))
```

Stata permits us to have multiple axes for the *x* variables and the *y* variables. In this example, we use the `yaxis(1)` option to place `faminc` on the first *y* axis and the `yaxis(2)` option to place `workers2` on the second *y* axis. To make the graph more readable, Stata moved the second *y* axis to the right side. The `yaxis(1)` option was not needed but was included for clarity.
Uses allstatesdc.dta & scheme vg_outc

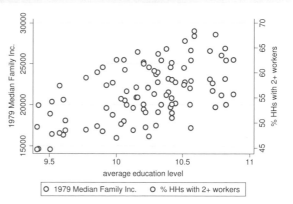

```
twoway (scatter faminc educ)
     (scatter workers2 educ, yaxis(2)), ylabel(40(5)80, axis(2))
```

To label `workers2` from 40 to 80, incrementing by 5, we would specify `ylabel(40(5)80, axis(2))`, because `workers2` is on the second axis. Without the `axis(2)` option, Stata would assume that we are referring to the first *y* axis and would change the scaling of `faminc`.
Uses allstatesdc.dta & scheme vg_outc

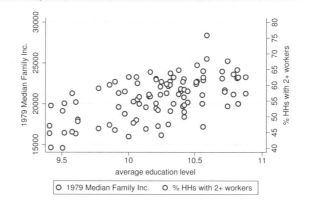

```
twoway (scatter faminc educ)
     (scatter workers2 educ, yaxis(2) ylabel(40(5)80))
```

We might be tempted to enter the `ylabel()` option as an option of the second `scatter` statement and expect the `ylabel()` to modify the scaling of `workers2`. However, we can see here that this does not work. This is explained further in the next example.
Uses allstatesdc.dta & scheme vg_outc

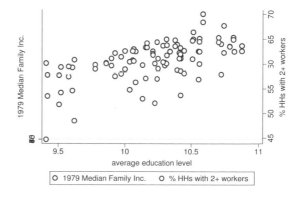

```
twoway (scatter faminc educ)
   (scatter workers2 educ, yaxis(2)), ylabel(40(5)80, axis(1))
```

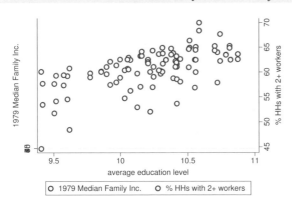

The `ylabel()` option is really an overall option, but Stata is willing to pretend that you specified this option globally, as though you had typed `ylabel()` as a global option as specified in this example. To make this clearer, we have added the default `axis(1)` to `ylabel()` to illustrate why this usage does not change the second y axis.

Uses allstatesdc.dta & scheme vg_outc

```
twoway (scatter faminc educ)
   (scatter workers2 educ, yaxis(2)),
   ytitle("Family income", axis(1)) ytitle("Two+ workers", axis(2))
```

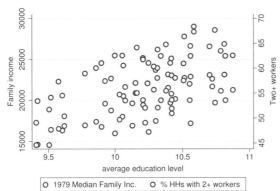

These same rules apply to modifying the axis titles and labels. In this example, we use the `ytitle()` option to change the titles for the first and second y axes.

Uses allstatesdc.dta & scheme vg_outc

9.8   Graphing by groups

This section provides more details about repeating graphs by using the `by()` option to show separate graphs for each by-group. For more information, see [G-3] *by_option*. This section uses the `vg_brite` scheme.

twoway scatter ownhome borninstate

We start by looking at a scatterplot of
`ownhome` and `borninstate`, and we see
a general positive relationship such that
the higher the percentage of those who
were born in the state, the higher the
percentage of homeowners in the state.
Uses allstatesdc.dta & scheme vg_brite

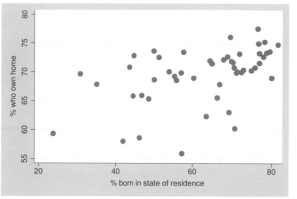

twoway scatter ownhome borninstate, by(north)

We use the by(north) option to look at
this relationship broken down by
whether the state is considered to be in
the North.
Uses allstatesdc.dta & scheme vg_brite

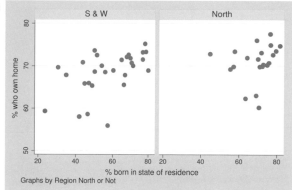

twoway scatter ownhome borninstate, by(north, total)

We use the `total` option to see the
overall relationship for all 50 states, as
well as the two plots separately, by the
levels of `north`.
Uses allstatesdc.dta & scheme vg_brite

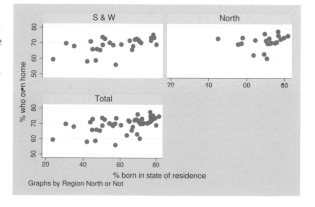

`twoway scatter ownhome borninstate, by(north, total colfirst)`

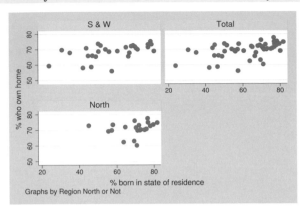

We add the `colfirst` option to show the graphs going down columns first rather than going across rows first, which is the default.

In the Object Browser, double-click on **Graph** to bring up the *Graph properties* dialog box. Choose the *Organization* tab and change the **Graph Sequence** to Down first.

Uses allstatesdc.dta & scheme vg_brite

`twoway scatter ownhome borninstate, by(north, total holes(2))`

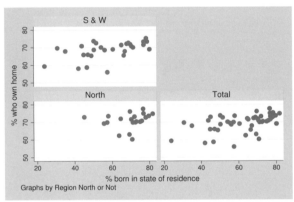

The `holes(2)` option leaves the second position empty. Here we specify one position to leave empty, but we can specify multiple positions within the `holes()` option.

In the Object Browser, double-click on **Graph**. In the *Organization* tab, change the **Positions to leave blank** to 2.

Uses allstatesdc.dta & scheme vg_brite

`twoway scatter ownhome borninstate, by(north, total rows(1))`

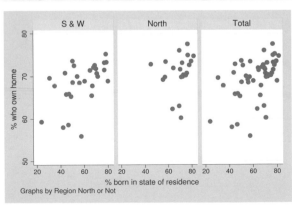

The `rows(1)` option indicates that the graph should be displayed in one row.

In the Object Browser, double-click on **Graph**. In the *Organization* tab, change the **Rows/Columns** to Rows and change the number of **Rows** to 1.

Uses allstatesdc.dta & scheme vg_brite

`twoway scatter ownhome borninstate, by(north, total cols(1))`

The `cols(1)` option shows the graph in
one column.

🖳 In the Object Browser, double-click
on **Graph**. In the *Organization* tab,
change the **Rows/Columns** to Cols
and change the number of **Cols** to 1.
Uses allstatesdc.dta & scheme vg_brite

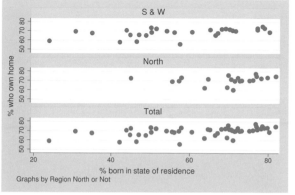

`twoway scatter ownhome borninstate, by(north, total iscale(*1.5))`

Sometimes with the `by()` option, the
graphs become small, making the text
and symbols difficult to see. The
`iscale()` option magnifies the size of
these elements. Here we increase the
size of these elements by a factor of 1.5.

🖳 In the Object Browser, double-click
on **Graph**. In the *Aspect/Size* tab,
change the **Scaling factor** to 1.5.
Uses allstatesdc.dta & scheme vg_brite

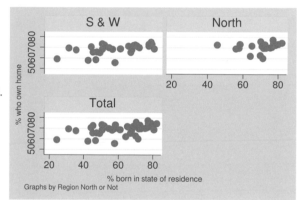

`twoway scatter ownhome borninstate, by(north, total compact)`

The `compact` option displays the graph
using a compact style, pushing the
graphs tightly together. This is almost
the same as specifying
`style(compact)`.
Uses allstatesdc.dta & scheme vg_brite

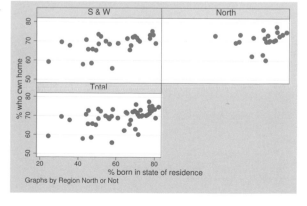

Introduction   Editor   Twoway   Matrix   Bar   Box   Dot   Pie   Options   Standard options   Styles   Apperdix

Markers   Marker labels   Connecting   Axis titles   Axis labels   Axis scales   Axis selection   By   Legend   Adding text   Textboxes   Text display

`twoway scatter ownhome borninstate, by(north, total noedgelabel)`

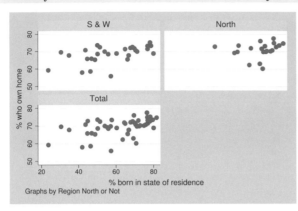

The `noedgelabel` option suppresses the display of the x axis for the graphs that do not appear on the bottom row, in this case, the graph for the North.

📖 Double-click on the North x axis, click on **Advanced**, and check **Hide Axis**.

Uses allstatesdc.dta & scheme vg_brite

`twoway scatter ownhome borninstate, by(north, yrescale)`

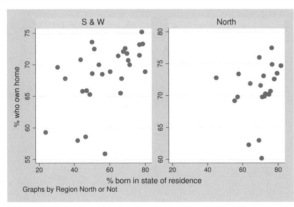

The `yrescale` option allows the y variables to be scaled independently for each by-group.

📖 In the Object Browser, expand **plotregion1**, right-click on **yaxis1[2]** (the first y axis for the second graph), and select **Show**. Double-click on each y axis and scale it as desired.

Uses allstatesdc.dta & scheme vg_brite

`twoway scatter ownhome borninstate, by(north, xrescale)`

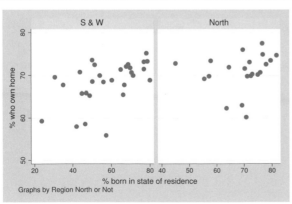

Likewise, the `xrescale` option allows the x variable to be scaled differently across all the by-groups.

📖 Double-click on each x axis and scale it as desired.

Uses allstatesdc.dta & scheme vg_brite

`twoway scatter ownhome borninstate, by(north, rescale)`

The `rescale` option scales both the x variable and y variable differently across the by-groups. Both axes are separately rescaled.

📖 Double-click on any x axis or y axis and scale it as desired.

Uses allstatesdc.dta & scheme vg_brite

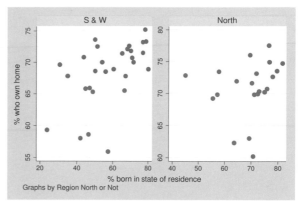

`twoway scatter ownhome borninstate, by(north, iyaxes)`

The `iyaxes` option displays the y axes for each individual graph.

📖 In the Object Browser, expand **plotregion1**. For any **yaxis1[ ]** that is dimmed, right-click on it and select **Show** to show the hidden axes.

Uses allstatesdc.dta & scheme vg_brite

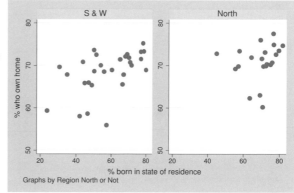

`twoway scatter ownhome borninstate, by(north, cols(1))`

Likewise, the `ixaxes` option displays the x axis for all graphs. Here we omit this option and instead use the `cols(1)` option to display two graphs in one column. Stata omits the x axis in the top graph.

Uses allstatesdc.dta & scheme vg_brite

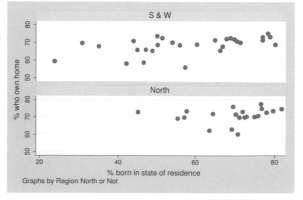

Introduction   Editor   Twoway   Matrix   Bar   Box   Dot   Pie   Options   Standard options   Styles   Appendix

Markers   Marker labels   Connecting   Axis titles   Axis labels   Axis scales   Axis selection   By   Legend   Adding text   Textboxes   Text display

`twoway scatter ownhome borninstate, by(north, ixaxes cols(1))`

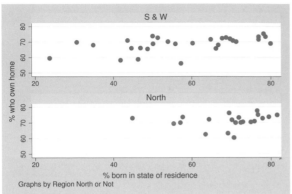

We now include the `ixaxes` option and see that the x axis is now displayed on the top graph.

In the Object Browser, expand **plotregion1**. For any **xaxis1[ ]** that is dimmed, right-click on it and select **Show** to show the hidden axes.

Uses allstatesdc.dta & scheme vg_brite

`twoway scatter ownhome borninstate, by(north, total iytitle)`

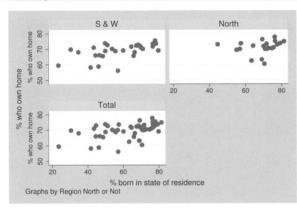

We display the title for each y axis by using the `iytitle` option.

In the Object Browser, expand **plotregion1**. For any dimmed **yaxis1[ ]**, right-click it to show it. For any **title** that is dimmed, right-click on it and select **Show** to show the hidden title.

Uses allstatesdc.dta & scheme vg_brite

`twoway scatter ownhome borninstate, by(north, total iyaxes iytitle)`

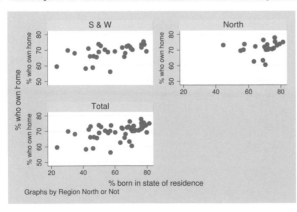

The y title was not displayed for the North because the y axis was omitted for that graph. We now include the `iyaxes` and `iytitle` options to display the y axis and y title for the North graph as well.

See the two previous examples.

Uses allstatesdc.dta & scheme vg_brite

```
twoway scatter ownhome borninstate, by(north, total ixaxes ixtitle)
```

Likewise, we can display the *x* title on each graph using the `ixaxes` and `ixtitle` options.

▣ In the Object Browser, expand **plotregion1**. For any dimmed **xaxis1[ ]**, right-click on it to show it. For any **title** that is dimmed, right-click on it and select **Show** to show the hidden title.

Uses allstatesdc.dta & scheme vg_brite

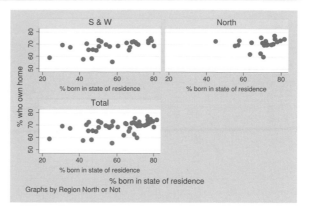

```
twoway scatter ownhome borninstate, by(north) title("My title")
```

If we include a `title()` option with `by()`, Stata creates each graph separately using the title we specify.

▣ In the Object Browser, expand **plotregion1**. Double-click on **title[1]** to change the title of the first graph and double-click on **title[2]** to change the title of the second graph. You can also change the note, caption, and subtitles for each graph here as well.

Uses allstatesdc.dta & scheme vg_brite

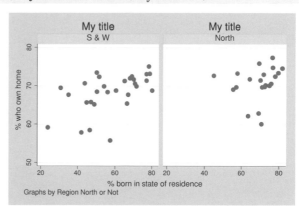

```
twoway scatter ownhome borninstate, by(north, title("My title"))
```

If we add `title()` within the `by()` option, Stata makes this an overall title for the graph.

▣ In the Object Browser, double-click on **Graph** and select the *Titles* tab. Change the title as desired.

Uses allstatesdc.dta & scheme vg_brite

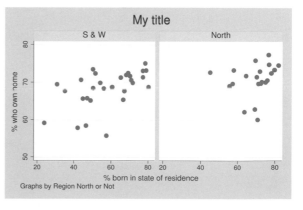

Introduction

Markers   Marker labels   Connecting

Editor   Twoway   Matrix   Bar   Box   Dot   Pie   Options   Standard options   Styles   Appendix

Axis titles   Axis labels   Axis scales   Axis selection   By   Legend   Adding text   Textboxes   Text display

```
twoway scatter ownhome borninstate, by(north, title("By title"))
    title("Regular title")
```

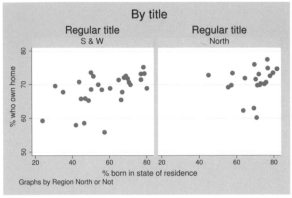

This example should help you to understand how these two types of titles work. When we use the `title()` option outside the `by()` option, it applies to all graphs that are created because it is repeated with the `by()` option. The `by(title())` option is applied after all smaller graphs are created, providing a title for the entire set of subgraphs.

Uses allstatesdc.dta & scheme vg_brite

```
twoway scatter ownhome borninstate, by(north)
    caption("Regular caption")
```

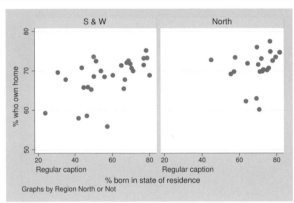

Stata treats the `caption()` option the same way that it treats titles. Here we include the `caption()` option outside the `by()` option. The caption is thus displayed with each graph.

In the Object Browser, expand **plotregion1**. Double-click on **caption[1]** to change the caption of the first graph and double-click on **caption[2]** to change the caption of the second graph.

Uses allstatesdc.dta & scheme vg_brite

```
twoway scatter ownhome borninstate, by(north, caption("By caption"))
```

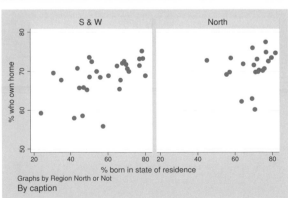

When we include the `caption()` option inside the `by()` option, it is displayed as a caption for the full graph.

In the Object Browser, double-click on **Graph**. Select the *Titles* tab and change the caption as desired.

Uses allstatesdc.dta & scheme vg_brite

```
twoway scatter ownhome borninstate, by(north)
    subtitle("This is a subtitle")
```

Stata treats the `subtitle()` option differently than the `title()` and `caption()` options. Here we include a `subtitle()` option, and we see that the subtitle has replaced the title above each graph that represented the names of the by-group.

Uses allstatesdc.dta & scheme vg_brite

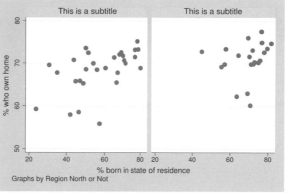

```
twoway scatter ownhome borninstate, by(north)
    subtitle("Region of state", prefix)
```

We can use the `subtitle()` option to add more labeling to the by-group names. Here we use the `prefix` option to insert text that appears in the subtitle before the name of the by-group.

📊 In the Contextual Toolbar, click on the subtitle and change the text.

Uses allstatesdc.dta & scheme vg_brite

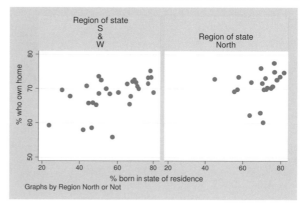

```
twoway scatter ownhome borninstate, by(north)
    subtitle("Region of state", suffix)
```

We can use the `suffix` option to insert text that appears in the subtitle after the name of the by-group.

📊 Click on the subtitle and, in the Contextual Toolbar, change the text.

Uses allstatesdc.dta & scheme vg_brite

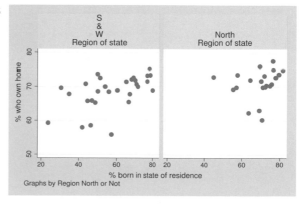

Introduction   Editor   Twoway   Matrix   Bar   Box   Dot   Pie   Options   Standard options   Styles   Appendix

Markers   Marker labels   Connecting   Axis titles   Axis labels   Axis scales   Axis selection   By   Legend   Adding text   Textboxes   Text display

```
twoway scatter ownhome borninstate, by(north)
   subtitle("State's location", prefix)
   subtitle("Based on Region", suffix)
```

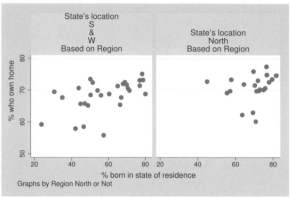

We can even combine the **prefix** and **suffix** options to insert text before and after the label of the by-group.

📖 Click on the subtitle and, in the Contextual Toolbar, change the text.
Uses allstatesdc.dta & scheme vg_brite

```
twoway scatter ownhome borninstate,
   by(north, subtitle("This is a subtitle"))
```

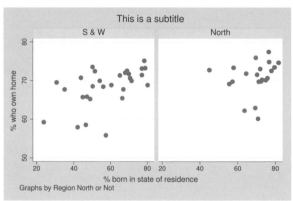

When used within the **by()** option, the **subtitle()** option works just like the **title()** and **caption()** options, placing a subtitle on the entire graph.

📖 In the Object Browser, double-click on **Graph**. Select the *Titles* tab and change the subtitle as desired.
Uses allstatesdc.dta & scheme vg_brite

```
twoway scatter ownhome borninstate, by(north) note("Regular note")
```

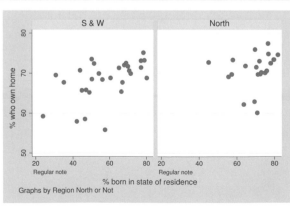

Stata treats the **note()** option much as it does the **title()**, **caption()**, and **subtitle()** options. Here we include a **note()** option and see that it is shown below both graphs.

📖 In the Object Browser, expand **plotregion1**. Double-click on **note[1]** to change the note of the first graph and double-click on **note[2]** to change the note of the second graph.
Uses allstatesdc.dta & scheme vg_brite

```
twoway scatter ownhome borninstate, by(north, note("By note"))
```

If we include `note()` within the `by()`
option, we see that our note overrides
the note that Stata provided to indicate
that the graphs were separated by the
variable `north`.

📖 Click on the overall note and, in the
Contextual Toolbar, change the text.
Uses allstatesdc.dta & scheme vg_brite

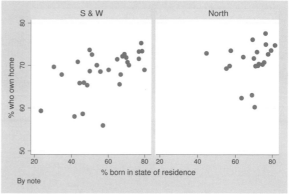

```
twoway scatter ownhome borninstate,
     by(north, note("North N=21, Not North N=29", suffix))
```

As shown with the `subtitle()` option,
the `prefix` or `suffix` option adds text
before or after the existing note.

📖 Click on the overall note and, in the
Contextual Toolbar, change the text.
Uses allstatesdc.dta & scheme vg_brite

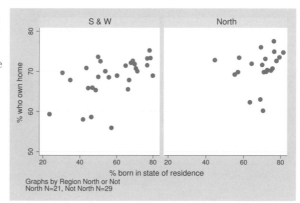

```
twoway scatter ownhome borninstate,
     by(north, total) subtitle(, position(11))
```

Previously, we saw that we could use
the `subtitle()` option to modify the
by-group names above each graph.
Here we use the `subtitle(,
position(11))` option to modify the
placement of this text to appear in the
eleven o'clock position.

📖 Double-click on a subtitle and, in
the *Format* tab, change the
justification to **Left**.
Uses allstatesdc.dta & scheme vg_brite

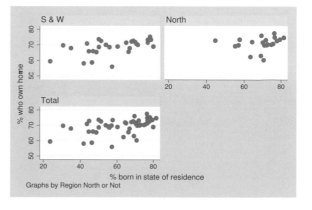

Introduction   Editor   Twoway   Matrix   Bar   Box   Dot   Pie   Options   Standard options   Styles   Appendix

Markers   Marker labels   Connecting   Axis titles   Axis labels   Axis scales   Axis selection   By   Legend   Adding text   Textboxes   Text display

```
twoway scatter ownhome borninstate,
    by(north, total) subtitle(, position(5) ring(0) nobexpand)
```

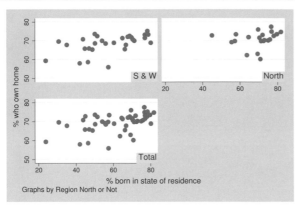

We can place the name of the by-group in the bottom right corner of each graph by using the `subtitle()` option. The options `position(5)` and `ring(0)` move the subtitle to the five o'clock position and inside the plot region. The `nobexpand` (no box expand) option prevents the by-group name from expanding to consume the entire plot region.

📈 See the next graph.

Uses allstatesdc.dta & scheme vg_brite

```
twoway scatter ownhome borninstate,
    by(north, total)
```

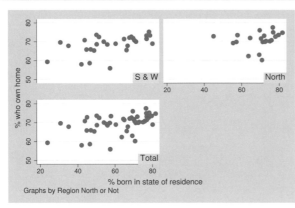

📈 Using the Graph Editor, select the Grid Edit 🔲 tool, click and hold *S & W*, and drag it into the middle of the graph. Using the Pointer ⬉ tool, double-click on *S & W* (which should be occupying the entire plot area), from the *Format* tab change the **Position** to Southeast, and uncheck **Expand area to fill cell**. Repeat this for the other subtitles.

Uses allstatesdc.dta & scheme vg_brite

```
twoway scatter ownhome borninstate,
    by(north, total title("My title", ring(0) position(5)))
```

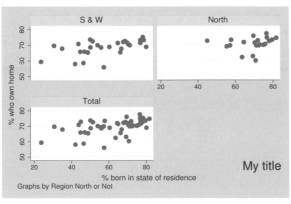

We can also use the `ring()` and `position()` options with `title()`, `note()`, and `caption()`. Here we use `position(5)` to put the title in the bottom right corner and `ring(0)` to locate it inside the plot region.

📈 Select the Grid Edit 🔲 tool and then point to the title and drag it into the center of the graph. Then, using the Pointer ⬉ tool, double-click on the title and, in the *Format* tab, change the **Position** to Southeast.

Uses allstatesdc.dta & scheme vg_brite

```
twoway scatter ownhome borninstate,
    by(north, total title("My title", position(5)))
```

We repeat the previous command without the `ring(0)` option to illustrate that without this option the title is placed outside the plot region.

⊞ Select the Grid Edit ⊞ tool and then point to the title and drag it to where the note is. Then, using the Pointer ↖ tool, double-click on the title and, in the *Format* tab, change the **Position** to **East**.

Uses allstatesdc.dta & scheme vg_brite

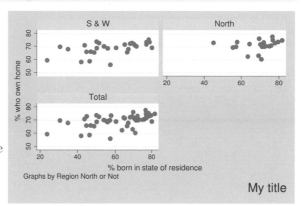

```
twoway scatter ownhome borninstate,
    by(north, total) l1title("left title") b1title("bottom title")
```

Including the `l1title()` option adds a title to the left (on the *y* axis) of each of the graphs. Likewise, the `b1title()` option adds a title to the bottom (on the *x* axis) of each of the graphs.

⊞ In the Object Browser, expand **positional titles** and double-click on **bottom 1** and **left 1** to add titles to the bottom and left.

Uses allstatesdc.dta & scheme vg_brite

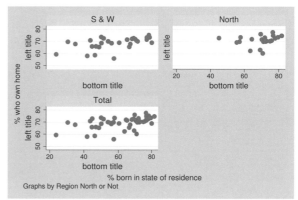

```
twoway scatter (borninstate propval100 ownhome), by(nsw)
    legend(label(1 "Born in state") label(2 "% > 100K"))
```

Here we use the `legend()` option to change the labels associated with the first two keys. These options modify the contents of the legend, so they should appear outside the `by()` option.

⊞ Simply double-click on items in the legend to change them.

Uses allstatesdc.dta & scheme vg_brite

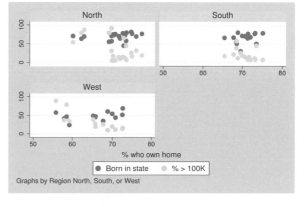

Introduction

Markers

Marker labels

Editor

Twoway

Connecting

Matrix

Bar

Axis titles

Box

Dot

Axis labels

Pie

Options

Axis scales

Standard options

Axis selection

Styles

By

Legend

Appendix

Adding text

Textboxes

Text display

```
twoway scatter (borninstate propval100 ownhome),
   by(nsw, legend(position(12)))
```

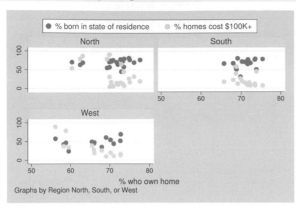

In this graph, we use the **position()** option to modify the position of the legend. Note that options that modify the position of the legend need to be placed within the **by()** option.

⌨ Using the Grid Edit 🖊 tool, in the Object Browser, click on **legend**. Then grab the legend and drag it to the top of the graph.

Uses allstatesdc.dta & scheme vg_brite

```
twoway scatter (borninstate propval100 ownhome),
   by(nsw, legend(position(12)))
   legend(label(1 "Born in state") label(2 "% > 100K"))
```

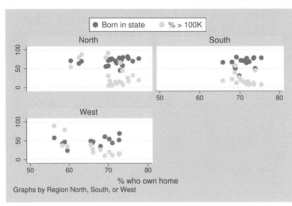

Here we use both options from the previous two graphs—the **legend()** option is used twice: inside the **by()** option to modify the legend's position and outside the **by()** option to modify the legend's contents. The use of **legend()** with the **by()** option is covered more in Options : Legend (361).

⌨ See the previous two examples.

Uses allstatesdc.dta & scheme vg_brite

```
twoway scatter ownhome borninstate,
   by(north, title("% own home" "by % born in state"))
   title("Region of state")
```

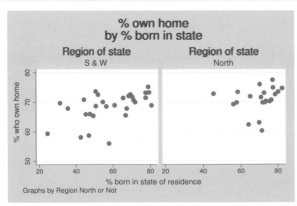

The **title()** option within the **by()** option makes a title for the entire graph, whereas the second **title()** option makes a title that is displayed for each graph.

⌨ See the previous two examples in this section regarding titles.

Uses allstatesdc.dta & scheme vg_brite

```
twoway scatter ownhome borninstate,
    by(north, total rescale ixtitle iytitle b1title("") l1title(""))
```

Here we obtain separate graphs for the
three groups, using `rescale` to obtain
different *x*- and *y*-axis labels and scales,
`ixtitle` and `iytitle` to title the
graphs separately, and `b1title()` and
`l1title()` to suppress the overall titles
for the *x* and *y* axes.

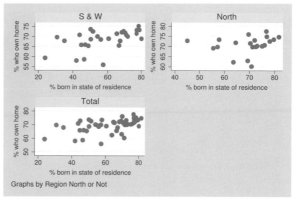

📈 See the previous examples in this
section.

Uses allstatesdc.dta & scheme vg_brite

9.9   Controlling legends

This section describes more details about using legends. Legends can be useful in several situations, and this section shows how to customize them. For more information about
legend options, see [G-3] *legend_options*. Also, for controlling the text and textbox of the
legend, see Options : Textboxes (379) and Options : Adding text (374). This section uses the
vg_s2c scheme.

```
twoway scatter ownhome propval100 urban
```

Legends can be created in many ways.
For example, here we have two *y*
variables, `ownhome` and `propval100`, on
the same plot, and Stata creates a
legend labeling the different points.
The default legend for this graph is
quite useful.

Uses allstatesdc.dta & scheme vg_s2c

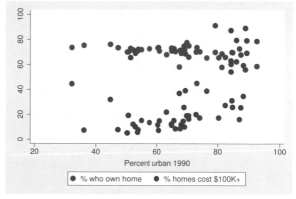

```
twoway (scatter ownhome urban) (lfit ownhome urban)
   (qfit ownhome urban)
```

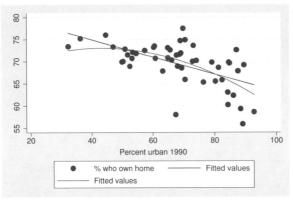

Legends are also created when we overlay plots. Here Stata adds a legend entry for each overlaid plot. The default legend is less useful because it does not help us differentiate between the kinds of fit values.

Uses allstatesdc.dta & scheme vg_s2c

```
twoway (scatter ownhome urban if north==0)
   (scatter ownhome urban if north==1)
```

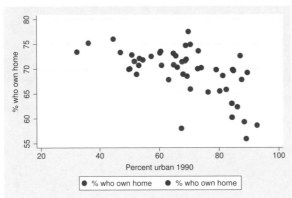

A third example is when we overlay two plots by using `if` to display the same variables but for different observations. Here we show the same scatterplot separately for states in the North and for those not in the North. This legend does not help us differentiate the markers.

Uses allstatesdc.dta & scheme vg_s2c

```
twoway (scatter ownhome urban) (lfit ownhome urban)
   (qfit ownhome urban)
```

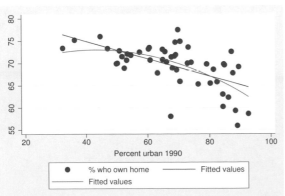

Regardless of the graph command(s) that generated the legend, it can be customized the same way. For many examples, we will use this graph for customizing the legend.

Uses allstatesdc.dta & scheme vg_s2c

```
twoway (scatter ownhome urban) (lfit ownhome urban)
    (qfit ownhome urban),
    legend(label(1 "% Own home") label(2 "Lin. Fit") label(3 "Quad. Fit"))
```

The label() option assigns labels for the keys. Note that we can use a separate label() option to modify each key.

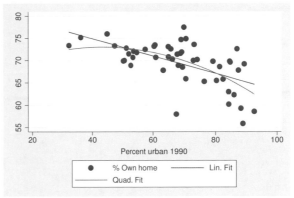 Click on any of the labels in the legend and, in the Contextual Toolbar, change the **Text**.

Uses allstatesdc.dta & scheme vg_s2c

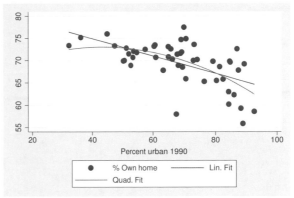

```
twoway (scatter ownhome urban) (lfit ownhome urban)
    (qfit ownhome urban),
    legend(label(2 "Lin. Fit") label(3 "Quad. Fit"))
```

We can also use the label() option to modify some of the keys. Here we modify only the second and third keys.

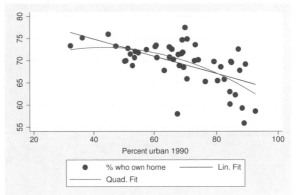 Click on the second label in the legend and, in the Contextual Toolbar, change the **Text**. Repeat the process for the third key.

Uses allstatesdc.dta & scheme vg_s2c

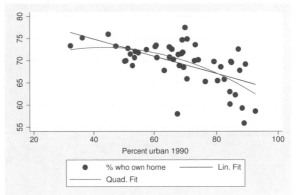

```
twoway (scatter ownhome urban) (lfit ownhome urban)
    (qfit ownhome urban),
    legend(label(1 "%own" "home") label(2 "Lin" "Fit") label(3 "Qd" "Fit"))
```

We can place the label on multiple lines by including multiple quoted strings.

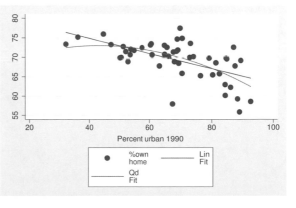 Use quotation marks in the same way in the Graph Editor to make titles appear on separate lines.

Uses allstatesdc.dta & scheme vg_s2c

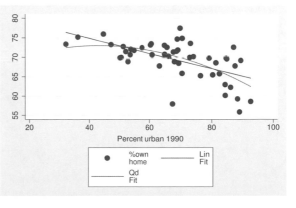

Introduction · Editor · Twoway · Matrix · Bar · Box · Dot · Pie · Options · Standard options · Styles · Appendix

Markers · Marker labels · Connecting · Axis titles · Axis labels · Axis scales · Axis selection · By · Legend · Adding text · Textboxes · Text display

```
twoway (scatter ownhome urban) (lfit ownhome urban)
   (qfit ownhome urban), legend(order(2 3 1))
```

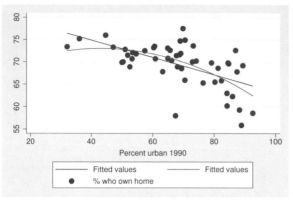

The `order()` option changes the order of the keys in the legend.

Although it is possible to change the order of the keys in the legend by using the Graph Editor, it is much simpler to do this with the `order()` option.

Uses allstatesdc.dta & scheme vg_s2c

```
twoway (scatter ownhome urban) (lfit ownhome urban)
   (qfit ownhome urban), legend(order(2 3))
```

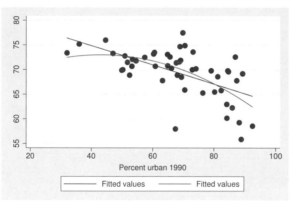

We can also omit keys from the `order()` option to suppress their display in the legend. Here we suppress the display of the first key.

Right-click on the key for the first key, `ownhome`, and choose **Hide**. Likewise, right-click on the label for `ownhome` and choose **Hide**.

Uses allstatesdc.dta & scheme vg_s2c

```
twoway (scatter ownhome urban) (lfit ownhome urban)
   (qfit ownhome urban), legend(order(2 "Lin. fit" 3 "Quad. fit"))
```

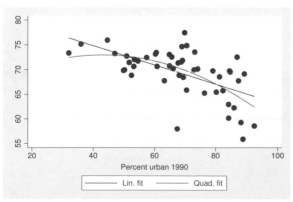

We can also insert and replace text for the keys with the `order()` option. Here we hide the first key and replace the text for keys 2 and 3.

See the previous example for how to hide the first key. Then, within the Contextual Toolbar, click on the text for the second and third key and change it.

Uses allstatesdc.dta & scheme vg_s2c

```
twoway (scatter ownhome urban) (lfit ownhome urban)
   (qfit ownhome urban),
   legend(order(- "Fitted" 2 "Lin. fit" 3 "Quad. fit" - "Observed" 1))
```

We use - "Fitted" to insert the word *Fitted* and - "Observed" to insert the word *Observed*. Because of the organization of the keys in the legend, the example of this option is hard to follow.

Uses allstatesdc.dta & scheme vg_s2c

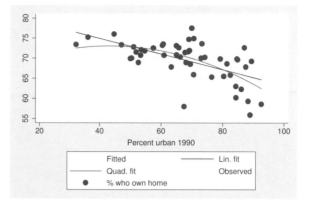

```
twoway (scatter ownhome urban) (lfit ownhome urban) (qfit ownhome urban),
   legend(order(- "Fitted" 2 "Lin. fit" 3 "Quad. fit" - "Observed" 1)
   cols(1))
```

The cols() option displays the legend in one column. Here the added text makes more sense, but the legend uses quite a bit of space.

📖 In the Object Browser, double-click on **legend** and, in the *Organization* tab, change **Rows/Columns** to Columns followed by 1.

Uses allstatesdc.dta & scheme vg_s2c

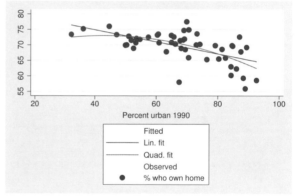

```
twoway (scatter ownhome urban) (lfit ownhome urban) (qfit ownhome urban),
   legend(order(- "Fitted" 2 "Lin. fit" 3 "Quad. fit" - "Observed" 1)
   rows(3))
```

Here we use the rows() option to display the legend in three rows. The next example shows how we can display the fitted keys in the left column and the observed keys in the right column.

📖 In the Object Browser, double-click on **legend** and, in the *Organization* tab, change **Rows/Columns** to Rows followed by 3.

Uses allstatesdc.dta & scheme vg_s2c

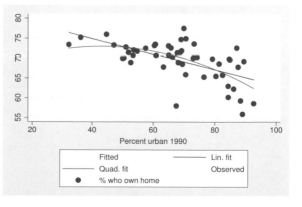

Introduction   Editor   Twoway   Matrix   Bar   Box   Dot   Pie   Options   Standard options   Styles   Appendix

Markers   Marker labels   Connecting   Axis titles   Axis labels   Axis scales   Axis selection   By   Legend   Adding text   Textboxes   Text display

```
twoway (scatter ownhome urban) (lfit ownhome urban) (qfit ownhome urban),
    legend(order(- "Fitted" 2 "Lin. fit" 3 "Quad. fit" - "Observed" 1)
    rows(3) colfirst)
```

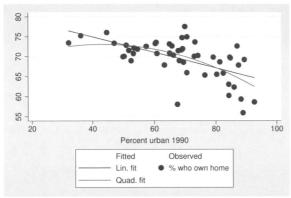

Adding the `colfirst` option displays the keys in column order instead of row order, with the *Fitted* keys in the left column and the *Observed* keys in the right column.

In the Object Browser, double-click on **legend** and, in the *Organization* tab, change **Key Sequence** to Down first.

Uses allstatesdc.dta & scheme vg_s2c

```
twoway (scatter ownhome urban) (lfit ownhome urban) (qfit ownhome urban),
    legend(order(- "Observed" 1 - "Fitted" 2 "Lin. fit" 3 "Quad. fit")
    rows(3) colfirst)
```

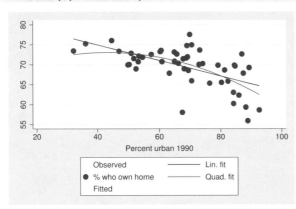

This legend is the same as the one in the previous example, but here we attempt to place the *Observed* keys in the left column and the *Fitted* keys in the right column. However, the word *Fitted* appears at the bottom of the first column instead of the top of the second column.

Uses allstatesdc.dta & scheme vg_s2c

```
twoway (scatter ownhome urban) (lfit ownhome urban) (qfit ownhome urban),
    legend(order(- "Observed" 1 - "Fitted" 2 "Lin. fit" 3 "Quad. fit")
    rows(3) holes(3) colfirst)
```

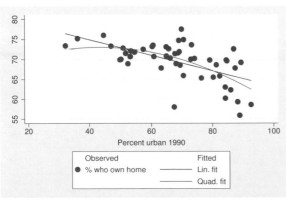

This legend is the same as the one in the previous two examples, but here we add the `holes(3)` option so a blank key is placed in the third position forcing the word *Fitted* in the fourth position at the top of the second column.

In the Object Browser, double-click on **legend** and, in the *Organization* tab, place a 3 in **Positions to leave blank**.

Uses allstatesdc.dta & scheme vg_s2c

```
twoway (scatter ownhome urban) (lfit ownhome urban) (qfit ownhome urban),
    legend(order(- "Observed" 1 - " " - "Fitted" 2 "Lin fit" 3 "Qd fit")
    rows(3) colfirst)
```

Instead of using `holes()` as we did in the previous example to insert a blank key, we use `- " "` in the `order()` option, which pushes the word *Fitted* to the next column.

Uses allstatesdc.dta & scheme vg_s2c

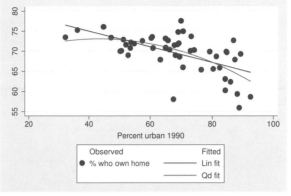

```
twoway (scatter ownhome urban) (lfit ownhome urban) (qfit ownhome urban),
    legend(order(- "Observed" 1 - " " - "Fitted" 2 "Lin fit" 3 "Qd fit")
    rows(3) colfirst textfirst)
```

With the `textfirst` option, the text for the key appears first, followed by the symbol.

In the Object Browser, double-click on **legend** and, in the *Organization* tab, change the **Symbol order** to Labels first.

Uses allstatesdc.dta & scheme vg_s2c

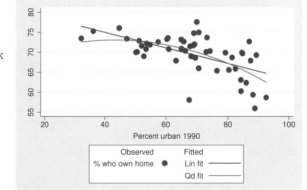

```
twoway (scatter ownhome urban) (lfit ownhome urban) (qfit ownhome urban),
    legend(order(2 "Linear" "Fit" 3 "Quadratic" "Fit")
    stack cols(1))
```

The `stack` option stacks the symbols above the labels. We use this option here to make a tall, narrow legend.

In the Object Browser, double-click on **legend** and, in the *Organization* tab, change **Stack symbols and text** to Yes.

Uses allstatesdc.dta & scheme vg_s2c

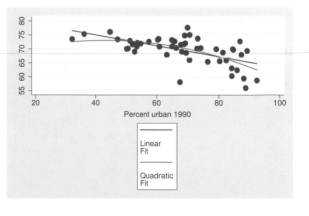

```
twoway (scatter ownhome urban) (lfit ownhome urban) (qfit ownhome urban),
    legend(order(2 "Linear" "Fit" 3 "Quadratic" "Fit")
    stack cols(1) position(3))
```

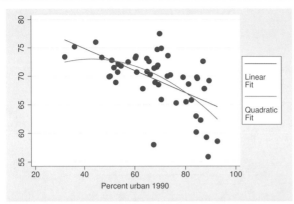

In this example, we use the `position()` option to move the narrow legend to the three o'clock position, right of the graph.

 Select the Grid Edit tool and, in the Object Browser, click on **legend**. The legend should be selected and gridlines should appear in the graph. Click and hold and drag the legend to the right of the plot.

Uses allstatesdc.dta & scheme vg_s2c

```
twoway (scatter ownhome urban) (lfit ownhome urban) (qfit ownhome urban),
    legend(order(2 "Linear" "Fit" 3 "Quadratic" "Fit")
    stack cols(1) ring(0) position(7))
```

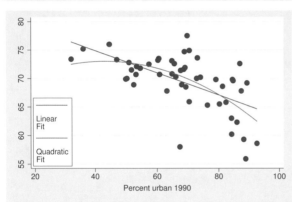

We now use the `ring(0)` option to place the legend inside the plot area and use `position(7)` to put it in the bottom left corner, using the empty space in the plot for the legend.

 Select the legend in the same way as the previous example and drag the legend to the center of the plot area. Then, using the Pointer tool, select the legend with the Object Browser and drag it to the bottom right corner.

Uses allstatesdc.dta & scheme vg_s2c

```
twoway (scatter ownhome urban) (lfit ownhome urban) (qfit ownhome urban),
    legend(order(1 "% Own Home" 2 "Linear" 3 "Quad")
    rows(1) position(12))
```

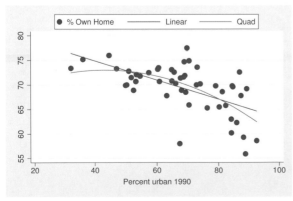

Here we make the legend a thin row at the top of the graph by using the `rows(1)` and `position(12)` options.

 As described in the previous examples, use the Grid Edit tool to move the legend to the top of the graph. In the Object Browser, double-click on **legend** and use **Rows/Columns** to make the legend 1 row.

Uses allstatesdc.dta & scheme vg_s2c

```
twoway (scatter ownhome urban) (lfit ownhome urban) (qfit ownhome urban),
    legend(order(1 "% Own Home" 2 "Linear" 3 "Quad")
    rows(1) position(12) bexpand)
```

We can expand the width of the legend
to the width of the plot area by using
the bexpand (box expand) option.

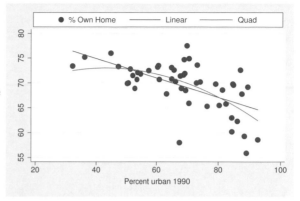

In the Object Browser, double-click
on **legend** and, in the *Advanced* tab,
check **Expand area to fill cell in the
X dimension**.

Uses allstatesdc.dta & scheme vg_s2c

```
twoway (scatter ownhome urban) (lfit ownhome urban) (qfit ownhome urban),
    legend(order(2 "Linear Fit" 3 "Quadratic Fit")
    rows(1) position(12) bexpand span)
```

The span option expands the legend to
the entire width of the graph area (not
just the plot area).

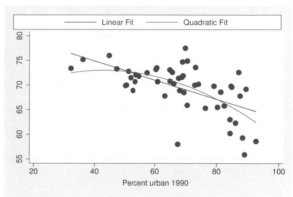

Choose the Grid Edit tool and,
in the Object Browser, click on **legend**.
In the Contextual Toolbar, choose
Expand Cell and then **Left 1 Cell**.

Uses allstatesdc.dta & scheme vg_s2c

```
twoway (scatter ownhome urban) (lfit ownhome urban) (qfit ownhome urban),
    legend( rows(1) title("Legend"))
```

We can add a title, subtitle, note, or
caption to the legend by using all the
features described in Standard
options · Titles (395). Here we add the
title() option to include a title in the
legend. A simple way to get a smaller
title is to use the subtitle() option.

In the Object Browser, double-click
on **legend** and choose the *Titles* tab,
where we can add a title as well as a
subtitle, caption, or note.

Uses allstatesdc.dta & scheme vg_s2c

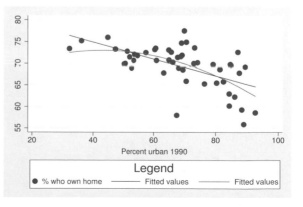

```
twoway (scatter ownhome urban) (lfit ownhome urban) (qfit ownhome urban),
    legend( rows(1) title("Legend", color(red) size(huge)))
```

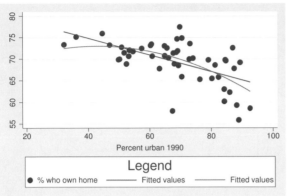

By using the `color()` and `size()` options, we make the legend title red and huge.

🖿 Click on the legend title and, from the Contextual Toolbar, change the **Color** to Red and the **Size** to Huge.

Uses allstatesdc.dta & scheme vg_s2c

```
twoway (scatter ownhome urban) (lfit ownhome urban) (qfit ownhome urban),
    legend(rows(1) title("Legend", color(red) size(huge) box bexpand))
```

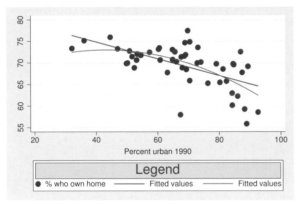

To emphasize the control we have, we put the title for the legend in a `box` and use `bexpand` to expand it the width of the legend.

🖿 Continuing from the previous graph, double-click on the legend title. In the *Box* tab, check **Place box around text** and, in the *Format* tab, check **Expand area to fill cell**.

Uses allstatesdc.dta & scheme vg_s2c

```
twoway (scatter ownhome urban) (lfit ownhome urban) (qfit ownhome urban),
    legend(note("Fit obtained with lfit and qfit"))
```

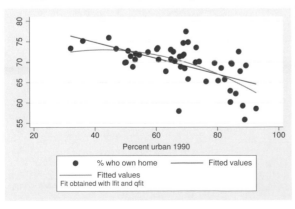

Here we use the `note()` option to show that a note can be added to the legend.

🖿 In the Object Browser, double-click on **legend** and, in the *Titles* tab, type in the note.

Uses allstatesdc.dta & scheme vg_s2c

```
twoway (scatter ownhome urban) (lfit ownhome urban) (qfit ownhome urban),
    legend(size(large) color(maroon) fcolor(eggshell) box)
```

We can also control the display of the labels for the keys with the `legend()` option. Here we request that those labels be large and maroon and be displayed with an eggshell background surrounded by a box.

📖 In the Object Browser, double-click on **legend** and, in the *Labels* tab, change the **size** to `Large`, change the **Color** to `Maroon`, check **Place box around text**, and change the **Fill color** to `Eggshell`.

Uses allstatesdc.dta & scheme vg_s2c

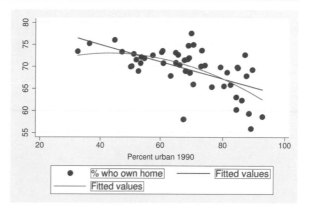

```
twoway (scatter ownhome urban) (lfit ownhome urban) (qfit ownhome urban),
    legend(region(fcolor(dimgray) lcolor(navy) lwidth(thick) margin(medium)))
```

The `region()` option controls the overall box in which the legend is placed. Here we specify the fill color dim gray and the line thick and navy with a medium-sized margin between the text and the box.

📖 In the Object Browser, double-click on **legend** and choose the *Region* tab. Change the **Margin** to `Medium`, the **Color** to `Dim Gray`, the **Outline Width** to `Thick`, and check **Different Outline Color** and change the **Outline color** to `Navy`.

Uses allstatesdc.dta & scheme vg_s2c

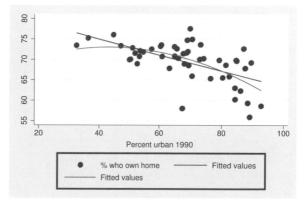

```
twoway (scatter ownhome urban) (lfit ownhome urban) (qfit ownhome urban),
    legend(rows(1) bmargin(t=10))
```

The `bmargin()` option adjusts the margin around the box of the legend. Here we make the margin 10 at the top, which increases the gap between the legend and the title of the *x* axis.

📖 In the Object Browser, double-click on **legend** and choose the *Advanced* tab. Change the **Outer Margin** to `Custom` and click on the **..** button, and change the top margin to `10` and the others to `0`.

Uses allstatesdc.dta & scheme vg_s2c

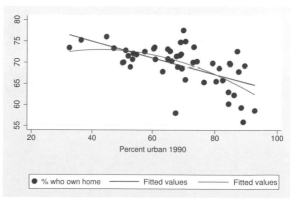

Introduction   Editor   Twoway   Matrix   Bar   Box   Dot   Pie   Options   Standard options   Styles   Appendix

Markers   Marker labels   Connecting   Axis titles   Axis labels   Axis scales   Axis selection   By   Legend   Adding text   Textboxes   Text display

```
twoway (scatter ownhome urban) (lfit ownhome urban) (qfit ownhome urban),
   legend(symxsize(30) symysize(20))
```

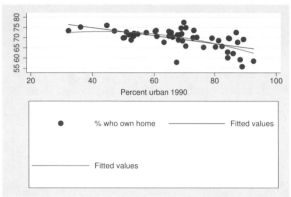

We control the width allocated to symbols with the symxsize() option and the height with the symysize() option.

In the Object Browser, double-click on **legend**. Next to **Key y size** enter 20 and next to **Key x size** enter 30.

Uses allstatesdc.dta & scheme vg_s2c

```
twoway (scatter ownhome urban) (lfit ownhome urban) (qfit ownhome urban),
   legend(colgap(25) rowgap(20))
```

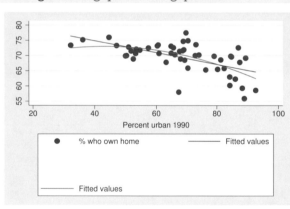

Here we control the space between columns of the legend with the colgap() option and the space between the rows with the rowgap() option. The rowgap() option does not affect the border between the top row and the box or the border between the bottom row and the box.

In the Object Browser, double-click on **legend**. Next to **Row gap** enter 20 and next to **Column gap** enter 25.

Uses allstatesdc.dta & scheme vg_s2c

```
twoway (scatter ownhome urban) (qfit ownhome urban),
   by(nsw)
```

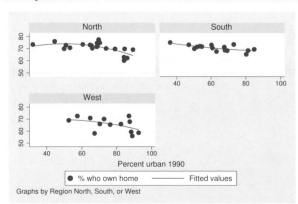

Consider this graph showing two overlaid scatterplots shown separately by the location of the state. We will now explore how to modify the legend for this kind of graph.

Uses allstatesdc.dta & scheme vg_s2c

```
twoway (scatter ownhome urban) (qfit ownhome urban),
    by(nsw) legend(position(12) label(2 "Quadratic Fit"))
```

Here we add a `legend()` option, but
the `position()` option has no effect
because the position of the legend did
not change.
Uses allstatesdc.dta & scheme vg_s2c

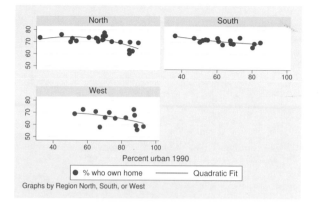

```
twoway (scatter ownhome urban) (qfit ownhome urban),
    by(nsw, legend(position(12))) legend(label(2 "Quadratic Fit"))
```

The previous command did not change
the position of the legend because
options for positioning the legend must
be placed within the `by()` option. Here
we place the `legend(position())`
option within the `by()` option, and the
legend is now placed above the graph.
📊 Using the Grid Edit 📊 tool, drag
the legend to the top of the graph.
Uses allstatesdc.dta & scheme vg_s2c

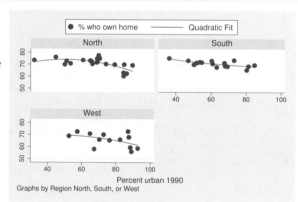

```
twoway (scatter ownhome urban) (qfit ownhome urban), by(nsw, legend(off))
```

Likewise, if we wish to turn the legend
off, we must place `legend(off)` within
the `by()` option.
📊 In the Object Browser, right-click on
legend and choose **Hide**. You can
reshow the legend by repeating this
step and instead choosing **Show**.
Uses allstatesdc.dta & scheme vg_s2c

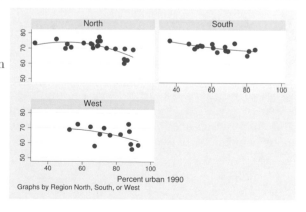

Introduction
Markers
  Marker labels
  Connecting
  Axis titles
  Axis labels
  Axis scales
  Axis selection
  By
  Legend
  Adding text
  Textboxes
  Text display
Editor
Twoway Matrix Bar Box Dot Pie Options Standard options Styles Appendix

```
twoway (scatter ownhome urban) (qfit ownhome urban),
   by(nsw, legend(at(4))) legend(cols(1))
```

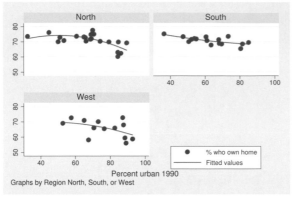

Here we place the legend in the fourth hole by using the at(4) option within the by() option. To display the legend in one column, we use the legend(cols(1)) option outside the by() option because this option does not effect the position of the legend.

Using the Grid Edit tool, drag the legend into the center of the graph. Then, using the Pointer tool, click and hold the legend and drag it to the desired position.

Uses allstatesdc.dta & scheme vg_s2c

```
twoway (scatter ownhome urban) (qfit ownhome urban),
   by(nsw, legend(position(center) at(4))) legend(cols(1))
```

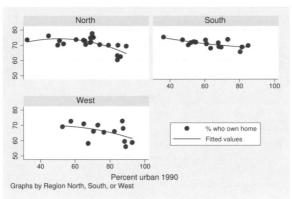

In this example, we add the position(center) option within the by() option to make the legend appear in the center of the fourth position.

See the previous example.

Uses allstatesdc.dta & scheme vg_s2c

9.10   Adding text to markers and positions

This section provides more details about the text() option for adding text to a graph. Although we can use added text in a wide variety of situations, this section will focus on how we can use it to label points and lines and to add descriptive text to our graph. For more information about the text() option, see [G-3] *added_text_options*. To learn more about how you can customize the text, see Options: Textboxes (379). This section uses the vg_teal scheme.

twoway scatter ownhome borninstate

In this scatterplot, one point appears to be an outlier. Because it is not labeled, we cannot tell from which state it originates.

Uses allstatesn.dta & scheme vg_teal

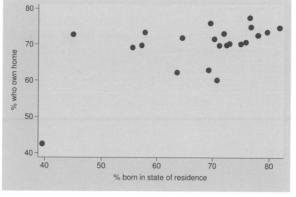

scatter ownhome borninstate, mlabel(stateab)

We use the `mlabel(stateab)` to label all points, which helps us see that the outlying point comes from Washington, DC. However, this plot is rather cluttered by all the labels.

Uses allstatesn.dta & scheme vg_teal

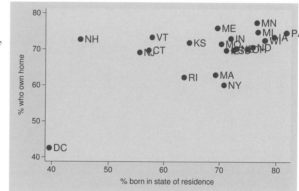

twoway (scatter ownhome borninstate)
(scatter ownhome borninstate if stateab == "DC", mlabel(stateab))

Here we repeat a second scatterplot just to label DC, but this is a bit cumbersome.

Simply select the Add Text **T** tool and click next to the marker for DC and add the desired text.

Uses allstatesn.dta & scheme vg_teal

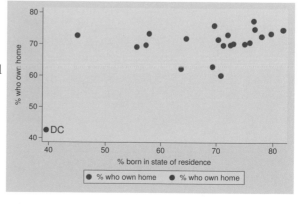

Introduction
Markers
Marker labels
Connecting
Axis titles
Axis labels
Axis scales
Axis selection
By
Legend
Adding text
Textboxes
Text display

Editor
Twoway
Matrix
Bar
Box
Dot
Pie
Options
Standard options
Styles
Appendix

`twoway scatter ownhome borninstate, text(43 40 "DC")`

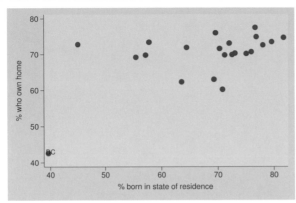

Instead, we use the `text()` option to add text to our graph. Looking at the values of `ownhome` and `borninstate` for DC, we see that their values are about 43 and 40, respectively. We use these as coordinates to label the point, but the `text()` option places the label at the center of the specified *y x* coordinate, sitting right over the point.

See the previous example.

Uses allstatesn.dta & scheme vg_teal

`twoway scatter ownhome borninstate, text(43 40 "DC", placement(ne))`

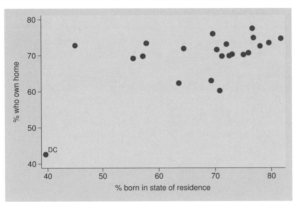

Adding the `placement(ne)` option within the `text()` option places the label above and to the right (northeast) of the point. Available placements include n, ne, e, se, s, sw, w, nw, and c (center); see Styles : Compassdir (415) for more details.

See the next graph.

Uses allstatesn.dta & scheme vg_teal

`twoway scatter ownhome borninstate`

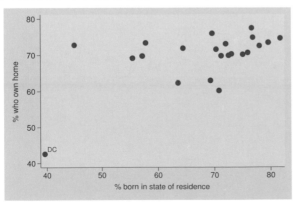

It is simple to add text to a graph using the Graph Editor. Select the Add Text **T** tool and click on the location where the text should be added (e.g., near the DC point). Then add DC to the **Text** and change the **Size** and **Color** as desired (here we choose `Medium` and `Navy`).

Uses allstatesn.dta & scheme vg_teal

```
twoway (scatter ownhome borninstate, text(43 40 "DC", placement(e)))
       (lfit ownhome borninstate) (lfit ownhome borninstate if stateab !="DC")
```

Consider this scatterplot showing a
linear fit between the two variables: one
including Washington, DC, and one
omitting Washington, DC. See the next
graph, which uses the `text()` option to
label the graph instead of the legend.
Uses allstatesn.dta & scheme vg_teal

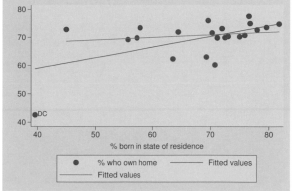

```
twoway (scatter ownhome borninstate, text(43 40 "DC", placement(ne)))
       (lfit ownhome borninstate) (lfit ownhome borninstate if stateab !="DC",
       text(72 50 "Without DC") text(60 50 "With DC")), legend(off)
```

This graph turns the legend off and
uses the `text()` option to label each
regression line to indicate which
regression line includes DC and which
one excludes DC.
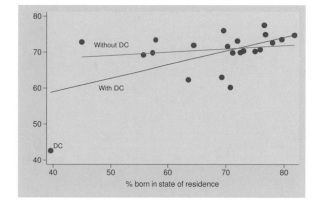 See the next graph.
Uses allstatesn.dta & scheme vg_teal

```
twoway (scatter ownhome borninstate, )
       (lfit ownhome borninstate)
       (lfit ownhome borninstate if stateab !="DC")
```

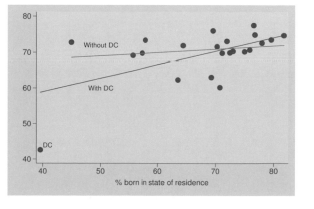 The customizations illustrated in
the previous graph can easily be done
with the Graph Editor. To hide the
legend, from the Object Browser,
right-click on **legend** and select **Hide**.
Then select the Add Text **T** tool and
click on the graph where the text
should be added. Add `With DC` to
Text. We can repeat this process to
label the other line `Without DC`.
Uses allstatesn.dta & scheme vg_teal

```
twoway (scatter ownhome borninstate, text(43 40 "DC", placement(ne)))
   (lfit ownhome borninstate) (lfit ownhome borninstate if stateab !="DC",
   text(71 50 "Without DC") text(60 50 "With DC")
   text(50 70 "Coef with DC .16" "Coef without DC .44")), legend(off)
```

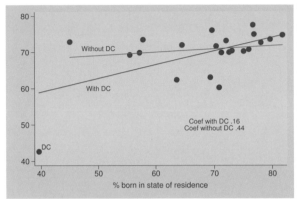

This graph adds explanatory text, showing the regression coefficient with and without DC.

⌨ Using the Add Text **T** tool, click where the explanatory text should be added. We can use double quotes as illustrated in the **text()** option to make the text appear on two lines.

Uses allstatesn.dta & scheme vg_teal

```
twoway (scatter ownhome propval100, xaxis(1) mlabel(stateab))
   (scatter ownhome borninstate, xaxis(2) mlabel(stateab))
```

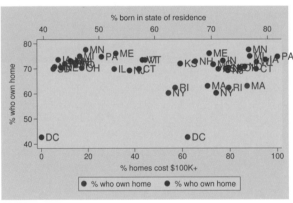

Consider this graph in which we overlay two scatterplots. We place **propval100** on the first *x* axis and **borninstate** on the second *x* axis.

Uses allstatesn.dta & scheme vg_teal

```
twoway (scatter ownhome propval100, xaxis(1))
   (scatter ownhome borninstate, xaxis(2)),
   text(43 66 "DC") text(43 42 "DC", xaxis(2))
```

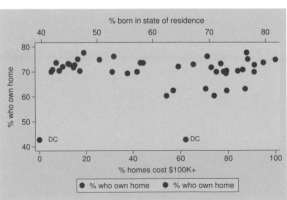

Rather than labeling all the points, we can label just the point for DC. We must be careful because we have two different *x* axes. The first **text()** option uses the first *x* axis, so no special option is required. The second **text()** option uses the second *x* axis, so we must specify the **xaxis(2)** option.

⌨ Using the Add Text **T** tool, add the text.

Uses allstatesn.dta & scheme vg_teal

9.11   Options for text and textboxes

This section describes more options for modifying textbox elements: titles, captions, notes, added text, and legends. Technically, all text in a graph is displayed within a textbox. We can modify the box's attributes, such as its size and color, the margin around the box, and the outline; and we can modify the attributes of the text within the box, such as its size, color, justification, and margin. These examples sometimes include the box option to see how both the textbox and its text are being displayed. This helps us to see if we should modify the attributes of the box containing the text or the text within the box. For more information, see [G-3] ***textbox_options*** and Options: Adding text (374). This section begins by showing examples illustrating how to control the placement, size, color, and orientation of text. This section begins using the vg_s1m scheme.

```
twoway scatter ownhome borninstate,
    text(43 40 "Washington, DC", placement(ne))
```

Consider this scatterplot, which has a dramatic outlying point. We have used the text() option to label that point, but, perhaps, we might want to control the size of the text for this label. See the next example for an illustration of how to do this.

Select the Add Text **T** tool and click where the text should begin and type in the desired text.

Uses allstatesn.dta & scheme vg_s1m

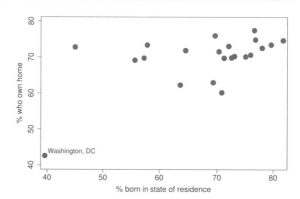

```
twoway scatter ownhome borninstate,
    text(43 40 "Washington, DC", placement(ne) size(vlarge))
```

We can alter the size of the text by using the size() option. Here we make the text very large. Available sizes include zero, miniscule, quarter_tiny, third_tiny, half_tiny, tiny, vsmall, small, medsmall, medium, modlarge, large, vlarge, huge, and vhuge; see Styles: Textsize (428) for more details.

With the Pointer ▸ tool, click on the text and, in the Contextual Toolbar, change the **Size** to v. Large.

Uses allstatesn.dta & scheme vg_s1m

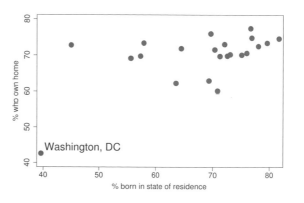

```
twoway scatter ownhome borninstate,
    text(43 40 "Washington, DC", placement(ne) color(gs9))
```

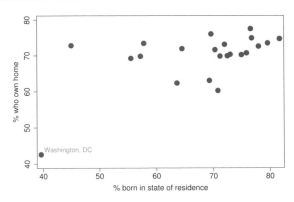

We can alter the color of the text by using the `color()` option. Here we make the text a middle-level gray. See Styles : Colors (412) for other colors.

🖼 Click on the text and, in the Contextual Toolbar, change the **Color** to **Gray 9**.

Uses allstatesn.dta & scheme vg_s1m

```
twoway scatter ownhome borninstate,
    text(43 40 "Washington, DC", placement(ne) orientation(vertical))
```

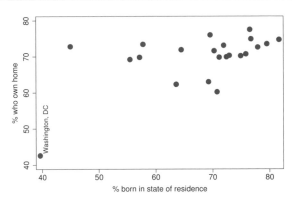

The `orientation()` option changes the direction of the text. Available orientations include **horizontal** for 0 degrees, **vertical** for 90 degrees, **rhorizontal** for 180 degrees, and **rvertical** for 270 degrees. See Styles : Orientation (426) for more details.

🖼 Double-click on the text and, in the *Format* tab, change the **Orientation** to **Vertical**.

Uses allstatesn.dta & scheme vg_s1m

This next set of examples considers options for justifying text within a box, sizing the box, and creating margins around the box. This is followed by options that control margins within the textbox. These examples use the **vg_rose** scheme.

```
twoway (scatter ownhome borninstate),
   title("% who own home by" "% that reside in state of birth", box)
```

Consider this example where we place a title on our graph. To help show how the options work, we put a box around the title.

📖 Double-click on the title, select the *Box* tab, and check **Place a box around text**.

Uses allstatesn.dta & scheme vg_rose

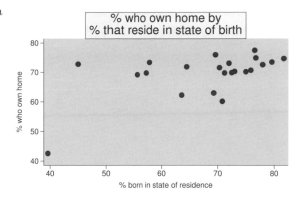

```
twoway (scatter ownhome borninstate),
   title("% who own home by" "% that reside in state of birth", box
   justification(left))
```

In this example, we left justify the text by using the justification() option. The title is justified within the textbox, not with respect to the entire graph area.

📖 Double-click on the title and, in the *Format* tab, change the **Justification** to Left.

Uses allstatesn.dta & scheme vg_rose

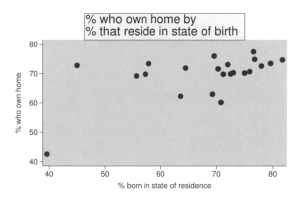

```
twoway (scatter ownhome borninstate),
   title("% who own home by" "% that reside in state of birth", box
   bexpand)
```

By using the bexpand (box expand) option, the textbox containing the title expands to fill the width of the plot area.

📖 Double-click on the title, select the *Format* tab, and check **Expand area to fill cell**.

Uses allstatesn.dta & scheme vg_rose

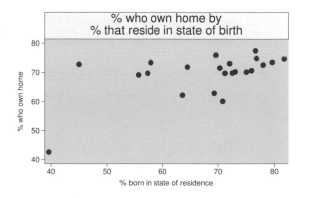

Introduction
Markers   Marker labels   Connecting
Editor   Twoway   Matrix   Bar   Box   Dot   Pie
Axis titles   Axis labels   Axis scales   Axis selection   By   Legend   Adding text   Textboxes   Text display
Options   Standard options   Styles   Appendix

```
twoway (scatter ownhome borninstate),
   title("% who own home by" "% that reside in state of birth", box
   bexpand justification(left))
```

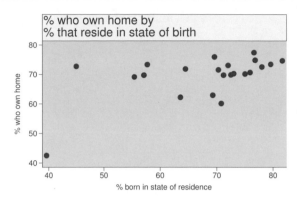

With the box expanded, the
justification(left) option now
makes the title flush left with the plot
area.

Double-click on the title, select the
Format tab, and change the
Justification to Left.
Uses allstatesn.dta & scheme vg_rose

```
twoway (scatter ownhome borninstate),
   title("% who own home by" "% that reside in state of birth",
   box bexpand justification(left) bmargin(medium))
```

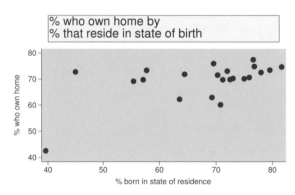

We can change the size of the margin
around the outside of the box by using
the bmargin(medium) (box margin)
option. Here we make the margin
medium-sized at all four edges: left,
right, top, and bottom.

Double-click on the title, select the
Box tab, and change the **Outer
margin** to Medium.
Uses allstatesn.dta & scheme vg_rose

```
twoway (scatter ownhome borninstate),
   title("% who own home by" "% that reside in state of birth",
   box bexpand justification(left) bmargin(0 0 3 3))
```

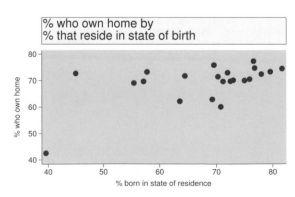

The margin around the title is 0 for the
left and right and 3 for the top and
bottom. The order of the margins is
bmargin($\#_{\text{left}}$ $\#_{\text{right}}$ $\#_{\text{top}}$ $\#_{\text{bottom}}$).

Double-click on the title, select the
Box tab, and change the **Outer
margin** to Custom. Then click on the ..
button, and specify the margin for the
top, bottom, left, and right.
Uses allstatesn.dta & scheme vg_rose

```
twoway (scatter ownhome borninstate),
   title("% who own home by" "% that reside in state of birth",
   box bmargin(b=3))
```

To make only the bottom margin 3, we specify `bmargin(b=3)`, where b=3 means to change the bottom margin to 3. The left, right, top, and bottom margins can be changed individually using l=, r=, t=, and b=, respectively.

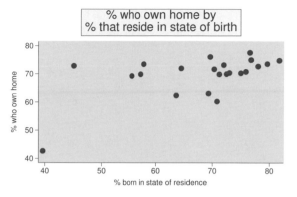

Double-click on the title, select the *Box* tab, and change the **Outer margin** to `Custom`. Then click on the `..` button and then specify 3 for the bottom margin and 0 for the rest.

Uses allstatesn.dta & scheme vg_rose

```
twoway (scatter ownhome borninstate),
   title("% who own home by" "% that reside in state of birth",
   box margin(medium))
```

We can expand the margin between the text and the box by using the `margin()` option. Note the difference between this and the `bmargin()` option (illustrated previously) is the increased margin around the text inside the box.

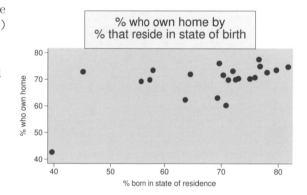

Double-click on the title, select the *Text* tab, and change the **Margin** to `Medium`.

Uses allstatesn.dta & scheme vg_rose

```
twoway (scatter ownhome borninstate),
   title("% who own home by" "% that reside in state of birth",
   box margin(5 5 2 2))
```

As with the `bmargin()` option, we can more precisely modify the margin around the text. Here we use the `margin()` option to make the size of the margin 5, 5, 2, and 2 for the left, right, top, and bottom, respectively.

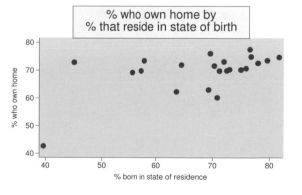

Double-click on the title and select the *Text* tab and change the **Margin** to `Custom`. Then click on the `..` button and then specify the margin for the left, right, top, and bottom.

Uses allstatesn.dta & scheme vg_rose

Introduction
Markers   Marker labels   Editor   Twoway   Matrix   Bar   Box   Dot   Pie   Options   Standard options   Styles   Appendix
Connecting   Axis titles   Axis labels   Axis scales   Axis selection   By   Legend   Adding text   Textboxes   Text display

```
twoway (scatter ownhome borninstate),
   title("% who own home by" "% that reside in state of birth",
   box linegap(4))
```

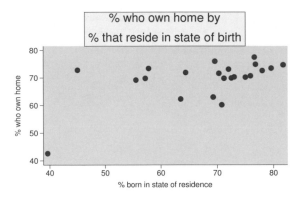

We can change the gap between the lines with the linegap() option. Here we make the gap larger than it would normally be. See Styles : Margins (423) for more details.

📖 Double-click on the title and select the *Format* tab and change the **Line gap** to 4.

Uses allstatesn.dta & scheme vg_rose

Let's now consider options that control the color of the textbox and the characteristics of the outline of the box (including the color, thickness, and pattern). This next set of examples uses the vg_past scheme.

```
twoway (scatter ownhome borninstate),
   title("% own home by % reside in state")
```

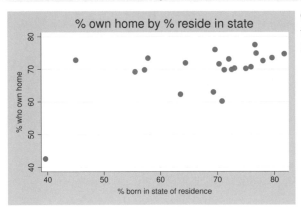

Consider this graph with a title at the top.

Uses allstatesn.dta & scheme vg_past

```
twoway (scatter ownhome borninstate),
    title("% own home by % reside in state", box)
```

We add the `box` option now for aesthetic purposes.

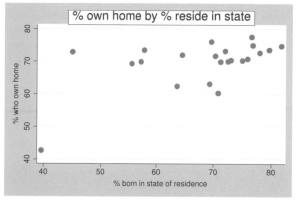 Double-click on the title, select the *Box* tab, and check **Place box around text**.

Uses allstatesn.dta & scheme vg_past

```
twoway (scatter ownhome borninstate),
    title("% own home by % reside in state",
    box fcolor(ltblue) lcolor(gray) lwidth(thick))
```

We can change the box fill color with the `fcolor()` option, the color of the line around the box with `lcolor()`, and the width of the surrounding box line with `lwidth()`. See Styles: Colors (412) for other values available with the `fcolor()` and `lcolor()` options and Styles: Linewidth (422) for other values available for `lwidth()`.

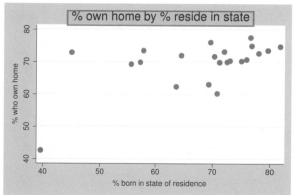 See the next graph.

Uses allstatesn.dta & scheme vg_past

```
twoway (scatter ownhome borninstate),
    title("% own home by % reside in state", box)
```

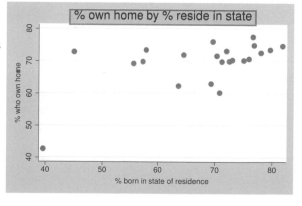 Using the Graph Editor, double-click on the title, select the *Box* tab, and change the **Fill color** to Light Blue, the **Outline color** to Gray, and the **Outline width** to Thick.

Uses allstatesn.dta & scheme vg_past

```
twoway (scatter ownhome borninstate),
   title("% own home by % reside in state", box bcolor(gold))
```

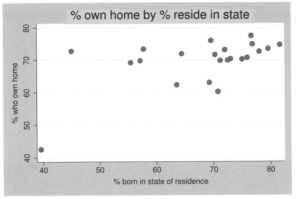

We can change the box color with the bcolor() option. Here we make the fill and outline color of the title box gold. ▥ Double-click on the title and select the *Box* tab and change the **Fill color** and the **Outline color** to Gold. *Uses allstatesn.dta & scheme vg_past*

Let's now use the `allstates` file and consider some examples that use the `by()` option to display multiple graphs broken down by the location of the state. We will look at options for placing and aligning text in graphs that use the `by()` option. These graphs use the `vg_s2c` scheme.

```
scatter ownhome borninstate,
   by(nsw, title("% own home" "by % born in state", box))
```

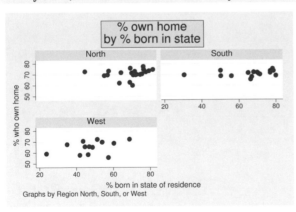

Consider this graph in which we use the by() option to show this scatterplot separately for states in the North, South, and West. We include the box option only to show the outline of the textbox, not for aesthetics. *Uses allstates.dta & scheme vg_s2c*

```
scatter ownhome borninstate,
    by(nsw, title("% own home" "by % born in state", box
    ring(0) position(5)))
```

Let's put the title in the open hole in
the right corner of the graph by using
the `ring(0)` and `position(5)` options.
In the Object Browser, click the
Grid Edit tool and select **title**.
Then drag the title to the open space
and release the mouse. Then click on
the Pointer ➤ tool and, with the title
still selected, drag it to the desired
position.
Uses allstates.dta & scheme vg_s2c

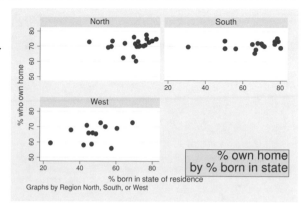

```
scatter ownhome borninstate,
    by(nsw, title("% own home" "by % born in state",
    ring(0) position(5) box width(65) height(40)))
```

We can make the area for the textbox
bigger by using the `width()` and
`height()` options. Here we change the
value to make the box approximately as
tall as the graph for the West and as
wide as the graph for the South.
Uses allstates.dta & scheme vg_s2c

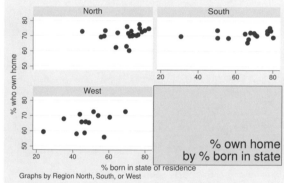

```
scatter ownhome borninstate,
    by(nsw, title("% own home" "by % born in state", ring(0) position(5)
    box width(65) height(40) justification(left) alignment(top)))
```

In this example, we left justify the text
and align it with the top by using the
`justification(left)` and
`alignment(top)` options. These
options make the title appear in the top
left corner of the empty hole.
Uses allstates.dta & scheme vg_s2c

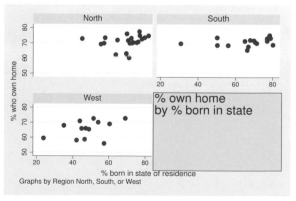

Introduction   Editor   Twoway   Matrix   Bar   Box   Dot   Pie   Options   Standard options   Styles   Appendix

Markers   Marker labels   Connecting   Axis titles   Axis labels   Axis scales   Axis selection   By   Legend   Adding text   Textboxes   Text display

```
scatter ownhome borninstate,
   by(nsw, title("% own home" "by % born in state", ring(0) position(5)
   width(65) height(40) justification(left) alignment(top)))
```

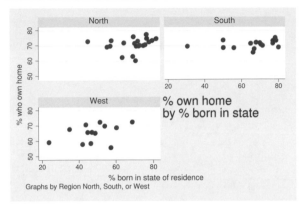

Now that we have aligned the text as we would like, we can remove the box by omitting the box option.

 Simply select the Add Text **T** tool and click any place in the graph to add text of your choosing.
Uses allstates.dta & scheme vg_s2c

9.12   More options controlling the display of text

This section describes additional options to control the display of text, such as how to display symbols (such as Greek letters), how to create subscripts and superscripts, and how to display text in bold or italics. This section also illustrates how to select from among different fonts. We begin by looking at examples illustrating the display of symbols.

```
graph twoway (scatter propval100 popk) (lfit propval100 popk),
   title(Property values by population)
```

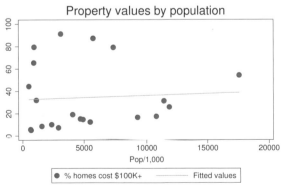

Consider this graph that includes a scatterplot with a fit line from a linear regression. We might want to include the regression equation for the linear regression as a title. We can do so as shown in the next example.
Uses allstatesn.dta & scheme vg_s1m

```
graph twoway (scatter propval100 popk) (lfit propval100 popk),
    title(y = {&alpha} + {&beta}*pop + {&epsilon})
```

Note how the {&alpha} in the title is replaced with lowercase α (alpha). Likewise, {&beta} is replaced with lowercase β (beta), and {&epsilon} is replaced with lowercase ϵ (epsilon). Stata can display the full Greek alphabet (uppercase and lowercase), as well as a variety of math symbols. You can see [G-4] *text* for more details. *Uses allstatesn.dta & scheme vg_s1m*

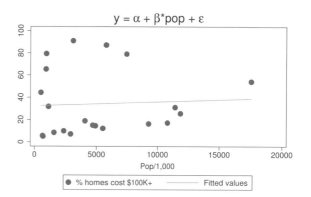

```
graph twoway (scatter propval100 popk) (lfit propval100 popk),
    title(y = {&Alpha} + {&Beta}*pop + {&Epsilon})
```

Although not customary, we could replace the Greek letters with their uppercase counterparts by making the first letter uppercase, e.g., replacing {&alpha} with {&Alpha}. *Uses allstatesn.dta & scheme vg_s1m*

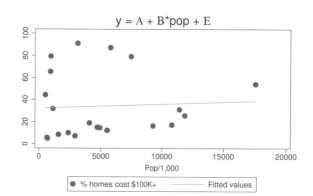

```
graph twoway (scatter propval100 popk) (lfit propval100 popk),
    title("p {&le} 0.150, y = {&function}(x), 1 {&ne} 2")
```

In addition to the full Greek alphabet, Stata can display a wide variety of symbols. For example, I have inserted some nonsense in the title of the graph to illustrate three symbols: {&le} (less than or equal to), {&function} (a function of), and {&ne} (not equal to). You can see [G-4] *text* for a complete list of the symbols Stata can display. *Uses allstatesn.dta & scheme vg_s1m*

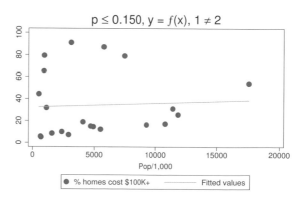

```
graph twoway (scatter propval100 popk) (qfit propval100 popk),
    title(y= {&Beta}{sub:0} + {&Beta}{sub:1}*Pop + {&Beta}{sub:2}*Pop{sup:2})
```

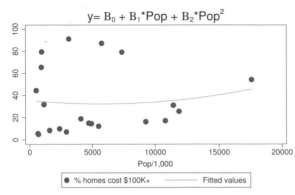

In this example, property values are modeled as a linear and quadratic function of population with the regression equation displayed in the title. Note how {sub:0} is used to display a 0 that is subscripted. Also note how {sup:2} displays squared (the number 2 superscripted).

Uses allstatesn.dta & scheme vg_s1m

```
graph twoway (scatter propval100 popk) (qfit propval100 popk),
    note({bf:Source of data: The 1990 US Census 5% PUMS household file})
    caption({it:Type of model: Quadratic regression})
```

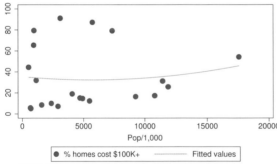

Suppose we want to display text as bold or as italics. In this example, the note() option specifies to display text in bold and the caption() option specifies to display text in italics. As you can see, the text that is enclosed between {bf: and } is displayed in bold. Likewise, the text that is enclosed between {it: and } is displayed in italics.

Uses allstatesn.dta & scheme vg_s1m

```
graph twoway (scatter propval100 popk) (qfit propval100 popk),
    note(Source of data: {bf:The 1990 US Census 5% PUMS household file})
    caption(Type of model: {it:Quadratic regression})
```

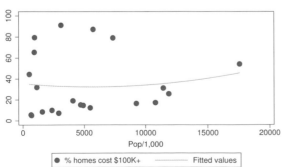

As shown in this example, you can also display just a portion of text using italics or bold. Part of the note is displayed in bold, and part of the caption is displayed in italics.

Uses allstatesn.dta & scheme vg_s1m

```
graph twoway (scatter propval100 popk) (qfit propval100 popk),
    note(Source of data: {it:The 1990 US Census} {bf:5% PUMS household file})
```

This example shows a note in which some of the text is displayed as normal text, some of the text is displayed in italics, and some of the text is displayed in bold.

Uses allstatesn.dta & scheme vg_s1m

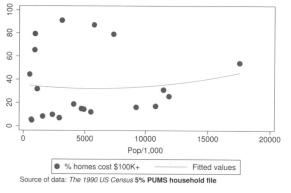

```
scatter propval100 popk,
    title(Property values by population)
```

Consider this basic graph that includes a title. In the following examples, we will see how we can specify different fonts.

Uses allstatesn.dta & scheme vg_s1m

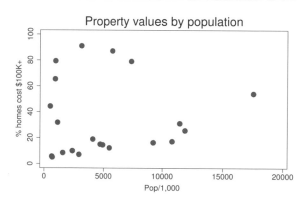

```
scatter propval100 popk,
    title({stSerif:This is the title in a serif font.})
```

The title in this example is displayed using a serif font. The text that is enclosed between {stSerif: and } is displayed using a serif font. The font name, **stSerif**, is a built-in font that is provided by Stata.

Uses allstatesn.dta & scheme vg_s1m

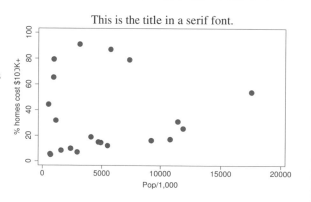

Introduction Editor Twoway Matrix Bar Box Dot Pie Options Standard options Styles Appendix

Markers Marker labels Connecting Axis titles Axis labels Axis scales Axis selection By Legend Adding text Textboxes Text display

```
scatter propval100 popk, title({stSans:This title uses a sans serif font.})
    subtitle({stSerif:This subtitle uses a serif font.})
    note({stMono:This note uses a monospace font})
    caption({stSymbol:ABCDEFG abcdefg})
```

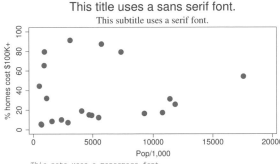

This example illustrates the four built-in fonts that are available within Stata. These fonts are named stSans, stSerif, stMono, and stSymbol. The title is displayed using the stSans font, the subtitle using stSerif, the note using stMono, and the caption using stSymbol. The default font for displaying text is stSans.
Uses allstatesn.dta & scheme vg_s1m

```
scatter propval100 popk,
    title(This is a {stSerif:serif font}. This is a {stMono:monospace font}.)
```

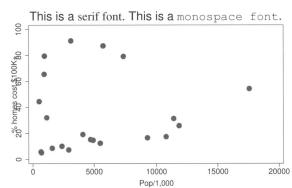

This example shows that you can alternate fonts within the title.
Uses allstatesn.dta & scheme vg_s1m

```
scatter propval100 popk,
    title(This is a {stSerif:serif font {it:displayed in italics}}.)
    subtitle(This is a {stMono:monospaced font {bf:displayed in bold}}.)
```

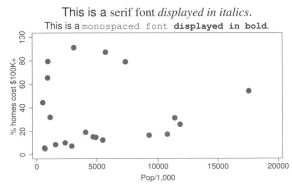

Note how this example both specifies a font type and displays some of the text in italics and some of the text in bold. This illustrates that you can specify a font type combined with italics or combined with bold.
Uses allstatesn.dta & scheme vg_s1m

```
scatter propval100 popk,
    title({fontface Gigi:This is the title in Gigi.})
```

In addition to the four built-in Stata fonts, you can select fonts that are available on your computer. For example, I was using my word processor on my Windows computer and saw that I have a font named `Gigi`. In this example, I specify that the title should be displayed using this font.
Uses allstatesn.dta & scheme vg_s1m

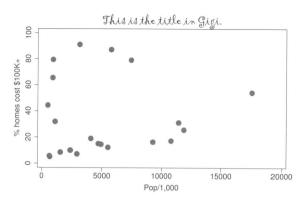

```
scatter propval100 popk, title({fontface Gigi:This is the title in Gigi.})
    subtitle({fontface Papyrus:This is the subtitle in Papyrus.})
    xtitle({fontface Batang:This is the x-axis title in Batang.})
    ytitle({fontface Elephant:This is the y-axis title in Elephant.})
```

Further looking at the list of fonts available in my word processor, I found that I have fonts named `Gigi`, `Papyrus`, `Batang`, and `Elephant`. I am going to go crazy in this example and display the title in `Gigi`, the subtitle using the `Papyrus` font, the *x*-axis title using the `Batang` font, and the *y*-axis title using the `Elephant` font.
Uses allstatesn.dta & scheme vg_s1m

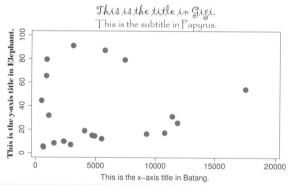

```
scatter propval100 popk,
    title('"{fontface "Segoe Script":This is the title in Segoe Script.}"')
```

Sometimes, the name of a font contains a space within it. For example, I have a font on my computer named `Segoe Script`. To specify this font, I need to enclose the name within double quotes. Because this introduces a double quotation within the `title` option, the title is surrounded by compound quotes; i.e., the title begins with `‘"` and ends with `"’` (see `help quotes`).
Uses allstatesn.dta & scheme vg_s1m

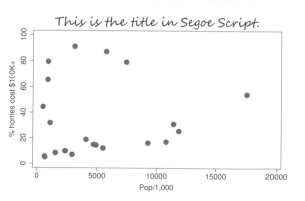

Introduction   Editor   Twoway   Matrix   Bar   Box   Dot   Pie   Options   Standard options   Styles   Appendix

Markers   Marker labels   Connecting   Axis titles   Axis labels   Axis scales   Axis selection   By   Legend   Adding text   Textboxes   Text display

I would like to underscore that these examples use custom fonts that were available on my computer. The fonts that you have available on your computer will likely be different than mine. There are several ways to discover the fonts that you have on your computer, but perhaps the easiest is to use your word processor and look at the font options that it provides.

10 Standard options available for all graphs

This chapter discusses a class of options Stata refers to as *standard options*, because these options can be used in all graphs. This chapter begins by discussing options that allow us to add or change the titles in the graph and then shows us how to use schemes to control the overall look and style of our graph. The next section demonstrates options for controlling the size of the graph and the scale of items within graphs. The chapter concludes by showing options that allow us to control the colors of the plot region, the graph region, and the borders that surround these regions. For more details, see [G-3] *std_options*.

10.1 Creating and controlling titles

Titles are useful for providing more information that explains the contents of a graph. Stata includes four standard options for adding explanatory text to graphs: `title()`, `subtitle()`, `note()`, and `caption()`. This section shows how to use these options to add titles and how to customize the title's content and placement. For more information about customizing the appearance of such titles (e.g., color, size, orientation), see Options: Textboxes (379). For more information about titles, see [G-3] *title_options*. This section uses the `vg_s1m` scheme.

`scatter propval100 ownhome, title("My title")`

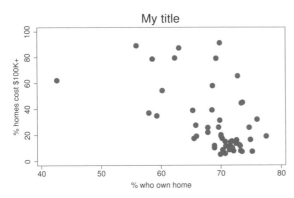

The `title()` option adds a title to a graph. Here we add a simple title to the graph. Although the title includes quotes, we could have omitted them. Later, we will see examples where the quotes become important.

In the Object Browser, double-click on **Graph**, choose the *Titles* tab, and add a title as desired.

Uses allstates.dta & scheme vg_s1m

`scatter propval100 ownhome, title("My title") subtitle("My subtitle")`

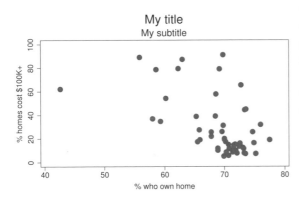

The `subtitle()` option adds a subtitle to a graph. The subtitle, by default, appears below the title in a smaller font.

In the Object Browser, double-click on **Graph**, choose the *Titles* tab, and add a subtitle as desired.

Uses allstates.dta & scheme vg_s1m

`scatter propval100 ownhome, subtitle("My smaller title")`

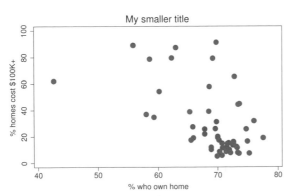

We do not have to specify `title()` to specify `subtitle()`. For example, we might want a title that is smaller than a regular title, so we could specify `subtitle()` alone.

See the previous example.

Uses allstates.dta & scheme vg_s1m

```
scatter propval100 ownhome, caption("My caption") note("My note")
```

Here the `caption()` option adds a
small-sized caption in the lower left
corner, and the `note()` option places a
smaller-sized note in the bottom left
corner. If we specify both options, the
note appears above the caption. Both
options do not need to be included in
the same graph.

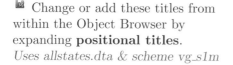 In the Object Browser, double-click
on **Graph**, choose the *Titles* tab, and
add a note and a caption as desired.
Uses allstates.dta & scheme vg_s1m

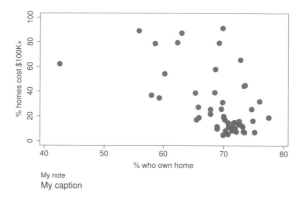

```
scatter propval100 ownhome, t1title("My t1title") t2title("My t2title")
    b1title("My b1title") b2title("My b2title") l1title("My l1title")
    l2title("My l2title") r1title("My r1title") r2title("My r2title")
```

Although these are not as commonly
used, Stata offers several other title
options for titling the top of the graph
(`t1title()` and `t2title()`), the
bottom of the graph (`b1title()` and
`b2title()`), the left side of the graph
(`l1title()` and `l2title()`), and the
right side of the graph (`r1title()` and
`r2title()`).

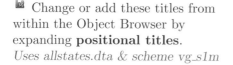 Change or add these titles from
within the Object Browser by
expanding **positional titles**.
Uses allstates.dta & scheme vg_s1m

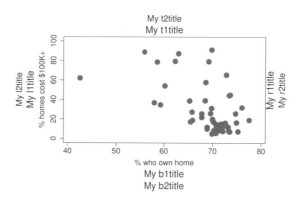

Stata gives you considerable flexibility in the placement of these titles, notes, and captions, as well as control over the size, color, and orientation of the text. This is illustrated
below by using the `title()` option, but the same options apply equally to the `subtitle()`,
`note()`, and `caption()` options.

Using the Graph Editor, you can add or modify the title, subtitle, note, or caption
by going to the Object Browser and double-clicking on **Graph** and then choosing the *Titles* tab. You can modify an existing title, subtitle, note, or caption by clicking on it and
making changes in the Contextual Toolbar or double-clicking on it to bring up a dialog box
that allows you to modify all its properties.

Introduction   Editor   Twoway   Matrix   Bar   Box   Dot   Pie   Options   Standard options   Styles   Appendix

Titles   Schemes   Sizing graphs   Graph regions

`scatter propval100 ownhome, title("My" "title")`

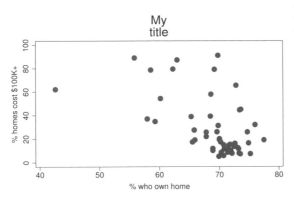

In this example, we use multiple sets of quotes in the `title()` option to tell Stata that we want the title to appear on two separate lines.

Use quotation marks in the same way when typing the title into the Graph Editor to make a title appear across two lines.

Uses allstates.dta & scheme vg_s1m

`scatter propval100 ownhome, title('"A "title" with quotes"')`

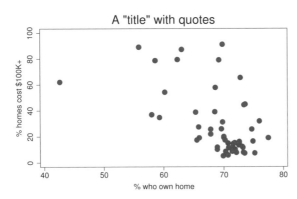

This example illustrates that we can have quotation marks in the `title()` option, as long as we open the title with '" and close it with "'. (The open single quote is often located below the tilde on your keyboard, and the close single quote is often located below the double quote on your keyboard.)

Use single quotes in the same way in the Graph Editor.

Uses allstates.dta & scheme vg_s1m

`scatter propval100 ownhome, title("My title", position(6))`

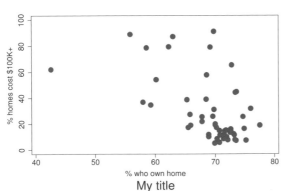

The `position()` option changes the position of the title. Here we place the title in the bottom of the graph by indicating that it should be at the six o'clock position. See Styles : Clockpos (414) for more details.

Select the Grid Edit tool and click on the title and drag it down to the bottom of the graph.

Uses allstates.dta & scheme vg_s1m

```
scatter propval100 ownhome, title("My title", position(7))
```

In this example, we place the title in the bottom left corner of the graph by indicating that it should be at the seven o'clock position.

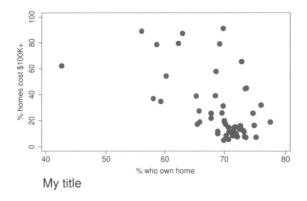

My title

Continuing from the previous example, now double-click on the title and, in the *Format* tab, change the **Position** to West.

Uses allstates.dta & scheme vg_s1m

```
scatter propval100 ownhome, title("My title", position(1) ring(0))
```

Here we not only use the `position()` option to place the title at the one o'clock position, but we also use the `ring(0)` option to place the title inside the plot region. Higher values for `ring()` place the item farther away from the plot region. Imagine concentric rings around the plot area with higher values corresponding to the rings that are farther from the center.

See the next graph.

Uses allstates.dta & scheme vg_s1m

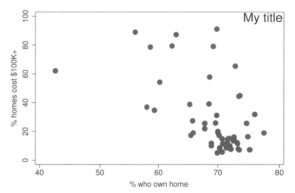

```
scatter propval100 ownhome, title("My title")
```

Using the Graph Editor, select the Grid Edit tool and click on the title and drag it right into the center of the plot. Then, using the Pointer Tool , double-click on the title and select the *Format* tab and change the **Position** to Northeast.

Uses allstates.dta & scheme vg_s1m

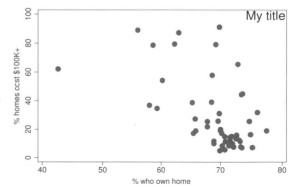

```
scatter propval100 ownhome, title("This is my" "title", box bexpand)
```

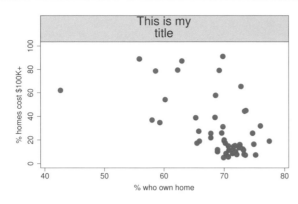

Because titles, subtitles, notes, and captions are considered textboxes, we can use the options associated with textboxes to customize their display. Here we add a box and expand it to fill the width of the plot region.

📊 Double-click on the title and, from the *Box* tab, check **Place box around text**. In the *Format* tab, check **Expand area to fill cell**.

Uses allstates.dta & scheme vg_s1m

```
scatter propval100 ownhome, title("This is my" "title", box bexpand
    justification(left))
```

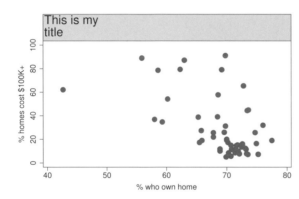

We can use the justification(left) option to left justify the text inside the box. Note the difference between the position() option (illustrated in previous examples), which positions the textbox, and the justification() option, which justifies the text within the textbox.

📊 Double-click on the title and, from the *Format* tab, change the **Justification** to Left.

Uses allstates.dta & scheme vg_s1m

```
scatter propval100 ownhome, title("This is my" "title", box bexpand
    justification(left) span)
```

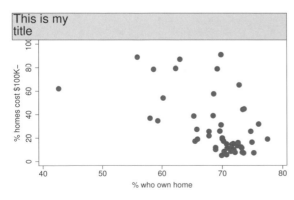

We can use the span option to make the box span the entire width of the graph. There are numerous other textbox options that we can use with titles; see Options : Textboxes (379) and [G-3] *textbox_options* for more details.

📊 Click on the Grid Edit 🔧 tool and click on the title. In the Contextual Toolbar, click on **Expand Cell** and then **Left 1 Cell**.

Uses allstates.dta & scheme vg_s1m

10.2   Using schemes to control the look of graphs

Schemes control the overall look of Stata graphs by providing default values for many graph options. You can accept these defaults or override them by using graph options. This section first examines the kinds of schemes available in Stata, discusses different methods for selecting schemes, and then shows how to obtain more schemes. For more information about schemes, see [G-4] **schemes intro**. Stata has two basic families of schemes, the s2 family and the s1 family, which each share similar characteristics. There are also other specialized schemes, including the **sj** scheme for making graphs like those in the *Stata Journal* and the **economist** scheme for making graphs like those that appear in *The Economist*. We will look at these schemes below.

```
twoway (scatter propval100 urban) (scatter rent700 urban)
  (lfit propval100 urban) (lfit rent700 urban), scheme(s2color)
```

This example uses the
`scheme(s2color)` option to create a
graph using the **s2color** scheme. By
using the `scheme()` option, we can
manually select which scheme to use for
displaying the graph we want to create.
The **s2color** scheme is the default
scheme for Stata graphs.
Uses allstates.dta & scheme s2color

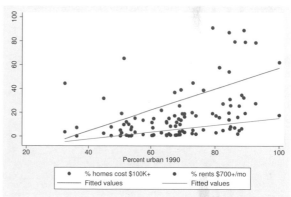

```
twoway (scatter propval100 urban) (scatter rent700 urban)
  (lfit propval100 urban) (lfit rent700 urban), scheme(s2mono)
```

The s2mono scheme is a
black-and-white version of the **s2color**
scheme. Here the symbols differ in gray
scale and size, and the lines differ in
their patterns.
Uses allstates.dta & scheme s2mono

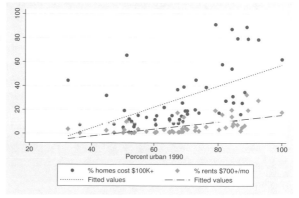

```
twoway (scatter propval100 urban) (scatter rent700 urban)
    (lfit propval100 urban) (lfit rent700 urban), scheme(s2manual)
```

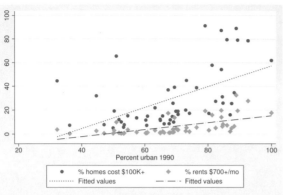

Here is an example using the s2manual scheme, which is similar to the s2mono scheme.

Uses allstates.dta & scheme s2manual

```
twoway (scatter propval100 urban) (scatter rent700 urban)
    (lfit propval100 urban) (lfit rent700 urban), scheme(s1color)
```

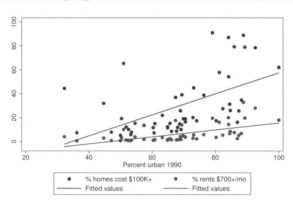

This is an example of a graph using the s1color scheme. Note how the lines and markers are only differentiated by their color. Both the plot area and the border around the plot are white. Also, note the absence of grid lines. (Stata also has an s1rcolor scheme, in which the plot area and border area are black. This is not shown because it would be difficult to read in print.)

Uses allstates.dta & scheme s1color

```
twoway (scatter propval100 urban) (scatter rent700 urban)
    (lfit propval100 urban) (lfit rent700 urban), scheme(s1mono)
```

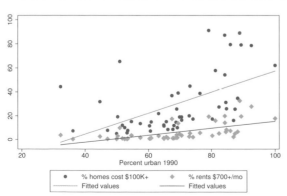

The s1mono scheme is similar to the s1color scheme in that the plot area and border are white and the grid is omitted. In a monochrome scheme, the markers differ in gray scale and size, and the lines differ in their pattern.

Uses allstates.dta & scheme s1mono

```
twoway (scatter propval100 urban) (scatter rent700 urban)
    (lfit propval100 urban) (lfit rent700 urban), scheme(s1manual)
```

The s1manual is similar to s1mono, but the sizes of the markers and text are increased. This is useful for a small graph, when these small elements must be magnified to be more easily seen.
Uses allstates.dta & scheme s1manual

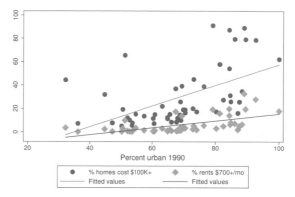

```
twoway (scatter propval100 urban) (scatter rent700 urban)
    (lfit propval100 urban) (lfit rent700 urban), scheme(sj)
```

The sj scheme is similar to the s2mono scheme. In fact, a comparison of this graph with the earlier graph that used the s2mono scheme shows no visible differences. The sj scheme is based on the s2mono scheme and only alters xsize() and ysize(). See Appendix: Customizing schemes (479) for more information about how to inspect (and alter) the contents of graph schemes.
Uses allstates.dta & scheme sj

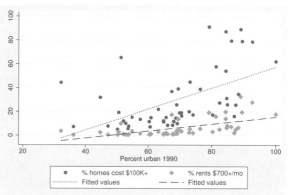

```
twoway (scatter propval100 urban) (scatter rent700 urban)
    (lfit propval100 urban) (lfit rent700 urban), scheme(economist)
```

The economist scheme is quite different from all the other schemes and is a good example of how much we can control with a scheme. Using this scheme modifies the colors of the plot area, border, markers, lines, the position of the y axis, and the legend. It also removes the line on the y axis and changes the angle of the labels on the y axis.
Uses allstates.dta & scheme economist

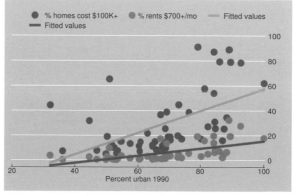

As these examples have shown, we can change the scheme of a graph by supplying the `scheme()` option on a graph command. If you want to use the same scheme over and over, you can use the `set scheme` command to set the default scheme. For example, if you typed

```
. set scheme economist
```

the default scheme would become `economist` until you quit Stata. Or, you could type

```
. set scheme economist, permanently
```

and the `economist` scheme would be your default scheme, even after you quit and restart Stata. If you will be creating a series of graphs that you want to have a common look, then schemes are a powerful tool for accomplishing this. Even though Stata has a variety of built-in schemes, you may want to obtain other schemes. You can use the `findit` command to search for information about schemes and to download schemes that others have developed. To search for schemes, type

```
. findit scheme
```

and Stata will list web pages and packages associated with the word *scheme*. See Intro : Schemes (15) for an overview of the schemes used in this book and Appendix : Online supplements (482) for instructions for obtaining the schemes for this book. Seeing how powerful and flexible schemes are, you might be interested in creating your own schemes. Stata gives you complete control over creating schemes. The section Appendix : Customizing schemes (479) provides tips for getting started.

10.3   Sizing graphs and their elements

This section illustrates how to use the `xsize()` and `ysize()` options to control the size of graphs, as well as how to use the `aspectratio()` option to control the aspect ratio of graphs. This section also illustrates the use of the `scale()` option for controlling the size of the text and markers. This section uses the `vg_s1c` scheme.

`scatter propval100 ownhome`

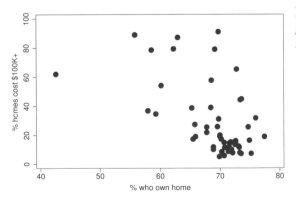

Let's first consider this graph. The graphs in this book have been sized to 3 inches wide by 2 inches tall.
Uses allstates.dta & scheme vg_s1c

`scatter propval100 ownhome, aspectratio(1.3)`

You can use the `aspectratio()` option
to change the aspect ratio of the graph
without changing the size. Numbers
greater than 1 create tall skinny graphs,
and numbers less than one create wide
fat graphs.

🔖 In the Object Browser, click on
Graph and, in the Contextual Toolbar,
change the **Aspect ratio** to 1.3.
Uses allstates.dta & scheme vg_s1c

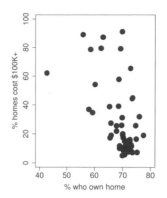

`scatter propval100 ownhome, xsize(3) ysize(1)`

Here we use `xsize()` and `ysize()` to
make the graph 3 inches wide and 1
inch tall. Although we can size the
graph on the screen and make it look
bigger, when we export it, the size will
be determined by the default `xsize()`
and `ysize()` or by the `graph export`
options.

🔖 In the Object Browser, double-click
on **Graph** and, in the *Aspect/Size* tab,
change the **Y size** to 1 and the **X size**
to 3.
Uses allstates.dta & scheme vg_s1c

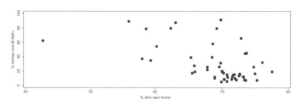

`scatter propval100 ownhome, xsize(2) ysize(2)`

Here is one more example where we create a graph that
is 2 inches wide by 2 inches tall.

🔖 In the Object Browser, double-click on **Graph** and,
in the *Aspect/Size* tab, change the **Y size** to 2 and the
X size to 2.
Uses allstates.dta & scheme vg_s1c

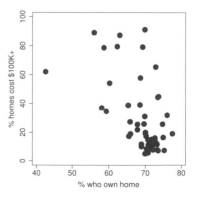

Introduction   Editor   Twoway   Matrix   Bar   Box   Dot   Pie   Options   Standard options   Styles   Appendix

Titles   Schemes   Sizing graphs   Graph regions

`scatter propval100 ownhome, scale(1.7)`

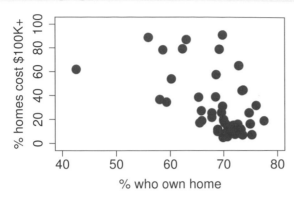

Here we add the `scale(1.7)` option to magnify the sizes of the text and markers in the graph, making them 1.7 times their normal sizes. This option can be useful when creating small graphs and increasing the sizes of the text and markers would make them easier to see.

In the Object Browser, click on **Graph** and, in the Contextual Toolbar, change the **Scale** to 1.7.

Uses allstates.dta & scheme vg_s1c

`scatter propval100 ownhome, scale(.5)`

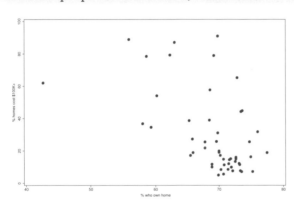

We can also use the `scale()` option to decrease the sizes of the text and markers. Here we make these elements half their normal size.

In the Object Browser, click on **Graph** and, in the Contextual Toolbar, change the **Scale** to .5.

Uses allstates.dta & scheme vg_s1c

10.4   Changing the look of graph regions

This section discusses the region options that we can control with the `plotregion()` and `graphregion()` options. These options allow us to control the plot region and graph region color, as well as the lines that border these regions. For more information, see [G-3] ***region_options***. This section uses the `vg_s2c` scheme.

```
scatter propval100 ownhome, title("My title")
```

Consider this scatterplot. In general, Stata sees this graph as having two overall regions. The area inside the x and y axes where the data are plotted is called the *plot region*. In this graph, the plot region is white. The area surrounding the plot region, where the axes and titles are placed, is called the *graph region*. Here the graph region is shaded light blue.

Uses allstates.dta & scheme vg_s2c

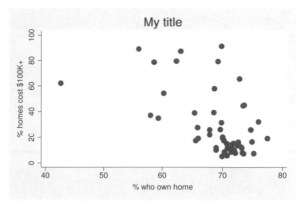

```
scatter propval100 ownhome, title("My title") plotregion(color(stone))
```

Here we use `plotregion(color(stone))` to make the color of the plot region stone. The `color()` option controls the color of the plot region.

📊 Click on an empty portion of the plot region (the inner region of the graph) and, in the Contextual Toolbar, change the **Color** to Stone.

Uses allstates.dta & scheme vg_s2c

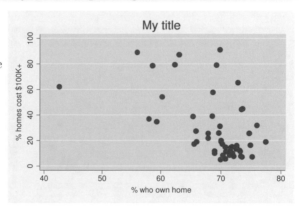

```
scatter propval100 ownhome, title("My title") plotregion(lcolor(navy)
    lwidth(thick) )
```

In this graph, we put a thick, navy blue line around the plot region by using the `lcolor()` and `lwidth()` options. This puts a bit of a frame around the plot region.

📊 Double-click on an empty portion of the plot region and, in the *Region* tab, change the **Outline width** to Thick, check **Different Outline Color**, and change the **Outline color** to Navy.

Uses allstates.dta & scheme vg_s2c

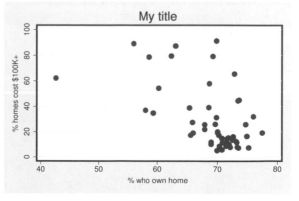

Introduction   Editor   Twoway   Matrix   Bar   Box   Dot   Pie   Options   Standard options   Styles   Appendix

Titles   Schemes   Sizing graphs   Graph regions

`scatter propval100 ownhome, title("My title") graphregion(color(erose))`

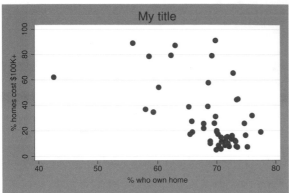

Here we use the `graphregion(color(erose))` option to modify the color of the graph region to be erose, a light rose color. The graph region is the area outside the plot region where the titles and axes are displayed.

In the Object Browser, double-click on **Graph** and, in the *Region* tab, change the **Color** to erose.

Uses allstates.dta & scheme vg_s2c

`scatter propval100 ownhome, title("My title")`
`    graphregion(ifcolor(erose) fcolor(maroon))`

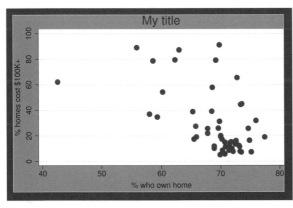

The graph region is actually composed of an inner part and an outer part. Here we use the `ifcolor(erose)` option to make the inner graph region light rose and the `fcolor(maroon)` option to make the outer graph region maroon.

Double-click on an empty portion of the periphery of the graph. In the *Region* tab, change the **Color** to maroon, and, in the *Inner Region* tab, change the **Color** to erose.

Uses allstates.dta & scheme vg_s2c

`scatter propval100 ownhome, title("My title")`
`    graphregion(lcolor(navy) lwidth(vthick))`

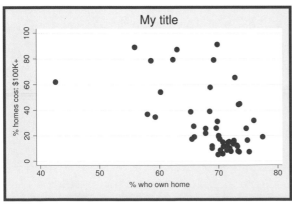

We can put a somewhat different frame around the graph by altering the size and color of the line that surrounds the graph region. Using the `lcolor(navy)` `lwidth(vthick)` options gives this graph a very thick, navy blue border.

Double-click on an empty portion of the plot region and, in the *Region* tab, change the **Outline width** to v Thick, check **Different Outline Color**, and change the **Outline color** to Navy.

Uses allstates.dta & scheme vg_s2c

You can control many details about the graph regions, but the commands can begin to get complicated, whereas some of these things can be easy with the Graph Editor. Here are some examples that solely rely on the Graph Editor.

`scatter propval100 ownhome, title("My title")`

In the Graph Editor's Object Browser, double-click on **Graph**. From the *Region* tab, play with the different settings to see what they do. This graph shows what is displayed with the **Margin** set to Vlarge, the **Color** set to Yellow, the **Outline width** set to vvv.Thick with a **Different outline color** checked, and the **Outline color** set to Green.
Uses allstates.dta & scheme vg_s2c

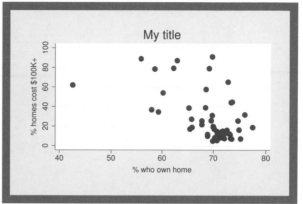

`scatter propval100 ownhome, title("My title")`

Continuing from the previous graph, double-click on an empty portion of the periphery of the graph. From the *Inner region* tab, play with the different settings to see what they do. This graph shows what is displayed with the **Color** set to Red, the **Outline width** set to Thick with a **Different outline color** checked, and the **Outline color** set to Blue.
Uses allstates.dta & scheme vg_s2c

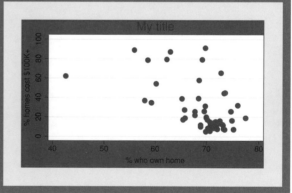

Introduction   Editor   Twoway   Matrix   Bar   Box   Dot   Pie   Options   Standard options   Styles   Appendix

Titles   Schemes   Sizing graphs   Graph regions

This section omitted many options that you could use to control the plot region and graph region, including more control of the inner and outer regions and more control of the lines that surround these regions. Stata gives you more control than you generally need, so rather than covering these options here, I refer you to [G-3] *region_options*.

11 Styles for changing the look of graphs

This chapter focuses on frequently used styles that arise in making graphs, such as *linepatternstyle*, *linewidthstyle*, or *markerstyle*. The chapter covers styles in alphabetical order, providing more details about the values we can choose. Each section refers to the corresponding section of the *Graphics Reference Manual*, which provides complete details on each style. We begin by using the `allstatesdc` file, which contains the `allstates` data with Washington, DC, omitted.

11.1 Angles

An *anglestyle* specifies the angle for displaying an item (or group of items) in the graph. Common examples include specifying the angle for marker labels with `mlabangle()` or the angle of the labels on the y axis with `ylabel(, angle())`. We can specify an *anglestyle* as a number of degrees of rotation (negative values are permitted, so for example, -90 can be used instead of 270). We can also use the keywords `horizontal` for 0 degrees, `vertical` for 90 degrees, `rhorizontal` for 180 degrees, and `rvertical` for 270 degrees. See [G-4] *anglestyle* for more information.

`scatter workers2 faminc, mlabel(stateab) mlabangle(45)`

Here we use the `mlabangle(45)` (marker label angle) to change the angle of the marker labels to 45 degrees.

🖳 Double-click on a marker label (e.g., VT) and change the **Angle** to 45 degrees.

Uses allstatesdc.dta & scheme vg_s2c

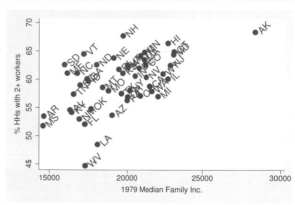

`scatter workers2 faminc, ylabel(, angle(0))`

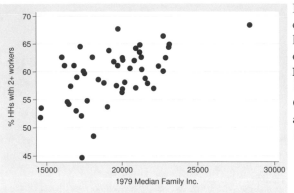

Here we change the angle of the labels of the y axis so that they read horizontally by using the `angle(0)` option. We could also have used `horizontal` to obtain the same effect.
📊 Click on the y axis and, in the Contextual Toolbar, change the **Label angle** to `Horizontal`.
Uses allstatesdc.dta & scheme vg_s2c

`scatter workers2 faminc, xlabel(15000(1000)30000, angle(45))`

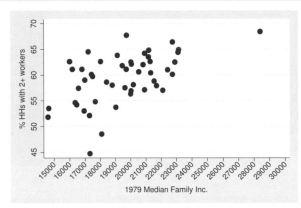

In this example, we label the x axis from 15,000 to 30,000, incrementing by 1,000. When we have so many labels, we can use the `angle(45)` option to display the labels at a 45-degree angle.
📊 Click on the x axis and, in the Contextual Toolbar, change the **Label angle** to `45 degrees`.
Uses allstatesdc.dta & scheme vg_s2c

11.2   Colors

A *colorstyle* allows us to modify the color of an object, be it a title, a marker, a marker label, a line around a box, a fill color of a box, or practically any other object in a graph. The two main ways to specify a color are either by giving a name of color (e.g., `red`, `pink`, `teal`) or by supplying an RGB value giving the amount of red, green, and blue to be mixed to form a custom color. See [G-4] ***colorstyle*** for more information.

`scatter workers2 faminc, mcolor(gs8)`

In this example, we use the `mcolor()` (marker color) option to make the marker a middle gray. Stata provides 17 levels of gray named `gs0` to `gs16`. The darkest is `gs0` (a synonym for `black`), and the lightest is `gs16` (a synonym for `white`).

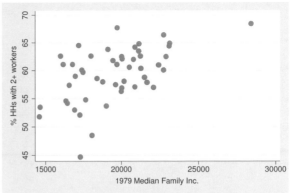

📉 Click on a marker and, from the Contextual Toolbar, change the **Color** to `Gray 8` (or any other color you desire).

Uses allstatesdc.dta & scheme vg_s2c

`scatter workers2 faminc, mcolor(lavender)`

Here we use the `mcolor(lavender)` option to make the markers lavender, one of the predefined colors created by Stata. The next example illustrates more of the colors from which you can choose.

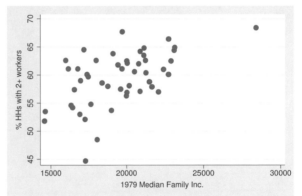

📉 Click on a marker and, in the Contextual Toolbar, change the **Color** to `Lavender`.

Uses allstatesdc.dta & scheme vg_s2c

`vgcolormap, quietly`

The author-written `vgcolormap` command shows the different standard colors available in Stata. We simply issue the command `vgcolormap`, and it creates a scatterplot that shows the colors we can choose and their names. See the list of colors available in [G-4] **colorstyle**, and see how to get `vgcolormap` in Appendix : Online supplements (482).

Uses allstatesdc.dta & scheme vg_s2c

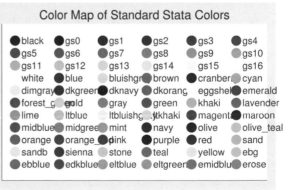

Introduction   Editor   Twoway   Matrix   Bar   Box   Dot   Pie   Options   Standard options   Styles   Appendix

Angles   Colors   Clockpos   Compassdir   Connect   Linepatterns   Linewidth   Margins   Markersize   Orientation   Symbols   Textsize

`scatter workers2 faminc, mcolor("255 255 0")`

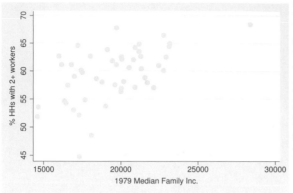

Despite all the standard color choices, we may want to mix our own colors by specifying how much red, green, and blue to mix together. We can mix between 0 and 255 units of each color. Mixing 0 units of each yields black and 255 units of each yields white. Here we mix 255 units of red, 255 units of green, and 0 units of blue to get a shade of yellow.

📊 See the next graph.

Uses allstatesdc.dta & scheme vg_s2c

`scatter workers2 faminc, mcolor("255 150 100")`

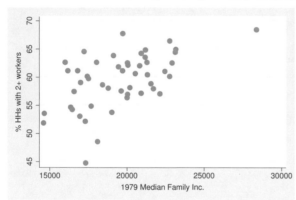

By mixing 255 parts red, 150 parts green, and 100 parts blue, we get a peach color. Because colors for web pages use this same principle of mixing red, green, and blue, we can do a web search using terms like *color mixing html* to learn more about mixing such colors.

📊 Double-click on any marker and change the **Color** to `Custom` and click on the color button that appears to the right to choose a custom color.

Uses allstatesdc.dta & scheme vg_s2c

11.3   Clock position

A clock position refers to a location using the numbers on an analog clock to indicate the location, with twelve o'clock being above the center, three o'clock to the right, six o'clock below the center, and nine o'clock to the left. A value of 0 refers to the center but may not always be valid. See [G-4] ***clockposstyle*** for more information.

```
scatter workers2 faminc, mlabel(stateab) mlabposition(5)
```

In this example, we add marker labels to a scatterplot and use the `mlabposition(5)` (marker label position) option to place the marker labels in the five o'clock position with respect to the markers.

📖 Click on one of the marker labels (e.g., ND) and, in the Contextual Toolbar, change the **Position** to 5 o'clock.

Uses allstatesdc.dta & scheme vg_s2c

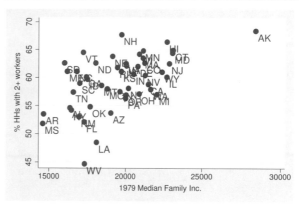

```
scatter workers2 faminc, mlabel(stateab) mlabposition(0) msymbol(i)
```

Here we place the markers in the center position by using the `mlabposition(0)` option. We also make the symbols invisible by using the `msymbol(i)` option. Otherwise, the markers and marker labels would be atop each other.

📖 Click on any marker and, in the Contextual Toolbar, change the **Symbol** to None. Then click on any marker label (e.g., ND) and, in the Contextual Toolbar, change the **Position** to Middle.

Uses allstatesdc.dta & scheme vg_s2c

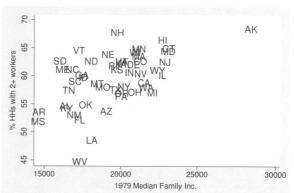

11.4   Compass direction

A *compassdirstyle* is much like *clockpos*, but where a *clockpos* has 12 possible outer positions, like a clock, the *compassdirstyle* has only 9 possible outer positions, like the major labels on a compass: `north`, `neast`, `east`, `seast`, `south`, `swest`, `west`, `nwest`, and `center`. These can be abbreviated as `n`, `ne`, `e`, `se`, `s`, `sw`, `w`, `nw`, and `c`. Stata permits you to use a *clockpos* even when a *compassdirstyle* is called for and makes intuitive translations; for example, 12 is translated to `north`, or 2 is translated to `neast`. See [G-4] ***compassdirstyle*** for more information.

Introduction

Angles   Colors   Clockpos   Compassdir   Connect   Linepatterns   Linewidth   Margins   Markersize   Orientation   Symbols   Textsize

Editor   Twoway   Matrix   Bar   Box   Dot   Pie   Options   Standard options   Styles   Appendix

```
scatter workers2 faminc, title("Work Status and Income",
    ring(0) placement(se))
```

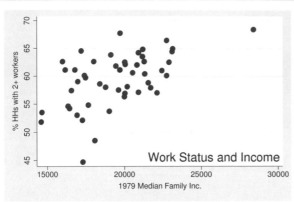

Here the `placement()` option positions the title in the southeast (bottom right corner) of the plot region. The `ring(0)` option moves the title inside the plot region.

Assuming the `ring(0)` option was included in the command, double-click on the title and select the *Format* tab and change the **Position** to Southeast.

Uses allstatesdc.dta & scheme vg_s2c

```
scatter workers2 faminc, title("Work Status and Income",
    ring(0) placement(4))
```

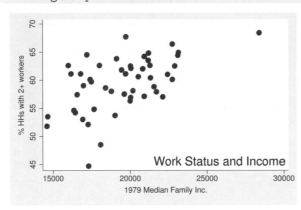

If we instead specify the `placement(4)` option (using a *clockpos* instead of *compassdir*), Stata makes a suitable substitution, and the title is placed in the bottom right corner.

Uses allstatesdc.dta & scheme vg_s2c

11.5   Connecting points

Stata supports a variety of methods for connecting points using different values for the *connectstyle*. These include l (lowercase L, as in line) to connect with a straight line, L to connect with a straight line only if the current *x* value is greater than the prior *x* value, J for stairstep, `stepstair` for step then stair, and i for invisible connections. The next few examples use the `spjanfeb2001` data file, keeping only the data for January and February of 2001. See [G-4] ***connectstyle*** for more information.

scatter close tradeday

Here we make a scatterplot showing the closing price on the y axis and the trading day (numbered 1–40) on the x axis. Normally, we would connect these points.

Uses spjanfeb2001.dta & scheme vg_s2c

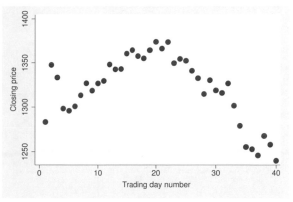

graph twoway connected close tradeday

We change to using the connected command to connect the points, but this is probably not the kind of graph we wanted to create. The problem is that the observations are in a random order, but the observations are connected in the same order as they appear in the data. We really want the points to be connected based on the order of tradeday.

Uses spjanfeb2001.dta & scheme vg_s2c

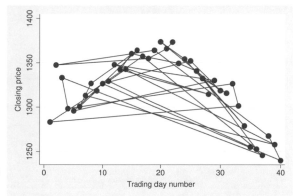

graph twoway connected close tradeday, sort

To fix the previous graph, we can either first use the sort command to sort the data on tradeday or, as we do here, use the sort option to tell Stata to sort the data on tradeday before connecting the points. We could have also specified sort(tradeday), and it would have had the same effect.

📈   Double-click on any marker and, in the **Line** tab, click on the **Sort** button.

Uses spjanfeb2001.dta & scheme vg_s2c

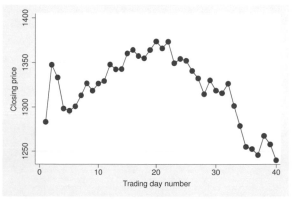

`graph twoway connected close predclose tradeday, sort`

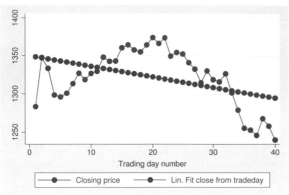

Say that we used the **regress** command to predict **close** from **tradeday** and generated a predicted value called **predclose**. Here we plot the actual closing prices and the predicted closing prices.

Uses spjanfeb2001.dta & scheme vg_s2c

`graph twoway connected close predclose tradeday, sort`
`   connect(i .) msymbol(. i)`

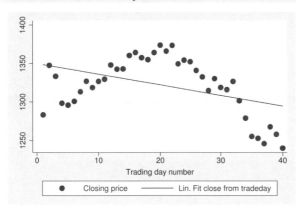

We use the `connect(i .)` option to suppress connecting the observed values while leaving the predicted values connected. The `i` option suppresses connecting the observed values whereas the `.` option indicates that the predicted values should be unchanged (i.e., remain connected). We also add `msymbol(. i)` to make the symbols for the observed values unchanged, but invisible for the fitted values.

📈 See the next graph.

Uses spjanfeb2001.dta & scheme vg_s2c

`graph twoway connected close predclose tradeday, sort`

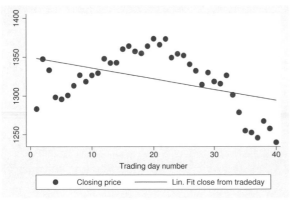

📈 Using the Graph Editor, double-click on one of the markers for the closing price and, in the *Line* tab, change the **Connecting method** to None. Then double-click on one of the markers for the fitted line and, in the *Markers* tab, change the **Symbol** to None.

Uses spjanfeb2001.dta & scheme vg_s2c

`graph twoway connected close tradeday, ` `connect(J) sort`

In other contexts (such as survival analysis), we might want to connect points using a stairstep pattern. Here we connect the observed closing prices with the J option (which can also be specified as **stairstep**) to get a stairstep effect.

🔳 Double-click on any of the observations and change the **Connecting method** to Stairstep.

Uses spjanfeb2001.dta & scheme vg_s2c

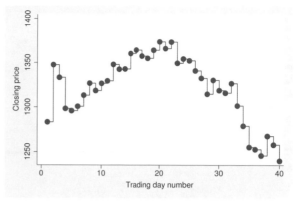

`graph twoway connected close tradeday, ` `connect(stepstair) sort`

In other contexts, we might want to connect points using a stepstair pattern. Here we connect the observed closing prices with the **stepstair** option to get a stepstair effect.

🔳 Double-click on any of the observations and change the **Connecting method** to Stairstep (up first).

Uses spjanfeb2001.dta & scheme vg_s2c

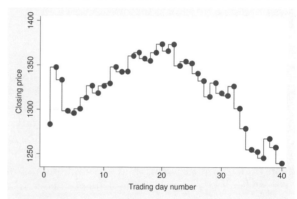

`graph twoway connected close dom, ` `connect(l) sort(date)`

Say that we created a variable called dom that represented the day of the month and wanted to graph the closing prices for January and February against the day of the month. By using the sort(date) option, we almost get what we want, but there is a line that swoops back connecting January 31 to February 1.

Uses spjanfeb2001.dta & scheme vg_s2c

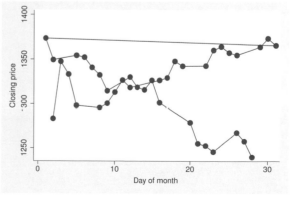

```
graph twoway connected close dom, connect(L) sort(date)
```

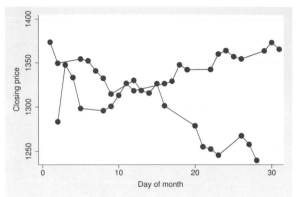

This kind of example calls for the `connect(L)` option, which avoids the line that swoops back by connecting points with a straight line, except when the *x* value (`dom`) decreases (e.g., goes from 31 to 1).

Uses spjanfeb2001.dta & scheme vg_s2c

11.6   Line patterns

We can specify the pattern we want for a line in three ways. We can specify a word that selects from a set of predefined styles, including `solid` (solid line), `dash` (a dashed line), `dot` (a dotted line), `shortdash` (short dashes), `longdash` (long dashes), and `blank` (invisible). There are also the combination styles `dash_dot`, `shortdash_dot`, and `longdash_dot`. We can also use a formula that combines the following five elements in any way that we want: `l` (letter l, solid line), `_` (underscore, long dash), `-` (hyphen, medium dash), `.` (period, short dash that is almost a dot), and `#` (small amount of space). You could specify `longdash_dot` or `"_."`, and they would be equivalent. See [G-4] *linepatternstyle* for more information.

```
twoway (line close tradeday, lpattern(solid) sort)
    (lfit close tradeday, lpattern(dash))
    (lowess close tradeday, lpattern(shortdash_dot))
```

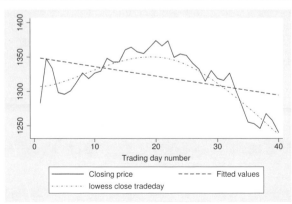

Here we make a line plot and use the `lpattern()` (line pattern) option to obtain a solid pattern for the observed data, a dash for the linear fit line, and a short dash and dot line for a lowess fit. Double-click on the line for the closing price and ensure the **Pattern** is `Solid`. Next double-click on the fitted line and change the **Pattern** to `Dash`. Finally, double-click on the lowess line and change the **Pattern** to `Short-dash dot`.

Uses spjanfeb2001.dta & scheme vg_s2c

```
twoway (line close tradeday, lpattern("l") sort)
   (lfit close tradeday, lpattern("._"))
   (lowess close tradeday, lpattern("-###"))
```

The `lpattern()` option specifies a formula to indicate the pattern for the lines. Here we specify a solid line for the `line` plot, a dot and dash for the `lfit` plot, and a dash and three spaces for the `lowess` fit.

Uses spjanfeb2001.dta & scheme vg_s2c

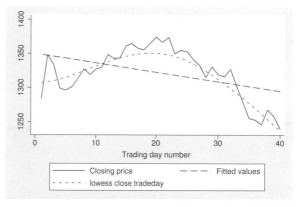

```
twoway (line close tradeday, lpattern("l") sort)
   (lfit close tradeday, lpattern("__##"))
   (lowess close tradeday, lpattern("-.#"))
```

This example shows other formulas to create, including `"__##"`, which yields long dashes with long breaks between, and `"-.#"`, which yields a medium dash, a dot, and a space. Using these formulas, we can create a wide variety of line patterns for those instances where we need to differentiate multiple lines.

Uses spjanfeb2001.dta & scheme vg_s2c

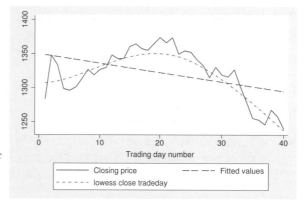

palette linepalette

The built-in Stata command `palette linepalette` shows the line patterns that are available within Stata to help us choose a pattern to our liking.

Uses spjanfeb2001.dta & scheme vg_s2c

Introduction Angles Colors Clockpos Compassdir Connect Linepatterns Linewidth Margins Markersize Orientation Symbols Textsize

Editor Twoway Matrix Bar Box Dot Pie Options Standard options Styles Appendix

11.7   Line width

We can indicate the width of a line in two ways. We can indicate a *linewidthstyle*, which allows us to use a word to specify the width of a line, including `none` (no width, invisible), `vvthin`, `vthin`, `thin`, `medthin`, `medium`, `medthick`, `thick`, `vthick`, `vvthick`, and even `vvvthick`. We can also specify a *relativesize*, which is a multiple of the line's normal thickness (e.g., *2 is twice as thick, or *.7 is .7 times as thick). See [G-4] ***linewidthstyle*** for more information.

```
twoway (line close tradeday, lwidth(vthick) sort)
   (lfit close tradeday, lwidth(thick))
   (lowess close tradeday, bwidth(.5) lwidth(thin))
```

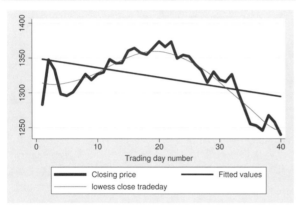

We now plot the same three lines from the previous section, but this time we differentiate them with line thickness by using the `lwidth()` (line width) option.

Change the line thickness from within the Contextual Toolbar. Click the closing price line and change the **Width** to v Thick. Next click on the fitted line and change the **Width** to Thick. Finally, click on the lowess line and change the **Width** to Thin.

Uses spjanfeb2001.dta & scheme vg_s2c

```
twoway (line close tradeday, lwidth(*4) sort)
   (lfit close tradeday, lwidth(*2))
   (lowess close tradeday, bwidth(.5) lwidth(*.5))
```

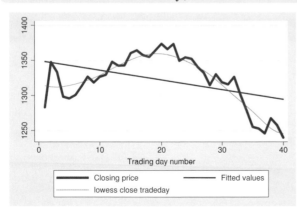

We could create a similar graph by using the `lwidth()` option and specifying the widths as relative sizes. Here we make the line for the `line` plot four times as wide, the line for the `lfit` plot twice as wide, and the line for the `lowess` plot half as wide.

Double-click on the line for the closing price and next to **Width** type in *4. Next double-click on the fitted line and change the **Width** to *2. Finally, double-click on the lowess line and change the **Width** to *.5.

Uses spjanfeb2001.dta & scheme vg_s2c

11.8 Margins

We can specify the size of a margin in three different ways. We can use a word that represents a predefined margin. Choices include `zero`, `vtiny`, `tiny`, `vsmall`, `small`, `medsmall`, `medium`, `medlarge`, `large`, and `vlarge`. They also include `top_bottom` to indicate a medium margin at the top and bottom and `sides` to indicate a medium margin at the left and right. A second method is to give four numbers giving the margins at the left, right, top, and bottom. A third method is to use expressions, such as `b=5` to modify one or more of the margins. These methods are illustrated below. See [G-4] ***marginstyle*** for more information.

```
scatter workers2 faminc, title("Overall title", margin(large) box)
```

We illustrate the control of margins by adding a title to this scatterplot and putting a box around it. The `margin()` option affects the distance between the title and the box. We specify a `large` margin, making the margin large on all four sides.

[icon] Click the title and, in the Contextual Toolbar, change the **Margin** to Large.
Uses allstatesdc.dta & scheme vg_s2c

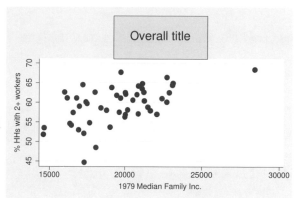

```
scatter workers2 faminc, title("Overall title", margin(top_bottom) box)
```

By using `margin(top_bottom)`, we obtain a margin that is medium on the top and bottom but zero on the left and right.

[icon] Click the title and, in the Contextual Toolbar, change the **Margin** to Top_bottom.
Uses allstatesdc.dta & scheme vg_s2c

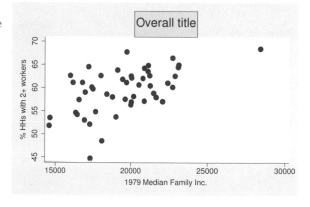

`scatter workers2 faminc, title("Overall title", margin(sides) box)`

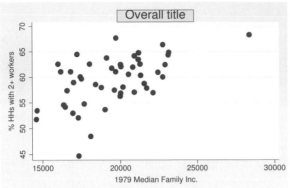

By using the `margin(sides)` option, we obtain a margin that is medium on the left and right but zero on the top and bottom.

📊 Click the title and, in the Contextual Toolbar, change the **Margin** to Sides.
Uses allstatesdc.dta & scheme vg_s2c

`scatter workers2 faminc, title("Overall title", margin(9 6 3 0) box)`

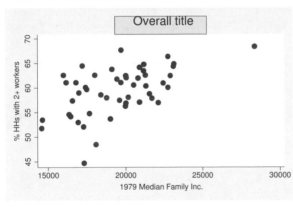

In addition to the words describing margins, we can manually specify the margins for the left, right, bottom, and top. Here we specify `margin(9 6 3 0)` and make the margin for the left 9, for the right 6, for the bottom 3, and for the top 0.

📊 Double-click on the title and, in the *Text* tab, change the **Margin** to Custom and then double-click on the `..` button. Change the left margin to 9, the right to 6, the bottom to 3, and the top to 0.
Uses allstatesdc.dta & scheme vg_s2c

`scatter workers2 faminc, title("Overall title", margin(l=9 r=9) box)`

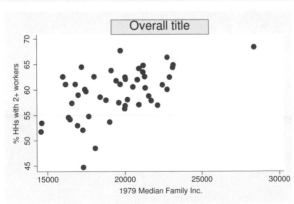

We can also manually change only some of the margins. By specifying `margin(l=9 r=9)`, we make the margins at the left and right 9 units, leaving the top and bottom unchanged. You can specify one or more of the expressions `l=`, `r=`, `t=`, or `b=` to modify the left, right, top, or bottom margins, respectively.

📊 See the previous example.
Uses allstatesdc.dta & scheme vg_s2c

11.9   Marker size

We can control the size of the markers by specifying a *markersizestyle* or a *relativesize*. The *markersizestyle* is a word that describes the size of a marker, including **vtiny**, **tiny**, **vsmall**, **small**, **medsmall**, **medium**, **medlarge**, **large**, **vlarge**, **huge**, **vhuge**, and **ehuge**. We could also specify the sizes as a *relativesize*, which is either an absolute size or a multiple of the original size of the marker (e.g., *2 is twice as large, or *.7 is .7 times as large). See [G-4] ***markersizestyle*** for more information.

```
twoway (scatter propval100 rent700 ownhome urban,
    msize(vsmall medium large))
```

Here we have an overlaid scatterplot where we graph three variables on the *y* axis (**propval100**, **rent700**, and **ownhome**) and use the **msize(vsmall medium large)** option to make the sizes of these markers very small, medium, and large, respectively.

Click on a marker and, in the Contextual Toolbar, change the **Size**.
Uses allstatesdc.dta & scheme vg_s2c

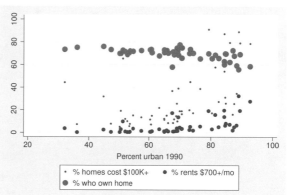

```
twoway (scatter propval100 rent700 ownhome urban, msize(*.5 *1 *1.5))
```

We can repeat the previous graph but use relative sizes within the **msize()** option to control the sizes of the markers, making them half the normal size, regular size, and 50% more than the normal size, respectively.

Double-click on a marker and change the **Size** to a multiple of the current size. For example, double-click on the marker for **propval100** and next to **Size** type in *.5.
Uses allstatesdc.dta & scheme vg_s2c

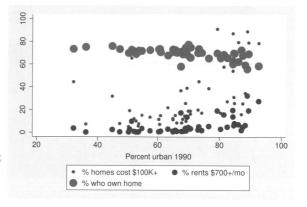

Introduction   Editor   Twoway   Matrix   Bar   Box   Dot   Pie   Options   Standard options   Styles   Appendix

Angles   Colors   Clockpos   Compassdir   Connect   Linepatterns   Linewidth   Margins   Markersize   Orientation   Symbols   Textsize

11.10   Orientation

We use an *orientationstyle* to change the orientation of text, such as a *y*-axis title, an *x*-axis title, or added text. An *orientationstyle* is similar to an *anglestyle*; see **Styles : Angles** (411). You can specify four different orientations using the keywords `horizontal` for 0 degrees, `vertical` for 90 degrees, `rhorizontal` for 180 degrees, and `rvertical` for 270 degrees. See [G-4] *orientationstyle* for more information.

```
scatter workers2 faminc,
    ytitle("Family" "Worker" "Status", orientation(horizontal))
```

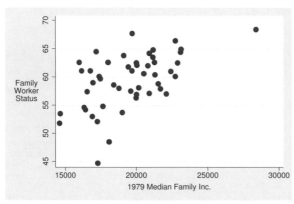

This example shows how we can rotate the title for the *y* axis using the `orientation(horizontal)` option to make the title horizontal.

📊 Double-click on the title for the *y* axis and, in the *Format* tab, change the **Orientation** to `Horizontal`.

Uses allstatesdc.dta & scheme vg_s2c

```
scatter workers2 faminc,
    xtitle("Family" "Income", orientation(vertical))
```

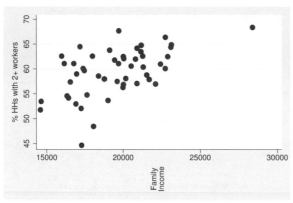

This example shows how we can rotate the title for the *x* axis to be vertical by using the `orientation(vertical)` option.

📊 Double-click on the title for the *x* axis and, in the *Format* tab, change the **Orientation** to `Vertical`.

Uses allstatesdc.dta & scheme vg_s2c

11.11   Marker symbols

Stata allows a wide variety of marker symbols. We can specify `O` (circle), `D` (diamond), `T` (triangle), `S` (square), `+` (plus sign), `X` (x), `p` (a tiny point), and `i` (invisible). We can also use lowercase letters `o`, `d`, `t`, `s`, and `x` to indicate smaller versions of these symbols. For circles, diamonds, triangles, and squares, we can append an `h` to indicate that the symbol should be displayed as hollow (e.g., `Oh` is a hollow circle). See [G-4] *symbolstyle* for more information.

`twoway (scatter propval100 rent700 ownhome urban, msymbol(S T O))`

We use the `msymbol(S T O)` (marker symbol) option to plot the three symbols in this graph using squares, triangles, and circles.

📈 Click on a marker for `propval100` and, in the Contextual Toolbar, change the **Symbol** to Square. Likewise, change the **Symbol** for `rent700` and `ownhome` to be Triangle and Circle, respectively.

Uses allstatesdc.dta & scheme vg_s2c

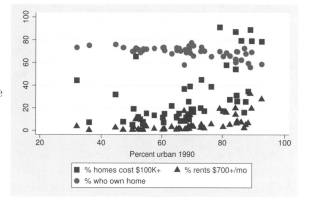

`twoway (scatter propval100 rent700 ownhome urban, msymbol(Sh Th Oh))`

We append an `h` to each marker symbol value to indicate that the symbol should be displayed as hollow.

📈 Click on a marker for `propval100` and use the Contextual Toolbar to change the **Symbol** to Hollow square. Likewise, change the **Symbol** for `rent700` and `ownhome` to be Hollow triangle and Hollow circle, respectively.

Uses allstatesdc.dta & scheme vg_s2c

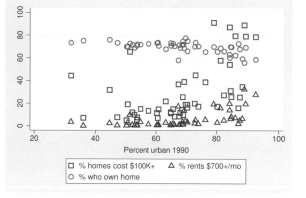

```
twoway (scatter propval100 rent700 ownhome urban, msymbol(s t o))
```

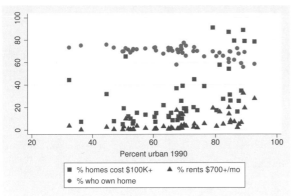

Here we use the `msymbol(s t o)` option to specify small squares, small triangles, and small circles.

▨ Click on a marker for `propval100` and use the Contextual Toolbar to change the **Symbol** to `Small square`. Likewise, change the **Symbol** for `rent700` and `ownhome` to be `Small triangle` and `Small circle`, respectively.

Uses allstatesdc.dta & scheme vg_s2c

11.12   Text size

We use the *textsizestyle* to control the size of text, either by specifying a keyword that corresponds to a particular size or by specifying a number representing a relative size. The predefined keywords include `zero`, `miniscule`, `quarter_tiny`, `third_tiny`, `half_tiny`, `tiny`, `vsmall`, `small`, `medsmall`, `medium`, `medlarge`, `large`, `vlarge`, `huge`, and `vhuge`. You could also specify the sizes as a relative size, which is a multiple of the original size of the text. See [G-4] *textsizestyle* for more information.

```
scatter workers2 faminc, mlabel(stateab) mlabsize(small)
```

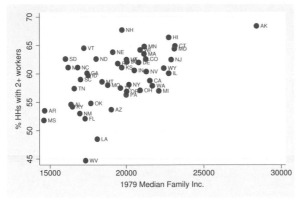

This example uses `mlabel(stateab)` to add marker labels with the state abbreviation labeling each point. We use the `mlabsize(small)` (marker label size) option to modify the size of the marker labels to make the labels small.

▨ Double-click on any marker label (e.g., AK) and change the **Size** to `Small`.

Uses allstatesdc.dta & scheme vg_s2c

`scatter workers2 faminc, mlabel(stateab) mlabsize(*1.5)`

In addition to using the keywords, we can specify a relative size that is a multiple of the current size. Here we use the `mlabsize(*1.5)` option to make the marker labels 1.5 times as large as they would normally be.

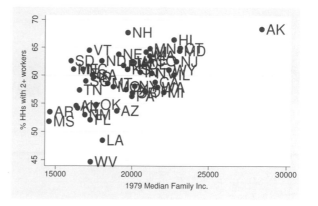

▨ Double-click on any marker label (e.g., AK) and next to **Size** type in *1.5.

Uses allstatesdc.dta & scheme vg_s2c

12 Appendix

The appendix contains a mixture of material that did not fit well in any of the previous chapters. The appendix shows some of the other kinds of statistical graphs Stata can produce that were not covered in the chapters and shows how to use the options illustrated in this book to make them. Next the `marginsplot` command is illustrated, focusing on how you can use options to customize such graphs. The next section looks at how to save graphs, redisplay graphs, and combine multiple graphs into one. This is followed by a section with more realistic examples that require a combination of multiple options or data manipulation to create the graph. The appendix reviews some common mistakes in writing graph commands and shows how to fix them, followed by a brief look at creating custom schemes. This chapter and the book conclude by describing the online supplements to the book and how to get them.

12.1 Overview of statistical graph commands

This section illustrates some of the Stata commands for producing specialized statistical graphs. Unlike other sections of this book, this section merely illustrates these kinds of graphs but does not further explain the syntax of the commands used to create them. The graphs are illustrated on the following six pages, with multiple graphs on each page. The title of each graph is the name of the Stata command that produced the graph. You can use the `help` command to find out more about that command, or you can find more information in the appropriate Stata manual. The figures are described below.

- Figure 12.1 illustrates several graphs used to examine the univariate distribution of variables.

- Figure 12.2 illustrates the `gladder` and `qladder` commands, which show the distribution of a variable according to the *ladder of powers* to help visually identify transformations for achieving normality.

- Figure 12.3 shows several graphs you can use to assess how your data meet the assumptions of linear regression.

- Figure 12.4 shows some plots that help to illustrate the results of a survival analysis.

- Figure 12.5 shows several different plots used to understand the nature of time-series data and to select among different time-series models.

- Figure 12.6 shows plots associated with receiver operating characteristic (ROC) analyses, which you can also use with logistic regression analysis.

431

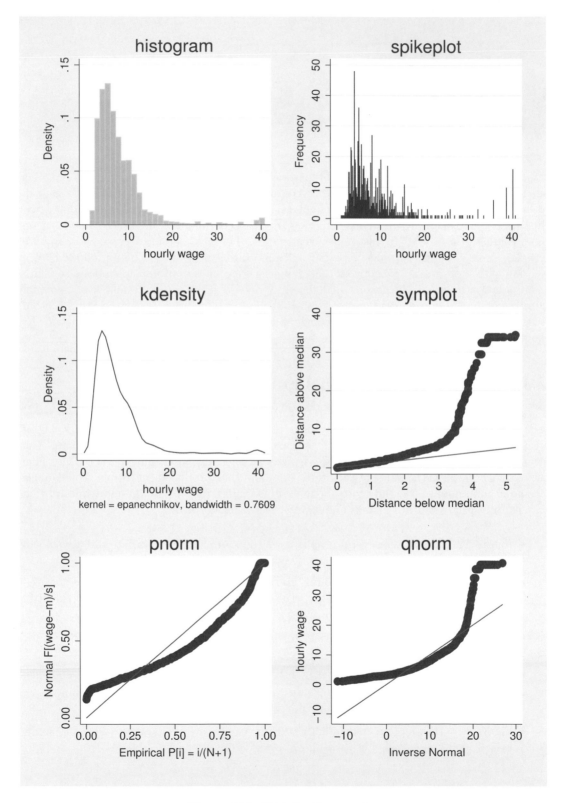

Figure 12.1: Distribution graphs

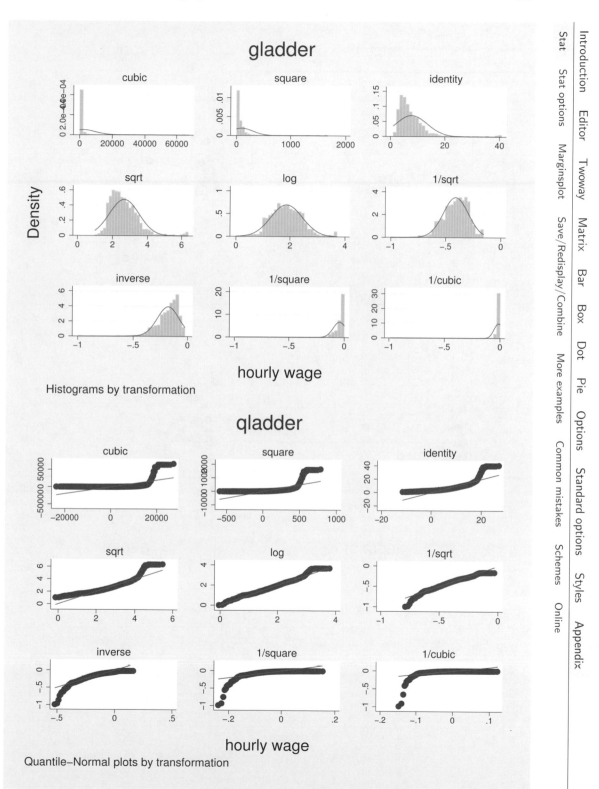

Figure 12.2: Ladder of powers graphs

Introduction   Editor   Twoway   Matrix   Bar   Box   Dot   Pie   Options   Standard options   Styles   Appendix

Stat   Stat options   Marginsplot   Save/Redisplay/Combine   More examples   Common mistakes   Schemes   Online

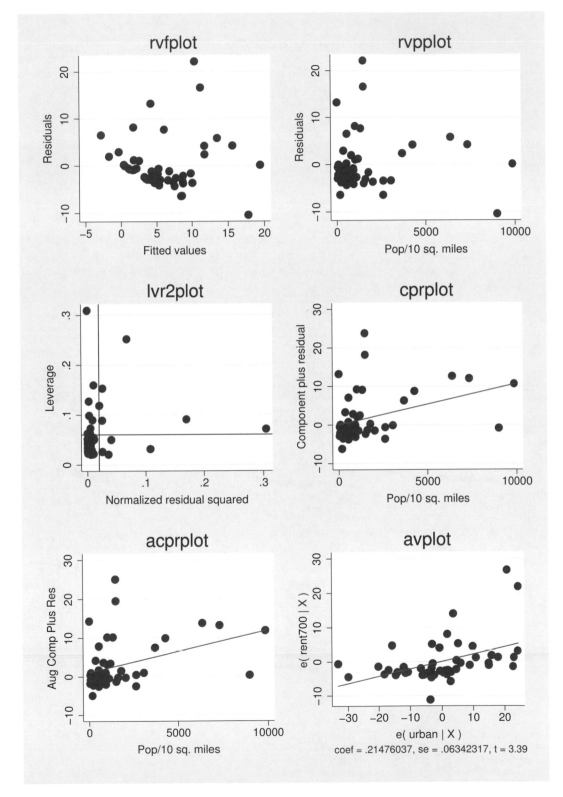

Figure 12.3: Regression diagnostics graphs

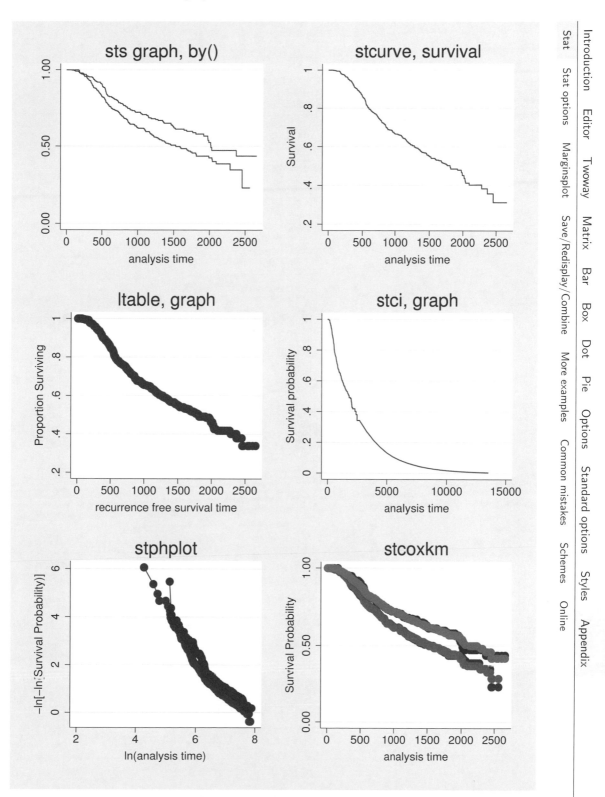

Figure 12.4: Survival graphs

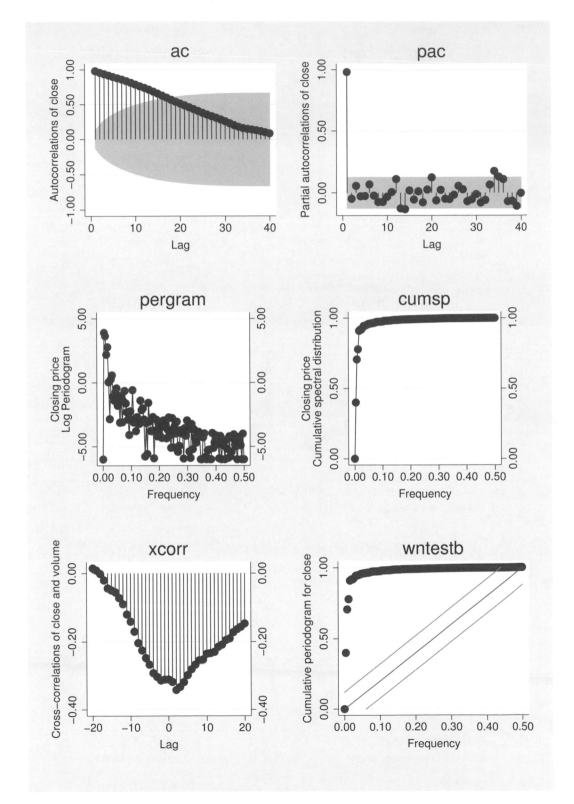

Figure 12.5: Time-series graphs

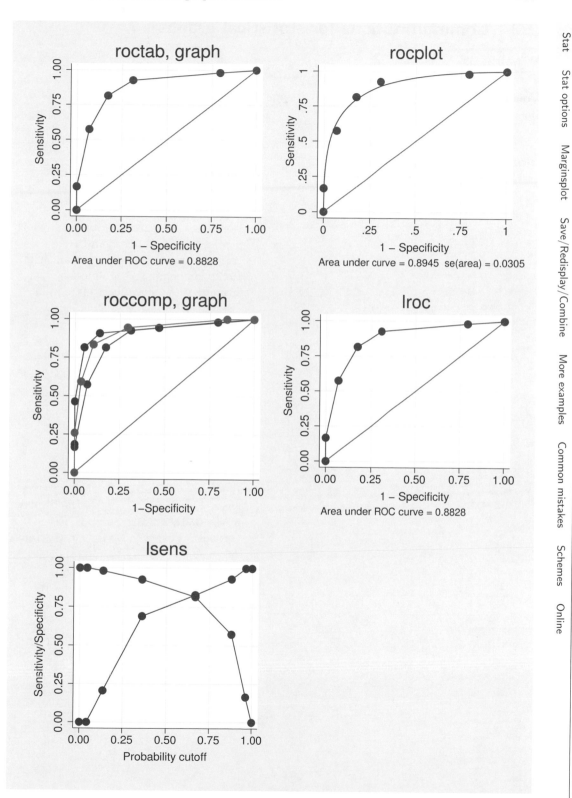

Figure 12.6: ROC graphs

12.2   Common options for statistical graphs

This section illustrates how to use Stata graph options with specialized statistical graph commands. Many of the examples will assume that you have run the command

```
. regress propval100 popden pcturban
```

and will illustrate subsequent commands with options to customize those specialized statistics graphs.

lvr2plot

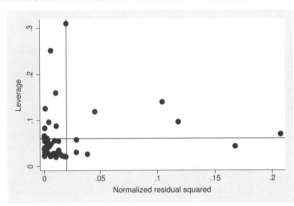

Consider this regression analysis, which predicts `propval100` from two variables, `popden` and `pcturban`. The `lvr2plot` command produces a leverage-versus-residual squared plot. *Uses allstates.dta & scheme vg_s2c Before running the graph command, type*
`reg propval100 popden pcturban`

lvr2plot, msymbol(Oh) msize(vlarge)

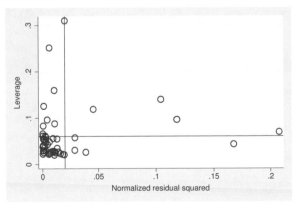

Here we add the `msymbol()` and `msize()` options to control the display of the markers in the graph. See Options: Markers (307) for more details.

📊 Click on any marker and, in the Contextual Toolbar, change the **Symbol** to `Hollow circle` and the **Size** to `v Large`. *Uses allstates.dta & scheme vg_s2c Before running the graph command, type*
`reg propval100 popden pcturban`

lvr2plot, mlabel(stateab)

The mlabel() option adds marker labels to the graph. We could also add more options to control the size, color, and position of the marker labels; see Options: Marker labels (320) for more details.

Uses allstates.dta & scheme vg_s2c
Before running the graph command, type
reg propval100 popden pcturban

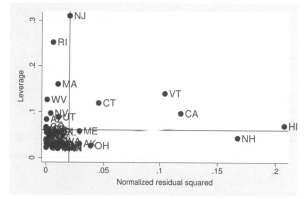

kdensity propval100

Consider this kernel density plot for the variable propval100. We could add options to control the display of the line. See the following example.
Uses allstates.dta & scheme vg_s2c

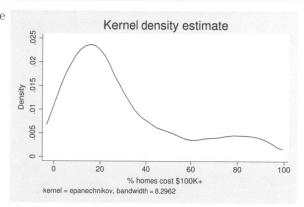

kdensity propval100, lwidth(thick) lpattern(dash)

The section Options: Connecting (323) shows several options we could add to control the display of the line. Here we add the lwidth() and lpattern() options to make the line thick and dashed.

▣ Click on the line and, in the Contextual Toolbar, change the **Width** to Thick and the **Pattern** to Dash.
Uses allstates.dta & scheme vg_s2c

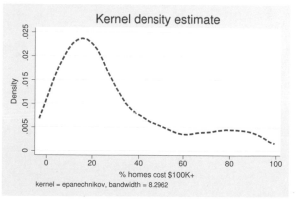

`avplot popden`

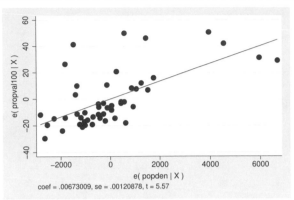

Consider this added-variable plot. We can modify the axes titles as illustrated in the following examples.

Uses allstates.dta & scheme vg_s2c
Before running the graph command, type
`reg propval100 popden pcturban`

`avplot popden, xtitle("popden adjusted for percent urban")`
`    ytitle("property value adjusted for percent urban")`

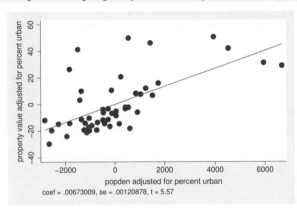

Here we use the `xtitle()` and `ytitle()` options to change the titles of the *x* and *y* axes. See Options : Axis titles (327) for more details.

📈 Click on the *x*-axis title and, in the Contextual Toolbar, change the **Text** to the desired title. Repeat these steps for the *y*-axis title.

Uses allstates.dta & scheme vg_s2c
Before running the graph command, type
`reg propval100 popden pcturban`

`avplot popden, note("Regression statistics for popden", prefix)`

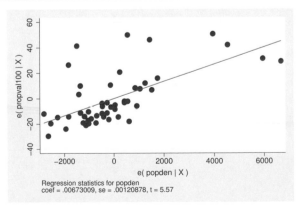

Here we use `prefix` within the `note()` option to add text before the existing note. We can do likewise for an existing title, subtitle, or caption. We could also use the `suffix` option to add information after an existing title.

📈 Click on the note and, in the Contextual Toolbar, make any desired changes to the **text**.

Uses allstates.dta & scheme vg_s2c
Before running the graph command, type
`reg propval100 popden pcturban`

avplot popden, xtitle(, size(huge))

We can modify the look of the existing title without changing the text. Here we add the size(huge) option to make the existing title huge. See Options : Axis titles (327) and Options : Textboxes (379) for more details.

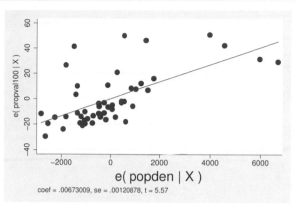

coef = .00673009, se = .00120878, t = 5.57

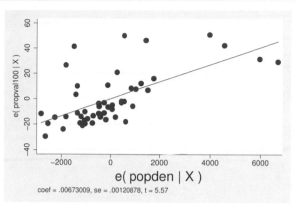 Click on the *x*-axis title and, in the Contextual Toolbar, change the **Size** to **Huge**.

Uses allstates.dta & scheme vg_s2c
Before running the graph command, type

```
reg propval100 popden pcturban
```

rvfplot

Consider this residual-versus-fit plot. We often hope to see an even distribution of points around zero on the *y* axis. To help evaluate this distribution, we might want to label the *y* axis identically for the values above 0 and below 0.

Uses allstates.dta & scheme vg_s2c
Before running the graph command, type

```
reg propval100 popden pcturban
```

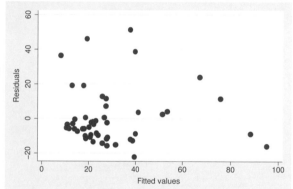

rvfplot, ylabel(-60(20)60, nogrid) yline(-20 20)

Here we add the ylabel() option to label the *y* axis from −60 to 60, incrementing by 20, and suppress the grid. Further, we use the yline() option to add a *y* line at 20 and −20. For more information about labeling and scaling axes, see Options : Axis labels (330) and Options : Axis scales (339).

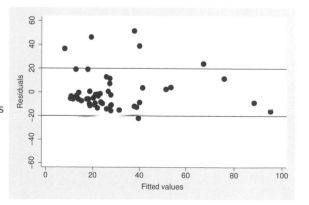 See the next graph.

Uses allstates.dta & scheme vg_s2c
Before running the graph command, type

```
reg propval100 popden pcturban
```

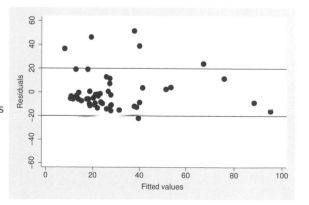

Introduction   Editor   Twoway   Matrix   Bar   Box   Dot   Pie   Options   Standard options   Styles   Appendix

Stat   Stat options   Marginsplot   Save/Redisplay/Combine   More examples   Common mistakes   Schemes   Online

`rvfplot, ylabel(-60(20)60, nogrid) yline(-20 20)`

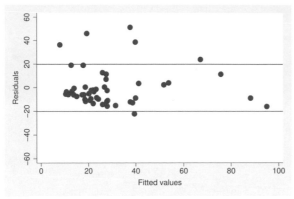

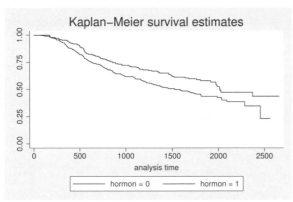 Using the Graph Editor, double-click on the *y*-axis label. Within **Axis rule**, select **Range/Delta** and enter **-60** for the **Minimum**, 60 for the **Maximum**, and 20 for the **Delta**. Click **Reference Line** and enter **-20 20** for the **Y axis value**. Finally, click on **Grid lines** and uncheck **Show grid**.

Uses allstates.dta & scheme vg_s2c
Before running the graph command, type
`reg propval100 popden pcturban`

`sts graph, by(hormon)`

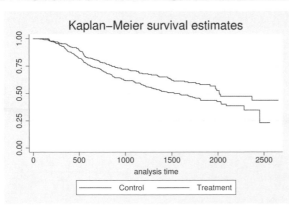

This graph shows survival-time estimates broken down by whether one is in the treatment group or the control group. The legend specifies the groups, but we might want to modify the labels as shown in the next example.

Uses hormone.dta & scheme vg_s2c

`sts graph, by(hormon) legend(label(1 Control) label(2 Treatment))`

We can use the `legend()` option to use different labels within the legend. See Options : Legend (361) for more details.

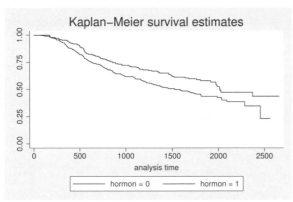 Within the legend, click on the label for the first group and, in the Contextual Toolbar, change the **Text** to `Control`. Likewise, for the second group change the label to `Treatment`.

Uses hormone.dta & scheme vg_s2c

```
sts graph, by(hormon) legend(off)
    text(.5 800 "Control") text(.8 1500 "Treatment")
```

We use the `legend(off)` option to suppress the display of the legend and use the `text()` option to add text directly to the graph to label the two lines; see **Options**: **Adding text** (374) for more information.

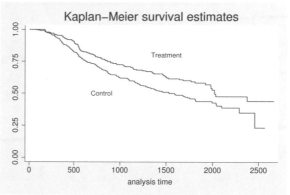

🖎 Select the Add Text **T** tool and click on a location above the top line and enter `Treatment` for the **Text**. Likewise, click on a location below the bottom line and enter `Control` for the **Text**.

Uses hormone.dta & scheme vg_s2c

```
avplot popden, title("Added-variable plot")
```

We return to the regression analysis predicting `propval100` from `popden` and `pcturban`. We add a title by using the `title()` option, but we could also add a `subtitle()`, `caption()`, or `note()`.

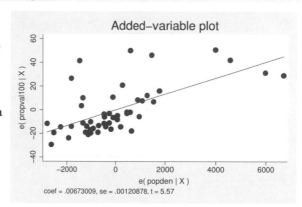

🖎 From the Main Menu, select **Graph** and then **Titles** and modify the **Title**.
Uses allstates.dta & scheme vg_s2c
Before running the graph command, type
```
reg propval100 popden pcturban
```

```
avplot popden, note("")
```

Here we add the `note("")` option, which suppresses the display of the note at the bottom showing the coefficients for the regression model.

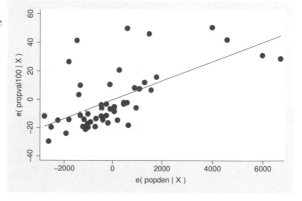

🖎 Right-click on the note and select **Hide**. You can redisplay the note by right-clicking it from the Object Browser and selecting **Show**.
Uses allstates.dta & scheme vg_s2c
Before running the graph command, type
```
reg propval100 popden pcturban
```

Introduction Editor Twoway Matrix Bar Box Dot Pie Options Standard options Styles Appendix

Stat Stat options Marginsplot Save/Redisplay/Combine More examples Common mistakes Schemes Online

`avplot popden, scheme(economist)`

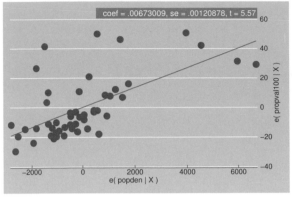

We can change the look of the graph by selecting a different scheme. Here we use scheme(economist) to display the graph using the economist scheme. See Standard options: Schemes (401) for more details.

📊 Before starting the Graph Editor, use the Main Menu to select **Edit** and then **Apply New Scheme** and change the **scheme** to economist.

Uses allstates.dta & scheme economist
Before running the graph command, type

`reg propval100 popden pcturban`

`avplot popden, xsize(3) ysize(1) scale(1.3)`

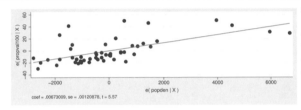

The section Standard options: Sizing graphs (404) describes options we can use to control the size of the graph and the scale of the contents of the graph. Here we show the xsize(), ysize(), and scale() options.

📊 Use the Main Menu to select **Graph** and then **Aspect Ratio** and change the **Y size** to 1, the **X size** to 3, and the **Scaling factor** to 1.3.

Uses allstates.dta & scheme vg_s2c
Before running the graph command, type

`reg propval100 popden pcturban`

12.3   The marginsplot command

This section illustrates how to use the `marginsplot` command, including options that are common to all graph commands as well as options that are specific to `marginsplot`. The following examples will assume that you have run the following `regress` and `margins` commands:

```
. use allstates
. regress propval100 popden pcturban80
. margins, at(pcturban80=(30(10)90))
```

The `regress` command predicts property values from population density and the percentage of the area considered to be urban. The `margins` command computes the adjusted means as a function of `pcturban80`.

marginsplot

This graph shows the adjusted means as a function of **pcturban80**. The following examples illustrate how you can use options to customize this graph.
Uses allstates.dta & scheme vg_s2c

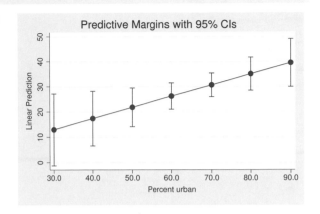

marginsplot, title(Title) subtitle(Subtitle) xtitle(X title)
ytitle(Y title) note(Note) caption(Caption)

You can add titles to the graph by using the `title()`, `subtitle()`, `xtitle()`, and `ytitle()` options. The `note()` and `caption()` options can also be used to annotate the graph. You can see Standard options : Titles (395) for more details about adding titles, notes, and captions.
Uses allstates.dta & scheme vg_s2c

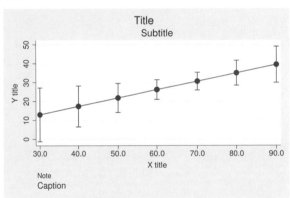

marginsplot, scheme(economist)

The `scheme()` option can be used to change the overall look of the graph. In this example, the `economist` scheme is used. See Standard options : Schemes (401) for more information about the selection of schemes.
Uses allstates.dta & scheme economist

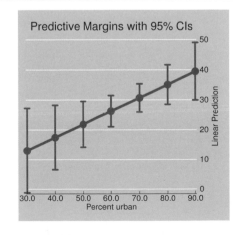

Stat   Stat options   Marginsplot   Save/Redisplay/Combine   More examples   Common mistakes   Schemes   Online

Introduction   Editor   Twoway   Matrix   Bar   Box   Dot   Pie   Options   Standard options   Styles   Appendix

`marginsplot, xtitle(Percent urban in 1980) ytitle(Property value)`

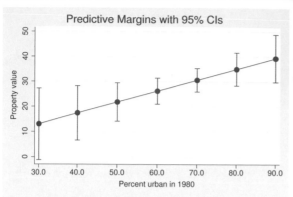

The `xtitle()` and `ytitle()` options are used to change the titles of the x and y axes. You can see Options : Axis titles (327) for more details about adding titles to axes.
Uses allstates.dta & scheme vg_s2c

`marginsplot, xlabel(30(5)90) ylabel(0(5)50, angle(0))`

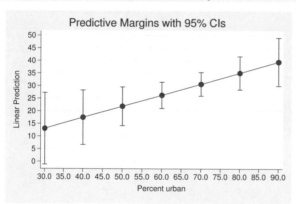

The labeling of the x and y axes can be controlled with the `xlabel()` and `ylabel()` options. You can find more information about labeling axes in Options : Axis labels (330).
Uses allstates.dta & scheme vg_s2c

`marginsplot, xscale(range(0 100)) yscale(range(0 60))`

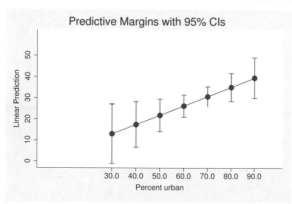

The `xscale()` and `yscale()` options can be used to expand the scale of the x and y axes. See Options : Axis scales (339) for more information about options that control the scale of the x and y axes.
Uses allstates.dta & scheme vg_s2c

marginsplot, plotopts(clwidth(thick))

The plotopts() option allows you to include options that control the look of the line and markers. This example uses the clwidth() suboption to make the fitted line thick. You can see Styles : Linewidth (422) for more details about controlling the thickness of lines.
Uses allstates.dta & scheme vg_s2c

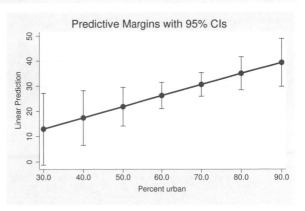

marginsplot, plotopts(msymbol(Oh) msize(large))

The plotopts() option is used in this example with the msymbol() and msize() suboptions to draw the markers as large hollow circles. For more information about selecting marker symbols, see Styles : Symbols (427).
Uses allstates.dta & scheme vg_s2c

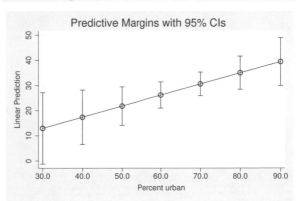

marginsplot, ciopts(lwidth(vthick) msize(huge))

The ciopts() option allows you to include options that control the look of the confidence interval. The lwidth() suboption makes the lines for the confidence intervals very thick, and the msize() suboption makes the cap of each confidence interval huge.
Uses allstates.dta & scheme vg_s2c

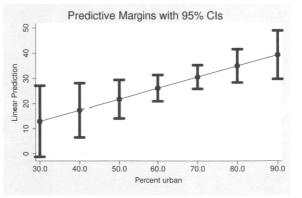

Introduction Editor Twoway Matrix Bar Box Dot Pie Options Standard options Styles Appendix

Stat Stat options Marginsplot Save/Redisplay/Combine More examples Common mistakes Schemes Online

`marginsplot, recast(line) recastci(rarea)`

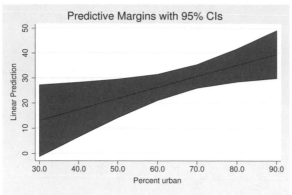

The recast(line) option specifies that the fitted line be drawn like a twoway line graph. The recastci(rarea) option specifies that the confidence interval be drawn like a twoway rarea graph.

Uses allstates.dta & scheme vg_s2c

`marginsplot, noci`

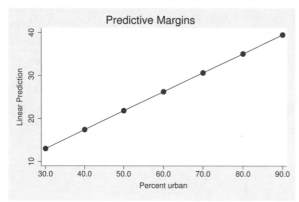

The noci option suppresses the display of the confidence interval.

Uses allstates.dta & scheme vg_s2c

`marginsplot, noci addplot(scatter propval100 pcturban80, msymbol(o))`

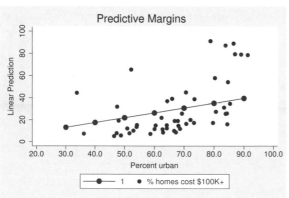

The addplot() option can be used to overlay a new graph onto the graph created by the margins command. In this case, it overlays a scatterplot of propval100 and pcturban80.

Uses allstates.dta & scheme vg_s2c

Let's now consider another example, this one based on a 5×2 analysis of variance (ANOVA). The commands below use the `nlsw.dta` dataset, run the `anova` command, and then use the `margins` command to obtain the adjusted mean of the outcome by `occ5` and `collgrad`. The output of these commands is suppressed to save space.

```
. use nlsw, clear
. anova wage i.occ5##i.collgrad
. margins occ5#collgrad
```

The `marginsplot` can then be used to graph the adjusted means computed by the `margins` command.

marginsplot

This example shows the graph created by the `marginsplot` command. The following examples illustrate how to customize this graph.
Uses nlsw.dta & scheme vg_s2c

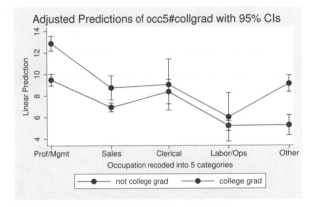

marginsplot, legend(subtitle("Education") rows(2))

This example includes the `legend()` option to customize the display of the legend, adding a subtitle and displaying the legend keys in two rows. You can see Options : Legend (361) for more details about customizing legends.
Uses nlsw.dta & scheme vg_s2c

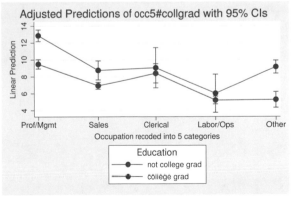

```
marginsplot, legend(subtitle("Education") rows(2) ring(0) pos(1))
```

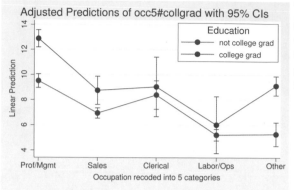

The ring() and pos() suboptions are added to the legend() option to display the legend within the graph in the 1 o'clock position. For more details about customizing the legend, see Options: Legend (361).

Uses nlsw.dta & scheme vg_s2c

```
marginsplot,
    xlabel(1 "Professional" 2 "Sales" 3 "Clerical" 4 "Labor" 5 "Other occ")
```

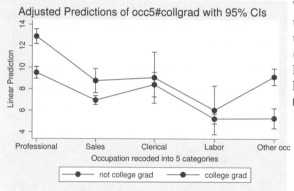

The xlabel() option is used to control the labeling of the x axis, as shown in this example. See Options: Axis labels (330) for more information about axis labels. The next example illustrates how to address the issue of the label Other occ being cut off.

Uses nlsw.dta & scheme vg_s2c

```
marginsplot,
    xlabel(1 "Professional" 2 "Sales" 3 "Clerical" 4 "Labor" 5 "Other occ")
    xscale(range(.75 5.25))
```

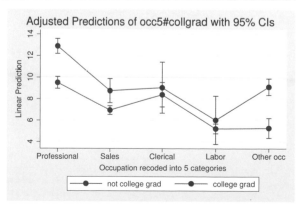

The xscale(range()) option is used to expand the range of the x axis to make additional room for longer x-axis labels. You can see Options: Axis scales (339) for more information about controlling axis scales.

Uses nlsw.dta & scheme vg_s2c

In the previous examples, the variable `occ5` was placed on the x axis, and the variable `collgrad` was graphed using separate lines. Suppose that instead we want to place `collgrad` on the x axis.

marginsplot, xdimension(collgrad)

The `xdimension()` option controls which variable is placed on the x axis. In this example, the variable `collgrad` is placed on the x axis. As a result, `collgrad` is graphed using separate lines.

Uses nlsw.dta & scheme vg_s2c

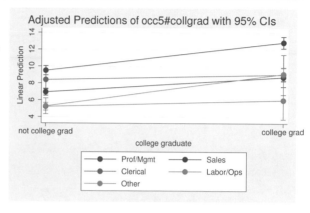

marginsplot, plotdimension(occ5)

The `plotdimension()` option controls which variable is graphed using the plot dimension, i.e., graphed using separate lines. In this example, `occ5` is graphed using separate lines, and thus `collgrad` is placed on the x axis.

Uses nlsw.dta & scheme vg_s2c

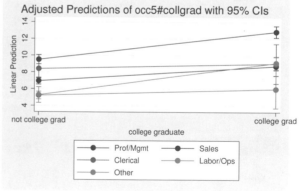

```
marginsplot,
    plotdim(occ5, labels("Professional" "Sales" "Clerical" "Labor" "Other"))
```

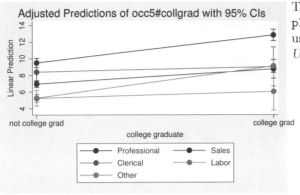

The labels() suboption within the plotdim() option changes the labels used for the plot dimension.
Uses nlsw.dta & scheme vg_s2c

```
marginsplot,
    plotdim(occ5, elabels(1 "Professional" 4 "Labor"))
```

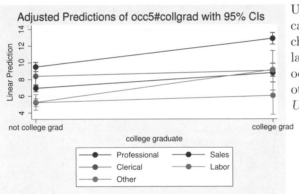

Using the elabels() suboption, you can selectively modify the labels of your choice. This example modifies the labels for the first and fourth occupations, leaving the labels for the other occupations unchanged.
Uses nlsw.dta & scheme vg_s2c

```
marginsplot, plotdim(occ5, nosimplelabels)
```

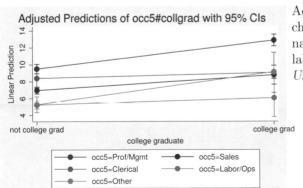

Adding the nosimplelabels suboption changes the plot label to the variable name, an equal sign, and the value label for the group.
Uses nlsw.dta & scheme vg_s2c

`marginsplot, plotdim(occ5, nolabels)`

Adding the `nolabels` suboption
changes the plot label to the variable
name, an equal sign, and the numeric
value for the group.
Uses nlsw.dta & scheme vg_s2c

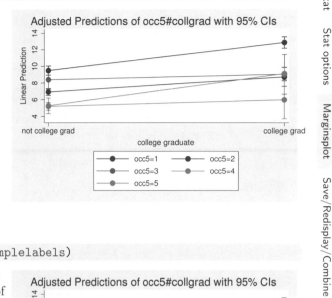

`marginsplot, plotdim(occ5, allsimplelabels)`

Using the `allsimplelabels` suboption
yields a label that is composed solely of
the value label for each group.
Uses nlsw.dta & scheme vg_s2c

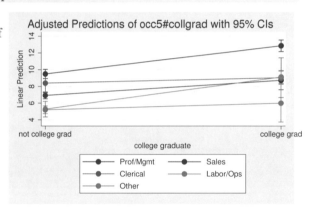

`marginsplot, plotdim(occ5, allsimplelabels nolabels)`

Using the `allsimplelabels` and
`nolabels` suboptions displays a label
that is composed solely of the numeric
value for each group.
Uses nlsw.dta & scheme vg_s2c

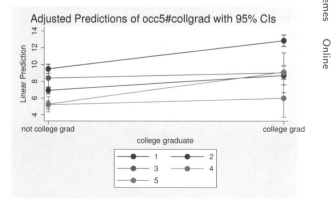

Stat · Stat options · Marginsplot · Save/Redisplay/Combine · More examples · Common mistakes · Schemes · Online

Introduction · Editor · Twoway · Matrix · Bar · Box · Dot · Pie · Options · Standard options · Styles · Appendix

Let's now consider an example where the `marginsplot` command involves by-groups as an additional dimension. This example predicts marital status from the interaction of education, whether one lives in the south, and whether one lives in an urban area. The variables `race` and `age` are included as covariates. The command for running this logistic regression is shown below:

```
. use nlsw
. logit married c.grade##i.south##i.urban2 i.race age
```

The `c.grade#south#urban2` interaction is significant. (The output is omitted to save space.) The `margins` command (below) is used to compute the predictive margin of the probability of being married as a function of education, whether one lives in the south, and whether one lives in an urban area, after adjusting for race and age.

```
. margins south#urban2, at(grade=(9(1)18))
```

marginsplot, noci

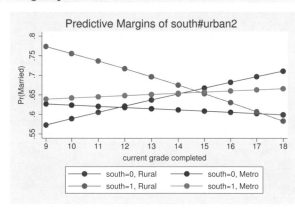

This example shows the graph created by the `marginsplot` command, graphing the predicted margins computed by the `margins` command. The `noci` option is used to suppress the confidence intervals. The following examples customize the graph, focusing on issues related to the by-dimension. *Uses nlsw.dta & scheme vg_s2c*

marginsplot, noci bydimension(south)

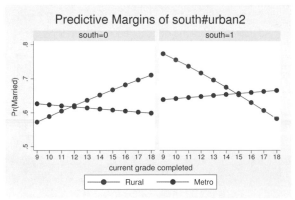

The `bydimension()` option is used to specify that separate graphs be created based on whether one lives in the south.

Uses nlsw.dta & scheme vg_s2c

`marginsplot, noci bydimension(south, label("Nonsouth" "South"))`

The `label()` suboption can be used to control the labeling of each of the graphs. You can control the labeling of the by-dimension by using the `bydimension()` option in the same manner that we controlled the plot dimension by using the `plotdimension()` option.

Uses nlsw.dta & scheme vg_s2c

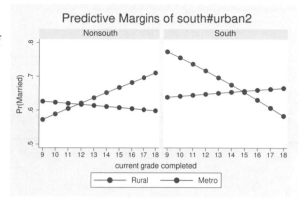

`marginsplot, noci bydimension(south) byopts(cols(1) ixaxes)`

The `byopts()` option allows you to specify suboptions that control the way the separate graphs are combined together. In this example, the `cols(1)` and `ixaxes` suboptions are specified to display the graphs in one column, each with its own x axis. You can see Options : By (346) for additional suboptions that you could supply within the `byopts()` option.

Uses nlsw.dta & scheme vg_s2c

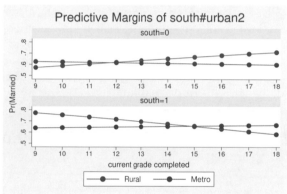

`marginsplot, noci bydimension(south)`
`    byopts(title(Title) subtitle(Subtitle) note(Note) caption(Caption))`

The `byopts()` option can be used to control the overall title, subtitle, note, and caption for the graph. By placing such options within the `byopts()` option, these options impact the overall title, subtitle, note, and caption for the graph. Contrast this with the next example.

Uses nlsw.dta & scheme vg_s2c

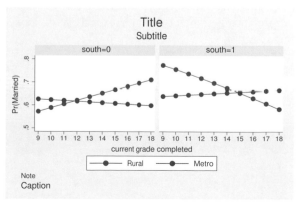

Introduction   Editor   Twoway   Matrix   Bar   Box   Dot   Pie   Options   Standard options   Styles   Appendix

Stat   Stat options   Marginsplot   Save/Redisplay/Combine   More examples   Common mistakes   Schemes   Online

```
marginsplot, noci bydimension(south)
    title(Title) subtitle(Subtitle) note(Note) caption(Caption)
```

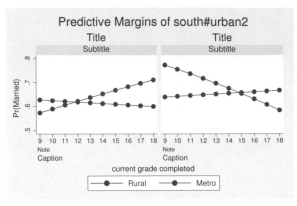

By placing these options outside the byopts() option, these options control the title, subtitle, note, and caption for each of the graphs. This is generally not the desired result.

Uses nlsw.dta & scheme vg_s2c

This section has illustrated some of the ways that you can customize graphs created by the marginsplot command. As we have seen, the marginsplot command supports standard graph options [as described in **Standard options** (395)]. The marginsplot command also supports options that you would apply to a twoway graph [as described in **Options** (307)]. In addition, this section illustrated options that are specific to the marginsplot command, such as the plotopts(), ciopts(), and plotdimension() options. For more information, you can see [R] **marginsplot**.

12.4   Saving, redisplaying, and combining graphs

This section shows how to save, redisplay, and combine Stata graphs.

The section begins by showing how to save graphs and use saved graphs. We can save graphs in one of two forms: *live* graphs or *as-is* graphs. We can edit a live graph by using the Graph Editor, and we can use and redisplay it with a different scheme. By contrast, an as-is graph can be displayed only exactly as it was saved—it can neither be edited nor be displayed with a different scheme. Such as-is graphs are generally smaller and will appear on another person's computer exactly as it looked when it was saved. Now let's look at a few examples.

twoway histogram urban

Let's start by creating a histogram. Suppose that we liked this graph so much that we wanted to save it for either sending to someone else (who owns Stata) or displaying it at a later time.

Uses allstates.dta & scheme vg_s2c

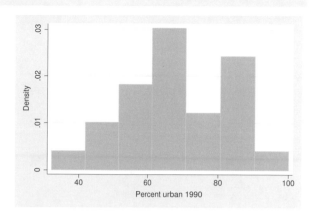

graph save hist1

The **graph save** command saves the currently displayed graph as a Stata .gph file. We save this graph in the current directory under the name hist1.gph. We will assume that in the following examples all graphs are stored in the current directory, but we can precede the filename with a directory name and store it wherever we want. This graph is stored as a live graph. If you added the **asis** option, the graph would have been stored as an as-is graph.

Uses allstates.dta & scheme vg_s2c

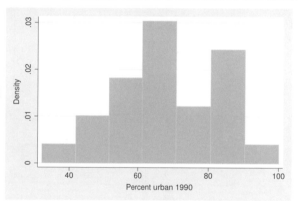

twoway histogram urban, saving(hist2, asis)

Most, if not all, Stata graph commands allow us to use the **saving()** option to save the graph as a Stata .gph file. This option allows us to create and save the graph in one step. Here we add the **asis** option to request the graph be stored as an as-is graph instead of a live graph. If hist2.gph existed, we would add the **replace** (next to the **asis** option), to overwrite the existing hist2.gph file.

Uses allstates.dta & scheme vg_s2c

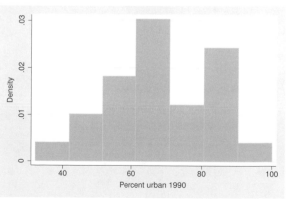

Introduction Editor Twoway Matrix Bar Box Dot Pie Options Standard options Styles Appendix

Stat Stat options Marginsplot Save/Redisplay/Combine More examples Common mistakes Schemes Online

`graph use hist1.gph`

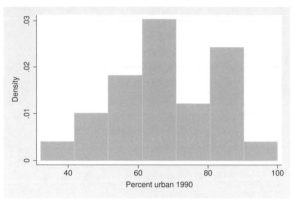

At a later time (including after quitting and restarting Stata), we can view a saved graph with the `graph use` command. Here we redisplay `hist1.gph`. If `hist1.gph` had been stored in a different directory, you would have to precede it with the directory where it was saved or use the `cd` command to change to that directory. Because this is a live graph, we can use the Graph Editor to edit this graph.

Uses allstates.dta & scheme vg_s2c

`graph use hist1.gph, scheme(vg_s1m)`

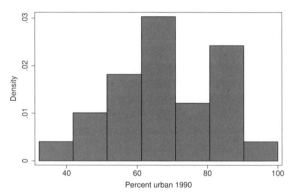

Also, because `hist1.gph` is a live graph, we can add the `scheme()` option to view the same graph using a different scheme. Here we view the last graph but use the `vg_s1m` scheme.

Uses allstates.dta & scheme vg_s1m

`graph use hist2.gph`

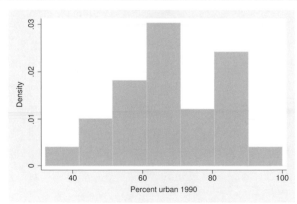

By contrast, `hist2.gph` is an as-is graph. We can display this graph with the `graph use` command, but we cannot display it with a different scheme, nor can we edit it by using the Graph Editor.

Uses allstates.dta & scheme vg_s1m

`twoway histogram propval100, name(hist2)`

The `name()` option is much like the `saving()` option, except that the graph is saved in memory instead of on disk. We can then view the graph later within the same Stata session, but once we quit Stata, the graph in memory is erased.

Uses allstates.dta & scheme vg_s1c

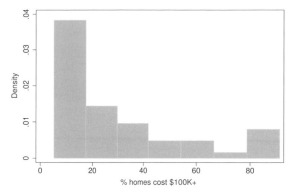

`graph display hist2`

The `graph display` command is similar to the `graph use` command, except that it redisplays graphs saved in memory. Here we redisplay the graph we created with the `name(hist2)` option.

Uses allstates.dta & scheme vg_s1c

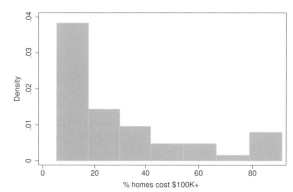

`graph display hist2, xsize(2) ysize(2)`

The `graph display` command allows us to use the `xsize()` and `ysize()` options to change the size and aspect ratio of the graph. Here we redisplay the graph we named `hist2` and make the graph 2 inches tall by 2 inches wide.

Uses allstates.dta & scheme vg_s1c

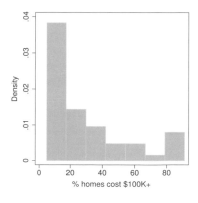

Introduction   Editor   Twoway   Matrix   Bar   Box   Dot   Pie   Options   Standard options   Styles   Appendix

Stat   Stat options   Marginsplot   Save/Redisplay/Combine   More examples   Common mistakes   Schemes   Online

`graph display hist2, scheme(s1mono)`

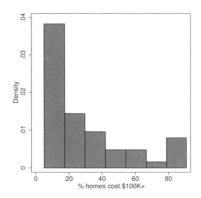

We can also use the `scheme()` option to view the same graph using a different scheme. Here we view the previous graph but with the **s1mono** scheme.

Uses allstates.dta & scheme s1mono

Let's now look at some examples that illustrate how to combine graphs once we have created and saved them. First, we will see how to show two scatterplots side by side rather than overlaying them.

`twoway scatter propval100 urban, name(scat1)`

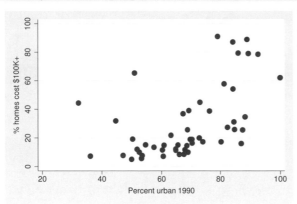

Using the `name(scat1)` option saves this scatterplot in memory with the name **scat1**.

Uses allstates.dta & scheme vg_s2c

`twoway scatter rent700 urban, name(scat2)`

We save this second scatterplot with the name `scat2`.

Uses allstates.dta & scheme vg_s2c

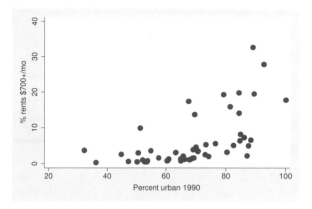

`graph combine scat1 scat2`

By using the `graph combine` command, we can see these two scatterplots side by side. In a sense, the *y* axis is on a different scale for these two graphs because they are different variables. However, in another sense, the scale for the two *y* axes is the same because they are both measured as percentages.

Uses allstates.dta & scheme vg_s2c

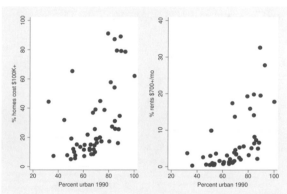

`graph combine scat1 scat2, ycommon`

This graph is the same as the last one, except that the *y* axes are placed on a common scale by using the `ycommon` option. This makes it easy to compare the two *y* variables by forcing them to be on the same metric. The `ycommon` option does not work when the graphs have been made by using different kinds of commands, e.g., `graph bar` and `graph box`.

Uses allstates.dta & scheme vg_s2c

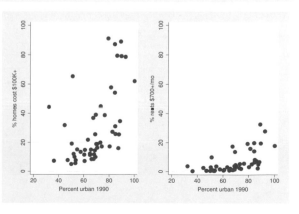

Introduction Editor Twoway Matrix Bar Box Dot Pie Options Standard options Styles Appendix

Stat Stat options Marginsplot Save/Redisplay/Combine More examples Common mistakes Schemes Online

`twoway scatter famsize urban, name(scat3)`

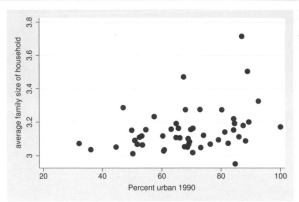

Let's make two more graphs. This graph is named **scat3**.
Uses allstates.dta & scheme vg_s2c

`twoway scatter popden urban, name(scat4)`

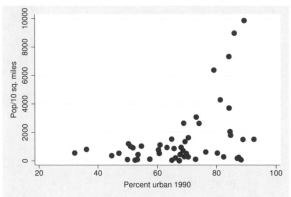

This example is named **scat4**. The next example will illustrate combining the four graphs **scat1**, **scat2**, **scat3**, and **scat4** into one graph.
Uses allstates.dta & scheme vg_s2c

`graph combine scat1 scat2 scat3 scat4`

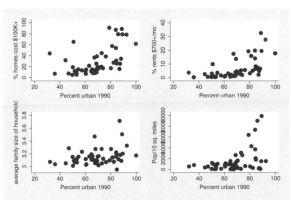

Note the default sizing of the text and markers in this combined graph. The next graph shows an alternate sizing we can select.
Uses allstates.dta & scheme vg_s2c

```
graph combine scat1 scat2 scat3 scat4, altshrink
```

Compare this graph with the previous
graph. Note how the `altshrink` option
reduces the size of the text and
markers. Sometimes this option might
be useful when combining graphs.
Uses allstates.dta & scheme vg_s2c

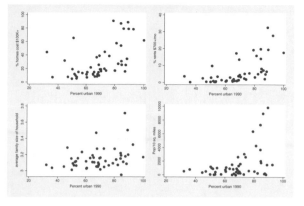

Here are more detailed examples showing how you can combine graphs and the options
to use in creating the graphs. The next set of examples uses the `sp2001ts` data file.

```
twoway rarea high low date, name(hilo)
```

We make a graph showing the high and
low closing price of the S&P 500 for
2001 and save this graph in memory,
naming it `hilo`.
Uses sp2001ts.dta & scheme vg_s2c

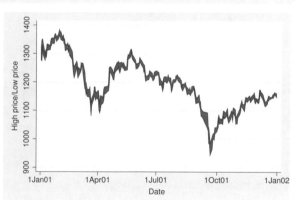

Introduction   Editor   Twoway   Matrix   Bar   Box   Dot   Pie   Options   Standard options   Styles   Appendix

Stat   Stat options   Marginsplot   Save/Redisplay/Combine   More examples   Common mistakes   Schemes   Online

`twoway spike volmil date, name(vol)`

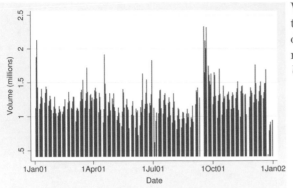

We can make another graph that shows the volume (millions of shares sold per day) for 2001 and save this graph in memory, naming it `vol`.

Uses sp2001ts.dta & scheme vg_s2c

`graph combine hilo vol`

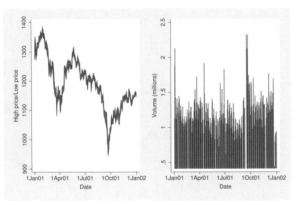

We can now use the `graph combine` command to combine these two graphs into one graph. The graphs are displayed as one row, but say that we would like to display them in one column.

Uses sp2001ts.dta & scheme vg_s2c

`graph combine hilo vol, cols(1)`

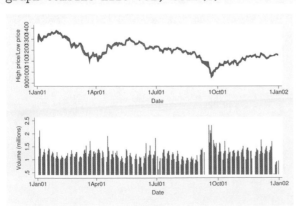

By using the `cols(1)` option, we can display the price above the volume. However, because the x axes of these two graphs are scaled the same, we could save space by removing the x-axis scale from the top graph.

Uses sp2001ts.dta & scheme vg_s2c

`twoway rarea high low date, xscale(off) name(hilo, replace)`

Here we use the `xscale(off)` option to suppress the display of the *x* axis, including the space that would be allocated for the labels. We name this graph `hilo` again, but we must use the `replace` option to replace the existing graph named `hilo`.

Uses sp2001ts.dta & scheme vg_s2c

`graph combine hilo vol, cols(1)`

We combine these two graphs; however, we might want to push the graphs a bit closer together.

Uses sp2001ts.dta & scheme vg_s2c

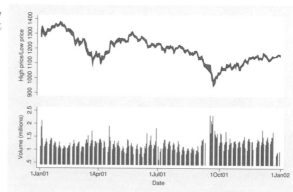

`graph combine hilo vol, cols(1) imargin(b=1 t=1)`

Here we use the `imargin(b=1 t=1)` option to make the margin at the top and bottom of the graphs to be small before combining them. However, we might want the lower graph of volume to be smaller.

Uses sp2001ts.dta & scheme vg_s2c

Introduction   Editor   Twoway   Matrix   Bar   Box   Dot   Pie   Options   Standard options   Styles   Appendix

Stat   Stat options   Marginsplot   Save/Redisplay/Combine   More examples   Common mistakes   Schemes   Online

```
twoway spike volmil date, ylabel(1 2) fysize(25) name(vol, replace)
```

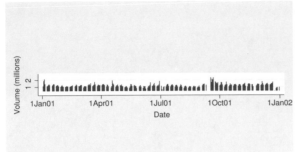

Using the `fysize()` (force y size) option makes the graph 25% of its normal size. We use this instead of `ysize()` because the `graph combine` command does not respect the `ysize()` or `xsize()` options. For aesthetics, we also reduce the number of labels. We save this graph in memory, replacing the existing graph named `vol`.

Uses sp2001ts.dta & scheme vg_s2c

```
graph combine hilo vol, cols(1) imargin(b=1 t=1)
```

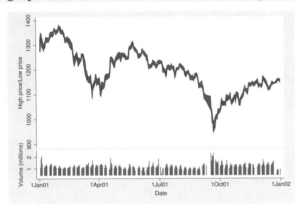

We combine these graphs again, and the combined graph looks pretty good. We might further tinker with the graph, changing the `xtitle()` for the volume graph to be shorter or modifying the `xlabel()` for the volume graph.

Uses sp2001ts.dta & scheme vg_s2c

12.5   More examples: Putting it all together

Most examples in this book have focused on the impact of one option or a few options, using datasets that required no manipulation before making the graph. In reality, many graphs use multiple options, and some require prior data management. This section addresses this issue by showing some examples that combine many options and require some data manipulation before making the graph.

```
twoway (scatter urban pcturban80) (function y=x, range(30 100)),
    xtitle(Percent Urban 1980) ytitle(Percent Urban 1990)
    legend(order(2 "Line where % Urban 1980 = % Urban 1990") pos(6) ring(0))
```

This graph shows the percentage of population living in an urban area of a state in 1990 against that of 1980. If there had been no changes from 1980 to 1990, the values would fall along a 45-degree line, where the value of y equals the value of x. Overlaying (`function y=x`), we can see any discrepancies from 1980 to 1990. The `range(30 100)` option makes the line span from 30 to 100 on the x axis.
Uses allstates.dta & scheme vg_s2c

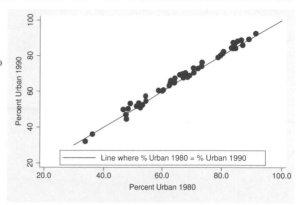

```
twoway (lfitci ownhome borninstate) (lfitci ownhome borninstate,
    ciplot(rline) lcolor(blue) lwidth(thick) lpattern(dash))
    (scatter ownhome borninstate), legend(off) ytitle("% Own Home")
```

This example shows how we can make a scatterplot, a regression line, and a confidence interval for the fit shown as an area. We also add a thick, blue, dashed line showing the upper and lower confidence limits. The first `lfitci` makes the fit line and area; the second `lfitci` makes a thick, blue, dashed outline for the area; and `scatter` overlays the scatterplot.
Uses allstates.dta & scheme vg_s2c

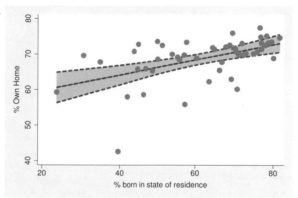

```
twoway scatter ownhome borninstate,
    by(nsw, hole(1) title("%Own home by" "%born in St." "by region",
    pos(11) ring(0) width(65) height(35) justification(center)
    alignment(middle)) note(""))
```

The `hole(1)` option leaves the first position empty when creating the graphs, and the title is placed there using `pos(11)` and `ring(0)`. We use `width()` and `height()` to adjust the size of the textbox and `justification()` and `alignment()` to center the textbox horizontally and vertically. The `note("")` option suppresses the note in the bottom corner of the graph.
Uses allstates.dta & scheme vg_s2c

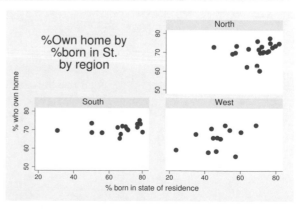

```
twoway (rspike hi low date) (rcap close close date, msize(medsmall)),
   tlabel(08jan2001 01feb2001 21feb2001) legend(off)
```

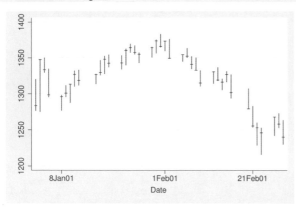

Before making this high/low/close graph, we first type `tsset date, daily` to tell Stata that `date` should be treated as a date in the `tlabel()` option. The `rcap` command uses `close` for both the high and the low values, making the tick line for the closing price, and the `legend(off)` option suppresses the legend. Using the `vg_samec` scheme makes the spikes and caps the same color.

Uses spjanfeb2001.dta & scheme vg_samec

```
twoway (rspike hi low date) (rcap close close date, msize(medsmall))
   (scatteri 1220 15027 1220 15034, recast(line) clwid(vthick) clcol(red)),
   tlabel(08jan2001 01feb2001 21feb2001) legend(off)
```

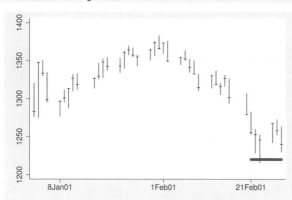

This example is the same as above, except that this one uses `scatteri()` to draw a support-level line. Two y x pairs are given after the `scatteri`, and the `recast(line)` option draws them as a line instead of two points. The x values were calculated beforehand by using `display d(21feb2001)` and `display d(28feb2001)` to compute the elapsed date values.

Uses spjanfeb2001.dta & scheme vg_samec

The rest of the examples in this section involve some data management before creating the graph. The next few examples use the `allstates` data file. First, run a regression command:

```
. vguse allstates
. regress ownhome propval100 workers2 urban
```

Then issue the

```
. dfbeta
```

command, creating DFBETAs for each predictor: `DFpropval100`, `DFworkers2`, and `DFurban`, which are used in the following graph. We also generated a sequential ID variable with the following command:

```
. generate id = _n
```

```
twoway dropline _dfbeta_1 _dfbeta_2 _dfbeta_3 statefips,
    mlabel(stateab stateab stateab)
```

Here we show each DFBETA as a
dropline plot. We add the `mlabel()`
option to label each point with the
state abbreviation.

Uses allstates.dta & scheme vg_s2c

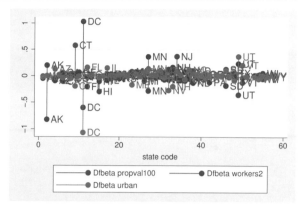

```
twoway (dropline _dfbeta_1 id if abs(_dfbeta_1)>.25, mlabel(stateab))
    (dropline _dfbeta_2 id if abs(_dfbeta_2)>.25, mlabel(stateab))
    (dropline _dfbeta_3 id if abs(_dfbeta_3)>.25, mlabel(stateab))
```

This example is similar to the one above
but simplifies the graph by showing
only the points where the DFBETA
exceeds .25. Note that we have taken
the example from above and converted
it into three overlaid dropline plots,
each of which has an `if` condition.

Uses allstates.dta & scheme vg_s2c

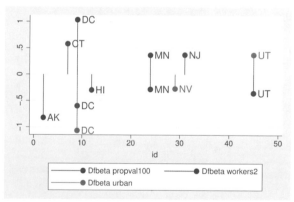

Before making the next graph, we must issue three `predict` commands to generate variables that contain the Cook's distance, the studentized residual, and the leverage from the
previous regression command:

```
. predict cd, cook
. predict rs, rstudent
. predict l, leverage
```

We are now ready to create the next graph.

Introduction   Editor   Twoway   Matrix   Bar   Box   Dot   Pie   Options   Standard options   Styles   Appendix

Stat   Stat options   Marginsplot   Save/Redisplay/Combine   More examples   Common mistakes   Schemes   Online

```
twoway (scatter rs id) (scatter rs id if abs(rs) > 2, mlabel(stateab)),
    legend(off)
```

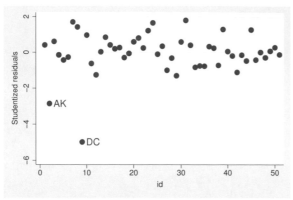

This graph uses `scatter rs id` to make an index plot of the studentized residuals. It also overlays a second `scatter` command with an `if` condition showing only studentized residuals that have an absolute value exceeding 2 and showing the labels for those observations. Using the `vg_samec` scheme makes the markers the same for both `scatter` commands.
Uses allstates.dta & scheme vg_samec

```
twoway (scatter rs id, text( -3 27 "Possible Outliers", size(vlarge)))
    (scatteri -3 18 -4.8 10, recast(line))
    (scatteri -3 18 -3 3, recast(line)), legend(off)
```

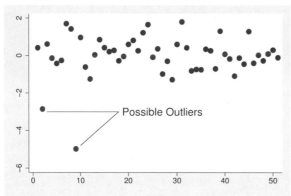

This graph is similar to the one above but uses the `text()` option to add text to the graph. We use two `scatteri` commands to draw a line from the text *Possible Outliers* to the markers for those points. The *y x* coordinates are given for the starting and ending positions, and `recast(line)` makes `scatteri` behave like a line plot, connecting the points to the text.
Uses allstates.dta & scheme vg_s2c

```
twoway (scatter rs l [aw=cd], msymbol(Oh))
    (scatter rs l if cd > .1, msymbol(i) mlabel(stateab) mlabpos(0))
    (scatter rs l if cd > .1, msymbol(i) mlabel(cd) mlabpos(6)), legend(off)
```

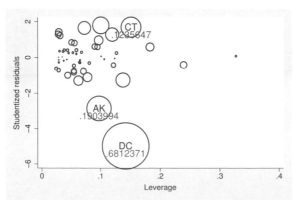

This graph shows the leverage-versus-studentized residuals, weighting the symbols by Cook's *D* (cd). We overlay it with a scatterplot showing the marker labels if `cd` exceeds .1, with the `cd` value placed below.
Uses allstates.dta & scheme vg_s2c

Say that we have a data file called `comp2001ts` that contains variables representing the stock prices of four hypothetical companies: `pricealpha`, `pricebeta`, `pricemu`, and `pricesigma`, as well as a variable `date`. To compare the performance of these companies, let's make a line plot for each company and stack them. We can do this by using `twoway tsline` with the `by(company)` option, but we first need to reshape the data into a `long` format. We do so with the following commands:

```
. vguse comp2001ts, clear
. reshape long price, i(date) j(compname) string
```

We now have variables `price` and `company` and can graph the prices by company.

```
twoway tsline price, by(compname, cols(1) yrescale note("") compact)
    ylabel(#2, nogrid) subtitle(, pos(5) ring(0) nobexpand nobox color(red))
    title(" ", box width(130) height(.001) bcolor(ebblue))
```

We graph `price` for the different companies with the `by()` option. The `cols(1)` option puts the graphs in one column. `yrescale` and `ylabel(#2)` allow the y axes to be scaled independently and labeled with about 2 values. The `subtitle()` option puts the name of the company in the bottom right corner of each graph. The `title()` option combined with the `compact` option creates a blue border between the graphs.
Uses comp2001ts.dta & scheme vg_s2c

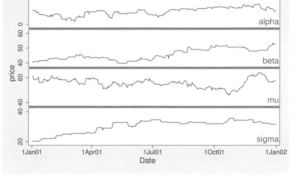

```
xtline price, i(compname) t(date) overlay
```

We can also use the `xtline` command to graph these same companies in one graph. The `i()` option specifies the variable that uniquely identifies an observation (e.g., the company), and the `t()` option provides the variable that represents time (plotted on the x axis). The `overlay` option places all the companies in the same graph.
Uses comp2001ts.dta & scheme vg_s2c

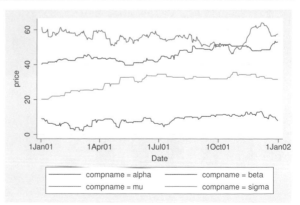

For the next graph, we want to create a bar chart that shows the mean of wages by occupation with error bars showing a 95% confidence interval for each mean. First, we collapse the data across the levels of occupation, creating the mean, standard deviation, and count. Next we create the variables `wageucl` and `wagelcl`, which are the upper and lower confidence limits, as shown below.

```
. vguse nlsw
. collapse (mean) mwage=wage (sd) sdwage=wage (count) nwage=wage, by(occ7)
. generate wageucl = mwage + invttail(nwage,0.025)*sdwage/sqrt(nwage)
. generate wagelcl = mwage - invttail(nwage,0.025)*sdwage/sqrt(nwage)
```

We are now ready to graph the data.

```
twoway (bar mwage occ7, barwidth(.5))
    (rcap wageucl wagelcl occ7, blwid(medthick) lcolor(navy) msize(large)),
    xlabel(1(1)7, valuelabel noticks) xscale(range(.5 7.5))
```

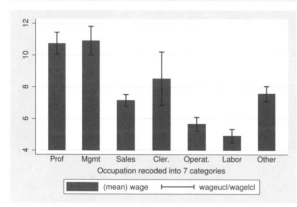

This bar chart is overlaid with a range plot showing the upper and lower confidence limits. The `xlabel()` option labels the values from 1 to 7, incrementing by 1. The `valuelabel` option indicates that the value labels for `occ7` will be used to label the x axis. The `xscale()` option adds a margin to the outer bars, and the `barwidth()` option creates the gap between the bars.
Uses nlsw.dta & scheme vg_s2c

```
twoway (rcap wageucl wagelcl occ7, lwidth(medthick) msize(large))
    (bar mwage occ7, barwidth(.5) bcolor(navy)),
    xlabel(1(1)7, valuelabel noticks) xscale(range(.5 7.5))
```

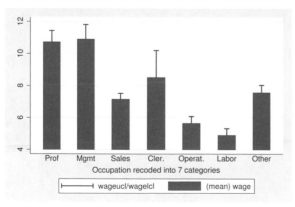

This graph is similar to the previous one, except that we have reversed the order of the commands, placing the `rcap` command first, followed by the `bar` command. As a result, only the top half of the error bar is shown. As in the previous example, the `xlabel()` option determines the labels on the x axis.
Uses nlsw.dta & scheme vg_s2c

Suppose that we wanted to show the mean wages with confidence intervals broken down by occupation and whether one graduated college. We can use the `collapse` command to create the mean, standard deviation, and count by the levels of `occ7` and `collgrad`. We can then create the upper and lower confidence limits. Finally, the `separate` command makes separate variables of `mwage` based on whether one graduated college, creating `mwage0` (wages for noncollege grad) and `mwage1` (wages for college grad). These commands are shown below, followed by the command to create the graph.

```
. vguse nlsw, clear
. collapse (mean) mwage=wage (sd) sdwage=wage
    (count) nwage=wage, by(occ7 collgrad)
. generate wageucl = mwage + invttail(nwage,0.025)*sdwage/sqrt(nwage)
. generate wagelcl = mwage - invttail(nwage,0.025)*sdwage/sqrt(nwage)
. separate mwage, by(collgrad)
```

```
twoway (line mwage0 mwage1 occ7) (rcap wageucl wagelcl occ7),
    xlabel( 1(1)7, valuelabel) xtitle(Occupation) ytitle(Wages)
    legend(order(1 "Not College Grad" 2 "College Grad"))
```

Here we make a line graph showing the mean wages for the noncollege graduates, `mwage0`, and the college graduates, `mwage1`, by occupation. We overlay that with a range plot showing the confidence interval. The `xlabel()` option labels the *x* axis with value labels, and the `legend()` option labels the legend.

Uses nlsw.dta & scheme vg_s2c

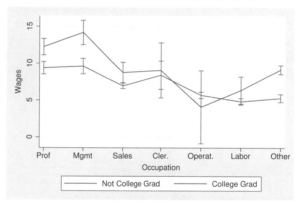

This next graph shows a type of scatterplot of the mean and confidence interval for `union` and `collgrad` for each level of `occ7`. To do this, collapse the data file by `occ7` and use those summary statistics to compute the confidence intervals below, followed by the command to create the graph.

```
. vguse nlsw, clear
. collapse (mean) pct_un=un pct_coll=collgrad
    (sd) sd_un=union sd_coll=collgrad
    (count) ct_un=union ct_coll=collgrad, by(occ7)
. gen lci_un = pct_un - sd_un/sqrt(ct_un)
. gen uci_un = pct_un + sd_un/sqrt(ct_un)
. gen lci_coll = pct_coll - sd_coll/sqrt(ct_coll)
. gen uci_coll = pct_coll + sd_coll/sqrt(ct_coll)
```

Stat | Stat options | Marginsplot | Save/Redisplay/Combine | More examples | Common mistakes | Schemes | Online

Introduction | Editor | Twoway | Matrix | Bar | Box | Dot | Pie | Options | Standard options | Styles | Appendix

```
twoway (rcap lci_coll uci_coll pct_un) (rcap lci_un uci_un pct_coll, hor)
   (sc pct_coll pct_un, msymbol(i) mlabel(occ7) mlabpos(10) mlabgap(5)),
   ylabel(0(.2).7) xtitle(% Union) ytitle(% Coll Grads) legend(off)
   title("% Union and % college graduates" "(with CIs) by occupation")
```

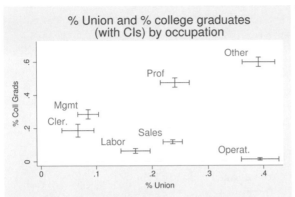

The overlaid `rcap` commands show the confidence intervals for both `union` and `collgrad` for each occupation. The `scatter` command uses an invisible marker and labels each occupation at the ten o'clock position with a larger gap than normal.

Uses nlsw.dta & scheme vg_s2c

This section concludes with a graph adapted from an example on the Stata web site. The graph combines numerous tricks, so rather than show it all at once, let's build it a piece at a time. Below is the ultimate graph we wish to create. It shows the population (in millions) for males and females in 17 different age groups, ranging from "Under 5" to "80–84". The blue bar represents the males, and the red bar represents the females.

`graph display`

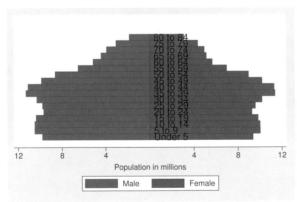

This is the graph that we wish to create. For now, we simply use the `graph display` command to display the graph. Because the `vg_s2c` scheme enhances the size of the text in the graph for readability in this book, some of the text may spillover but would not do so when using a scheme like `s2color`.

Uses pop2000mf.dta & scheme vg_s2c

To build this graph, we first use the data file `pop2000mf`, which contains 17 observations corresponding to 17 age groups (for example, "Under 5", "5–9", "10–14", and so forth). The variables `femtotal` and `maletotal` contain the number of females and males in each age group. After loading the data into Stata, we create `femmil`, which is the number of females per million, and `malmil`, which is the number of males per million, but this is made negative so that the male (blue) bar will be scaled in the negative direction. We also generate a variable called `zero`, which contains 0 for all observations.

```
. vguse pop2000mf, clear
. gen femmil = femtotal/1000000
. gen malmil = -maletotal/1000000
. gen zero = 0
```

We now take the first step in making this graph.

`twoway (bar malmil agegrp) (bar femmil agegrp)`

This is our first attempt to make this graph by overlaying the bar chart for the males with the bar chart for the females. The `agegrp` variable ranges from 1 to 17 and forms the x axis, but we can rotate this as shown in the next example.

Uses pop2000mf.dta & scheme vg_s2c

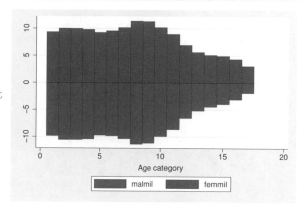

`twoway (bar malmil agegrp, horizontal) (bar femmil agegrp, horizontal)`

Adding the `horizontal` option to each bar chart, we can see the graph taking shape. However, we would like the age categories to appear inside the red (female) bars.

Uses pop2000mf.dta & scheme vg_s2c

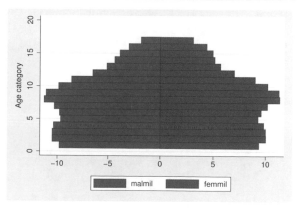

Introduction   Editor   Twoway   Matrix   Bar   Box   Dot   Pie   Options   Standard options   Styles   Appendix

Stat   Stat options   Marginsplot   Save/Redisplay/Combine   More examples   Common mistakes   Schemes   Online

```
twoway (bar malmil agegrp, horizontal) (bar femmil agegrp, horizontal)
    (scatter agegrp zero, msymbol(i) mlabel(agegrp) mlabcolor(black))
```

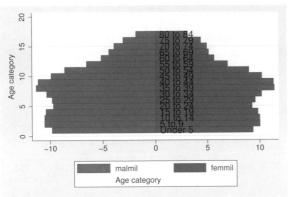

This `scatter` command uses `agegrp` (ranging from 1 to 17) as the y value and `zero` (0) for the x value, leading to the stack of 17 observations. Using the `msymbol()` and `mlabel()` options suppresses the symbol but displays the name of the age group from the labeled value of `agegrp`. Next we will fix the label and title for the x axis.
Uses pop2000mf.dta & scheme vg_s2c

```
twoway (bar malmil agegrp, horizontal) (bar femmil agegrp, horizontal)
    (scatter agegrp zero, msymbol(i) mlabel(agegrp) mlabcolor(black)),
    xlabel(-12 "12" -8 "8" -4 "4" 4 8 12) xtitle("Population in millions")
```

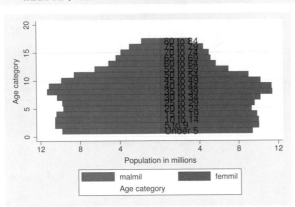

We use the `xlabel()` to change -12 to 12, -8 to 8, -4 to 4, and to label the positive side of the x axis as 4, 8, and 12. We also add a title for the x axis. Next let's fix the y axis and the legend.
Uses pop2000mf.dta & scheme vg_s2c

```
twoway (bar malmil agegrp, horizontal) (bar femmil agegrp, horizontal)
    (scatter agegrp zero, msymbol(i) mlabel(agegrp) mlabcolor(black)),
    xlabel(-12 "12" -8 "8" -4 "4" 4 8 12) xtitle("Population in millions")
    yscale(off) ylabel(, nogrid) legend(order(1 "Male" 2 "Female"))
```

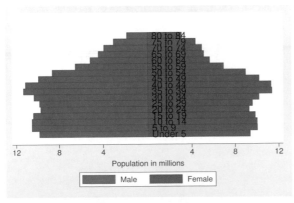

We suppress the display of the y axis by using the `yscale(off)` option and suppress the grid lines by using the `ylabel(, nogrid)` option. Finally, we use the `legend()` option to label the bars and suppress the display of the third symbol in the legend.
Uses pop2000mf.dta & scheme vg_s2c

12.6   Common mistakes

This section discusses mistakes that are frequently made when creating Stata graphs.

Commas with graph options

Graph options can accept their own options (sometimes referred to as *suboptions*); for example,

```
. twoway scatter propval100 popden rent700, xtitle("My Title", box)
```

The `xtitle()` option allows us to specify the x-axis title followed by a comma and a suboption that places a box around the x-axis title. If we had been content with the existing x-axis title and only wanted to add the box around the title, we could have issued this command:

```
. twoway scatter propval100 popden rent700, xtitle( , box)
```

Note the comma before the `box` option.

Now suppose that we are content with the existing legend but want to make the legend display in one column.

```
. twoway scatter propval100 popden rent700, xtitle( , box) legend(cols(1))
```

Based on the syntax from the `title()` option, we might have been tempted to type `legend( , cols(1))`, but that would have led to an error. Some options, like the `legend()` option, simply take a list of options with no comma permitted.

Using options in the wrong context

Consider the example below. Our goal is to move the labels for the x axis from their default position at the bottom of the graph to the alternate position at the top of the graph.

```
. twoway scatter propval100 rent700, xlabel( , alternate)
```

This command executes, but it does not have the desired effect. Instead, it staggers the labels of the x axis, alternating between the upper and lower positions. In this context, the `alternate` option means something different than we had intended. We really wanted to specify the `xscale(alternate)` option:

```
. twoway scatter propval100 rent700, xscale(alternate)
```

This command moves the entire scale of the x axis to the alternate position and has the desired effect. Another mistake we might have made was to put the `alternate` option as an overall option. This command is shown below with the result:

```
. twoway scatter propval100 rent700, alternate
option alternate not allowed
invalid syntax
r(198);
```

Here we are half right. There is an option `alternate`, but we have used it in the wrong context, yielding the syntax error. The option we are specifying may be right, but we just need to put it into the right context.

Introduction   Editor   Twoway   Matrix   Bar   Box   Dot   Pie   Options   Standard options   Styles   Appendix

Stat   Stat options   Marginsplot   Save/Redisplay/Combine   More examples   Common mistakes   Schemes   Online

Options appear to have no effect

When we add an option to a graph, we generally expect to see the effect of adding the option. However, sometimes adding an option has no effect. Consider this example:

```
. twoway scatter propval100 rent700, mlabpos(12)
```

This command executes, but nothing changes as a result of including the `mlabpos(12)` option, which would change the position of the marker labels to the twelve o'clock position. There are no marker labels in the graph, so adding this option has no effect. We would have to use the `mlabel()` option to add marker labels before we would see the effect of this option.

Consider another example, which is a bit more subtle. We would like to make the line (periphery) of the marker thick. When we run the following command, we do not see any effect from adding the `mlwidth(thick)` option:

```
. twoway scatter propval100 rent700, mlwidth(thick)
```

The reason for this is that the marker has a `line` color and a `fill` color, and by default, they are the same color, so it is impossible to see the effect of changing the thickness of the line around the marker. However, if we make the `line` and `fill` colors different, as in the following example, we can see the effect of the `mlwidth()` option:

```
. twoway scatter propval100 rent700, mlwidth(thick)
        mlcolor(black) mfcolor(gs13)
```

Options when using by()

Using the `by()` option changes the meaning of some options. Consider the following example:

```
. twoway scatter propval100 rent700, by(north) title(My title)
```

We might think that the `title()` option will provide a title for the entire graph, as it would when the `by()` option is not included. However, each graph will have "My title" as the title; the graph as a whole will not. To provide a title for the whole graph, we would specify the command this way:

```
. twoway scatter propval100 rent700, by(north, title(My title))
```

When using the `legend()` option combined with the `by()` option, we should place options that affect the position of the legend within the `by()` option. Consider this example:

```
. twoway scatter propval100 popden rent700,
        by(north, legend(pos(12))) legend(cols(1))
```

Here the `legend(pos(12))` option controls the position of the legend, placing it at the twelve o'clock position, so we place it within the `by()` option. On the other hand, the `legend(cols(1))` option does not affect the position of the legend, so we place it outside the `by()` option. For more details on this, see **Options : By** (346).

Altering the wrong axis

When we use multiple x or y axes, it is easy to modify the wrong axis. Consider this example:

```
. twoway (scatter  propval100 ownhome)
         (scatter rent700 ownhome, yaxis(2) ytitle(Rents over 700))
```

We might think that the `ytitle()` option will change the title for the second y axis, but it will actually change the first axis. Because `ytitle()` is an option that concerns the overall graph, we should place it at the end of the graph command, as shown below.

```
. twoway (scatter  propval100 ownhome)
         (scatter rent700 ownhome, yaxis(2)), ytitle(Rents over 700, axis(2))
```

We use the `axis(2)` option to indicate that `ytitle()` should be modified for the second y axis.

When all else fails

I hope that by describing these errors you can avoid some common problems. Here are more ideas and resources to help you when you are struggling:

- Build graphs slowly. Rather than trying to make a final graph at once, try building the graph slowly by adding one option at a time. This is illustrated in Intro : Building graphs (29), where we took a complex graph and built it one piece at a time. Building slowly helps isolate problems for a particular option, which can then be investigated.

- When possible, model graphs from existing examples. This book strives to provide examples from which to model. For more online examples, see Appendix : Online supplements (482) for the companion web site for the book, which links to more examples.

- For more information about the syntax of Stata graphics, see [G-2] **graph**. Some of the graph commands available in Stata may have been added after the printing of [G-2] **graph** but will be documented in the online help; type `help graph` in Stata. See also Appendix : Online supplements (482), which has links to the online help that are organized according to the table of contents of this book.

- Reach out to fellow Stata users: colleagues, friends on Statalist (visit http://www.stata.com/statalist/), or Stata technical support (visit http://www.stata.com/support/tech-support/).

12.7 Customizing schemes

This section shows how to customize your own schemes. Although schemes can look complicated, it is possible to easily create some simple schemes on your own. Let's look at the `vg_lgndc` scheme as an example. This scheme is based on the `s2color` scheme but changes the legend to display at the nine o'clock position, in one column, with the keys stacked on top of the symbols. Here are the contents of that scheme:

Stat Stat options Marginsplot Save/Redisplay/Combine More examples Common mistakes Schemes Online

Introduction Editor Twoway Matrix Bar Box Dot Pie Options Standard options Styles Appendix

```
#include s2color // start with the s2color scheme

clockdir legend_position  9   // put the legend in the nine o'clock position
numstyle legend_cols      1   // make the legend display in 1 column
yesno legend_stacked      yes // stack the keys & symbols on top of each other

gsize legend_key_gap      half_tiny // very, very small gap between key and label
gsize legend_row_gap      small     // somewhat larger gap between key/label pairs
```

Rather than creating the **vg_lgndc** scheme from scratch, which would be laborious, I used the **#include s2color** statement to base this new scheme on the **s2color** scheme. The subsequent statements changed the position of the legend and the number of columns in the legend and stacked the legend keys and symbols upon each other.

Say that we liked the **vg_lgndc** scheme but wanted to make our own version in which the legend is in the three o'clock position instead of the nine o'clock position, naming our version **legend3**. To do this, we would start the Stata Do-file Editor, for example, by typing **doedit**, and then type the following into it: (Of course, the scheme will work fine if you omit comments after the double slashes.)

```
#include s2color // start with the s2color scheme

clockdir legend_position  3   // put the legend in the three o'clock position
numstyle legend_cols      1   // make the legend display in 1 column
yesno legend_stacked      yes // stack the keys & symbols on top of each other

gsize legend_key_gap      half_tiny // very, very small gap between key and label
gsize legend_row_gap      small     // somewhat larger gap between key/label pairs
```

We can then save the file as **scheme-legend3.scheme**. We can use the **scheme(legend3)** option at the end of a graph command or type **set scheme legend3**, and Stata will use that scheme for displaying our graph. Below we show an example using this scheme. (The **legend3** scheme is not included among the downloadable schemes.)

```
twoway (scatter propval100 rent700) (lfit propval100 rent700),
   scheme(legend3)
```

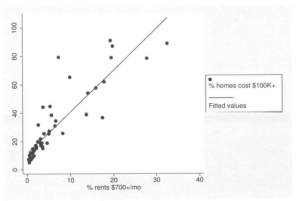

This is an example using the newly created **legend3** scheme. We see the legend in the three o'clock position, in one column, with the legend stacked. *Uses allstates.dta & scheme legend3*

So far, things are going great. However, Stata will only know how to find the newly created scheme-legend3.scheme while you are working in the directory where you saved that scheme. If you change to a different directory, Stata will not know where to find scheme-legend3.scheme. If, however, we save the scheme into your PERSONAL directory, Stata will know where to find it regardless of your current directory. For example, on my computer, I typed the sysdir command, which produced

```
. sysdir
    STATA:  C:\Program Files (x86)\Stata12\
  UPDATES:  C:\Program Files (x86)\Stata12\ado\updates\
     BASE:  C:\Program Files (x86)\Stata12\ado\base\
     SITE:  C:\Program Files (x86)\Stata12\ado\site\
     PLUS:  c:\ado\plus\
 PERSONAL:  c:\ado\personal\
 OLDPLACE:  c:\ado\
```

From this, I see that my PERSONAL directory is located in c:\ado\personal\, so if I store either .ado files or .scheme files there, Stata will find them. So, instead of saving scheme-legend3.scheme into the current directory, you may want to save it into your PERSONAL directory. (If you have already saved scheme-legend3.scheme to the current directory and also save it to the PERSONAL directory, you should remove the copy from the current directory.)

This section has really focused on the mechanics of creating a scheme but has not said much about the possible content that you could place inside a scheme. This is beyond the scope of this introduction, but here are three places where you can find more advanced information:

First, the help file for schemes, help schemes, provides information on schemes in general. Also, help scheme files contains documentation about scheme files and what can be changed with schemes.

Second, look at the downloaded schemes from this book to get ideas for your own schemes. See Appendix: Online supplements (482). Say that you wanted to look at the vg_rose scheme. You could type which scheme-vg_rose.scheme to find out where that scheme is located. Then you could use any editor (including the Do-file Editor) to view that scheme for ideas.

Third, look at the built-in Stata schemes, such as s1color, s2color, or economist. Looking at these schemes shows you the menu of items that you can fiddle with in your own schemes, but these schemes should never be modified directly. You can use the strategy outlined above where you made your own scheme and used #include to read in a scheme. You can then add your own statements to modify the scheme as desired.

Schemes that other people have created and the schemes built into Stata will contain statements that control some aspect of a graph, but you may not know which aspect they control. For example, in the vg_rose scheme there is the statement

```
    color     background eggshell
```

which obviously controls the color of some kind of background element, but you might not know which element it controls. You can find out by making a copy of the scheme and then changing eggshell to some other nonsubtle value, such as red, and then make a graph using this new scheme (using scheme(*schemename*), not set scheme *schemename*). The part of the graph that becomes red will indicate the part that is controlled by the `color background` statement.

Of course, I have just scratched the surface of how to create and customize schemes. However, this should provide the basic tools needed for making a basic scheme, storing it in the `PERSONAL` directory, and then playing with the scheme. Because schemes are so powerful, they can appear complicated, but if built slowly and methodically, the process can be straightforward, logical, and, actually, quite a bit of fun.

12.8   Online supplements

This book has several online resources associated with it. I encourage all readers to take advantage of these online additions by visiting the web site for the book at

http://www.stata-press.com/books/vgsg3.html

Resources on the web site include

- Programs and help files. You can easily download and install the programs and help files associated with this book. To install these programs and help files, just type

  ```
  . net from http://www.stata-press.com/data/vgsg3/
  . net install vgsg3
  ```

  After installing the programs and help, type `help vgsg` for an overview of what has been installed.

  If you have the first edition and have installed the programs and help files associated with that edition, you should instead type `adoupdate vgsg`.

- Data files. All the data files used in the book are available at the web site for downloading. I encourage you to download the data files used in this book, play with these examples, and try variations on your own to solidify and extend your understanding. You can quickly download all the datasets into your current working directory from within Stata by typing

  ```
  . net from http://www.stata-press.com/data/vgsg3/
  . net get vgsg3
  ```

  If you prefer, you can obtain any of the data files over the Internet with the `vguse` command. Each example concludes by indicating the data file and scheme that was used to make the graph, for example,

    Uses allstates.dta & scheme vg_s2c

  This statement indicates that you can type `vguse allstates`, and Stata will download and use the data file over the Internet for you (assuming that you have installed the programs).

- Schemes. This book uses a variety of schemes, and when you download the programs and help files (see above), the schemes used in this book are downloaded as well, allowing you to use them to reproduce the look of the graphs in this book.

- Hopefully, a short or empty *Errata* will be found at the web site. Although I have tried hard to make this book true and accurate, I know that some errors will be found, and they will be listed there.

- Other resources that may be placed on the site after this book goes to press, so visit the site to see what else may appear.

Introduction   Editor   Twoway   Matrix   Bar   Box   Dot   Pie   Options   Standard options   Styles   Appendix

Stat   Stat options   Marginsplot   Save/Redisplay/Combine   More examples   Common mistakes   Schemes   Online

Subject index